The Elements of Environmental Pollution

Environmental pollution is one of humanity's most pressing issues and will remain so for the foreseeable future. Anthropogenic activity is disturbing natural cycles and generating pollutants that are altering the atmosphere, accumulating in the food chain and contaminating the world's soils, rivers and oceans. Human health and ecosystems continue to be damaged by toxic metals, persistent organic pollutants, radionuclides and other hazardous materials. *The Elements of Environmental Pollution* provides comprehensive coverage of this essential subject. It explains the key principles of pollution science, assesses human disturbances of natural element cycles and describes local and global pollution impacts, from smoggy cities, polluted lakes and toxic soils to climate change, ocean acidification and marine dead zones. The book is informed by the latest pollution research and benefits from numerous real-world examples and international case studies. A comprehensive glossary provides clear and concise explanations of key concepts.

This textbook will support teaching and learning in environment-related university courses and will be vital reading for anyone with an interest in environmental protection.

John Rieuwerts is Associate Professor in Environmental Science at Plymouth University, UK.

"This is an excellent primer that provides a refreshing approach to understanding contemporary pollution issues. It provides a useful systematic approach to examining the role of key elements that chiefly affect human and environmental health; combining their chemistry with relevant environmental processes. This is a very useful text book which should be on hand to all Environmental Science, Chemistry and Geography students."

Crispin Halsall, Lancaster University, UK

"As our generation contemplates conditions being created for future generations, the topic of anthropogenic pollution of air, waters and soils will need to be closely evaluated and considered in order to protect and preserve the Earth's lifeforms, including humans. This book is an important extension of the literature on the role of human activities in changing the environment of the Earth and the topic is handled exceptionally well."

Howard W. Mielke, Tulane University, USA

"I am very impressed by this book. It is clearly written, accurate and places pollution in its correct context. The book fulfils an important need and it is comprehensive and readable."

J.N.B. Bell, Imperial College London, UK

"This book offers a fresh look at the interaction between the building blocks of our planet and human activities. Travelling through the periodic table, it examines the way we affect and may disrupt the life cycles of the most important and abundant elements necessary for life, and as a result influence our health and the health of the ecosystem around us. It is written in an accessible style whilst at the same time is underpinned by clear scientific principles."

Hemda Garelick, Middlesex University, UK

"A highly readable text that is both interesting and informative. I recommend this book as a valuable resource for undergraduates undertaking modules in environmental pollution and, more broadly, as background reading for students taking environmental degrees."

Peter J. Shaw, University of Southampton, UK

"Identifying and comprehending the phenomenon of pollution requires that we understand the difference between natural material fluxes and those caused by mankind. This book, which classifies pollution processes according to individual elements, should be part of the library of all those concerned with environmental geochemistry and pollution."

Martin Mihaljevič, Charles University, Czech Republic

The Elements of Environmental Pollution

John Rieuwerts

LONDON AND NEW YORK

First published 2015
by Routledge
2 Park Square, Milton Park, Abingdon, Oxon OX14 4RN

and by Routledge
711 Third Avenue, New York, NY 10017

Routledge is an imprint of the Taylor & Francis Group, an informa business

British Library Cataloguing-in-Publication Data
A catalogue record for this book is available from the British Library

Library of Congress Cataloging-in-Publication Data
Rieuwerts, John.
The elements of environmental pollution / John Rieuwerts.
pages cm
Includes bibliographical references and index.
1. Pollutants. 2. Environmental chemistry. 3. Pollution--Environmental aspects. I. Title.
TD174.R55 2015
628.5'2--dc23
2014038175

ISBN: 978-0-415-85919-6 (hbk)
ISBN: 978-0-415-85920-2 (pbk)
ISBN: 978-0-203-79869-0 (ebk)

Typeset in Sabon
by Saxon Graphics Ltd, Derby

Printed and bound in the United States of America by Publishers Graphics, LLC on sustainably sourced paper.

This book is dedicated to my father, Dr James H. Rieuwerts,
for inspiring a love of the Earth (inside and out) and a passion for
academic study.

Contents

Pollution issues

Pollution issue	Element chapter(s)
Acid mine drainage	S
Acid rain	S, N
Air pollution (local)	C, N, S
Biochemical oxygen demand	C
Biomass burning and haze	C, N, S
Classic smog	C, S
Endocrine disruption	Br, C, Cl
Eutrophication and marine dead zones	N, P
Global dimming and brightening	C, S
Global warming	C, N
Mineral extraction	Most elements
Nitrates in drinking water	N
Ocean acidification	C
Oceanic eutrophication and 'dead zones'	N, P
Oil spills	C
Organic waste and biochemical oxygen demand	C
Particulate pollution and human health	C, S
Persistent organic pollutants	Br, C, Cl
Pesticides	C, Cl, P, Sn
Photochemical smog	N
Plant and crop damage	N, S, metal(loid)s*
Radioactive waste	U
Radioactivity and human health	U
Stratospheric ozone depletion	Br, Cl, Fl
Suspended solids	C
Toxic elements and ecological impacts	Br, C, Cl, CN^-, F, metal(loid)s*
Toxic elements and human health	Br, C, Cl, F, metal(loid)s*
Water pollution	Most elements

*Metals and metalloids: Al, As, B, Cd, Cr, Cu, Mn, Mo, Pb, Hg, Ni, Se, Ag, Tl, Sn, V, Zn

Figures

Tables

Boxes

Preface

The environmental impacts of human activities can largely be classified into two broad categories: 'resource depletion' and 'pollution'. Resource issues, including energy choices, food and water scarcity, natural landscape damage and biodiversity loss, are covered extensively by a number of excellent texts, but there have been fewer text books devoted to environmental pollution in recent years, with the clear exception of those about carbon emissions and global warming. This is surprising; environmental pollution, on global, regional and local scales, is a priority issue for humanity in the modern era and will remain so for the foreseeable future. Global population growth and mass consumerism continue unabated and we are seeing agricultural, industrial and urban expansion in many parts of the world, in order to satisfy the human race's requirements for food, housing, energy and material goods. Human activity is disturbing natural cycles and generating pollution that is altering the atmosphere and contaminating the world's soils and waters.

The main aim of this book is to show how local and global pollution problems arise from anthropogenic disturbances of natural elemental cycles and reservoirs and from our introduction of hazardous materials into the natural world. Chapter 1 explains the key concepts in pollution science, including the linking of pollutant sources and receptors by specific environmental pathways. The final section of chapter 1 focuses on the main types of response to pollution; some specific examples are also given in the book's other chapters. The first chapter is followed, in chapters 2 to 5, by consideration of the four major elemental cycles (carbon, nitrogen, sulfur and phosphorus) and the consequences of their disturbance by human activities. Similarly, each of the following chapters (6 to 19) details the natural reservoirs and fluxes of a specific pollutant element (and its compounds), the reasons for our disturbance of these reservoirs/fluxes and the consequences of the pollution caused. To complete the picture, a number of additional chemical, biological and physical pollutants are covered in the final chapter (20). Pollutants do not occur in isolation and within the text there are clearly indicated cross-references to other element chapters where appropriate, using the {→N} format (where the cross-reference in this example is to the nitrogen chapter).

The Elements of Environmental Pollution is aimed primarily at first- and second-year undergraduate students taking courses in environment-related studies, but final year and postgraduate students should also benefit from its contents. The reader would benefit from an understanding of some chemistry, but it is not essential; chemical equations are used for illustrative purposes in approximately half of the chapters, but a reading of the adjacent text should be sufficient to understand the processes in question. Key terms or concepts in pollution science are introduced in chapter 1, indicated in bold type. A glossary is also included at the end of the book to aid understanding of other important terms used in the text.

Acknowledgements and image credits

I would like to express my thanks to Helen Bell at Routledge-Earthscan for her advice and guidance. Thanks also to the various (and anonymous) peer reviewers who commented on early drafts. I am very grateful to Kate Sardella of Plymouth University for bringing my rough sketches for Chapter 1 so brilliantly to life. Last but not least, thanks to Gill, Caitlin and Beth for their support and for tolerating a frequently absent (and absent-minded) husband/father during the writing of this book.

Uncredited photographs and diagrams are the author's, but all other photographs, and the diagrams of molecular structures, were dependent on the work of others and I am grateful to them for either putting their images in the public domain or allowing their work to be used under Creative Commons (CC) Attribution 1.0/2.0/2.5/3.0 Generic licenses (details of individual images below). Public domain and CC images were accessed via the websites Flickr (www.flickr.com), Wikimedia Commons (http://commons.wikimedia.org), pdpphoto (www.pdpphoto.org) and some US government websites. Under each image, the original creator/owner and web source is credited. In most CC license cases, the photograph used is a derivative of the original, as allowed under the license; i.e. the original colour image has been converted to monochrome during publication. Exceptions are Figs 1.15 and 2.4, which were monochrome images. Some additional images were reproduced by kind permission of academic publishers and, in such cases, the author(s) and publisher are credited below the figure (e.g. Fig. 12.4).

The following image is licensed under CC-BY-1.0: Fig. 12.1.
Details available at: http://creativecommons.org/licenses/by/1.0/

The following images are licensed under CC-BY-2.0: Figs 1.3, 1.4, 1.13, 1.15, 1.17, 1.18, 3.2, 3.4, 3.6, 3.10, 3.11, 4.1, 5.9, 6.1, 6.2, 6.4, 6.6, 6.9, 7.1, 7.4, 9.1, 9.6, 9.7, 10.1, 13.1, 13.2, 14.3, 15.1, 18.2, 18.3, 19.1.
Details available at: http://creativecommons.org/licenses/by/2.0/

The following image is licensed under CC-BY-2.5: Fig. 2.7d.
Details available at: http://creativecommons.org/licenses/by/2.5/

The following image is licensed under CC-BY-3.0: Fig. 2.4.
Details available at: http://creativecommons.org/licenses/by/3.0/

The following are public domain images: Figs 1.6, 1.16, 2.7a, 2.7b, 2.7c, 2.11, 2.15, 2.17, 2.18, 4.2, 4.4, 4.5, 5.4, 7.2, 7.3, 8.3, 9.4, 9.11, 9.14, 9.15, 11.1, 11.3, 12.2, 12.3, 14.1, 15.2, 16.1, 16.2, 17.1, 18.1.

1 Pollution

What is pollution?

When a substance occurs in a location or organism at higher levels than normal, we say that the environment or body has been 'contaminated' by the substance. 'Pollution' is generally considered as an extension of this; it implies that environmental harm will, or might, be caused by the contamination. Furthermore, pollution is typically (although not always) associated with human activity; for example, significant quantities of polluting substances are lost to the environment during the extraction and processing of raw materials and the manufacture, use and disposal of final products. The major categories of anthropogenic pollutants are listed in Table 1.1.

Table 1.1 The main anthropogenic pollutants of concern.

Atmospheric pollutants
Greenhouse gases: carbon dioxide, CO_2; methane, CH_4; nitrous oxide, N_2O; ozone, O_3
Oxides of nitrogen (NO_X): nitric oxide, NO and nitrogen dioxide, NO_2
Oxides of sulfur (SO_X): sulfur dioxide, SO_2 and sulfur trioxide, SO_3
Stratospheric ozone depleting compounds: halogenated hydrocarbons (e.g. CFC); nitrous oxide, N_2O
Ammonia, NH_3
Carbon monoxide, CO
Volatile organic compounds, VOCs
Particulate matter
Freshwater and seawater pollutants
Nutrients: nitrate and phosphate
Acidifying substances
Respirable organic matter
Toxic trace elements
Persistent organic pollutants
Crude oil and petrochemicals
Litter, including plastics

Table 1.1 continued

Soil pollutants
Nutrients: nitrate and phosphate
Toxic trace elements
Persistent organic pollutants
Petrochemicals
Radionuclides

Natural pollutants

There are many natural forms of pollution (Fig. 1.1). The Earth's crust is a major reservoir of most elements, some of them concentrated in relatively small volumes – in fossil fuel and metal sulfide deposits, for example. Over geological timescales, tectonic processes and erosion bring toxic elements, like arsenic and lead, and major elements, like carbon, to the surface environment where they can have adverse impacts. Elsewhere, volcanoes and natural forest fires emit gases and particles into the atmosphere. Some elements are emitted directly into the atmosphere from the Earth's crust in gaseous form; for example, radon is a radioactive gas that is naturally emitted by granite and other rocks and can cause lung cancer if inhaled over a period of time. Mercury vapour also diffuses into the atmosphere from the Earth's crust. Some naturally occurring organic compounds are very toxic. An example of natural

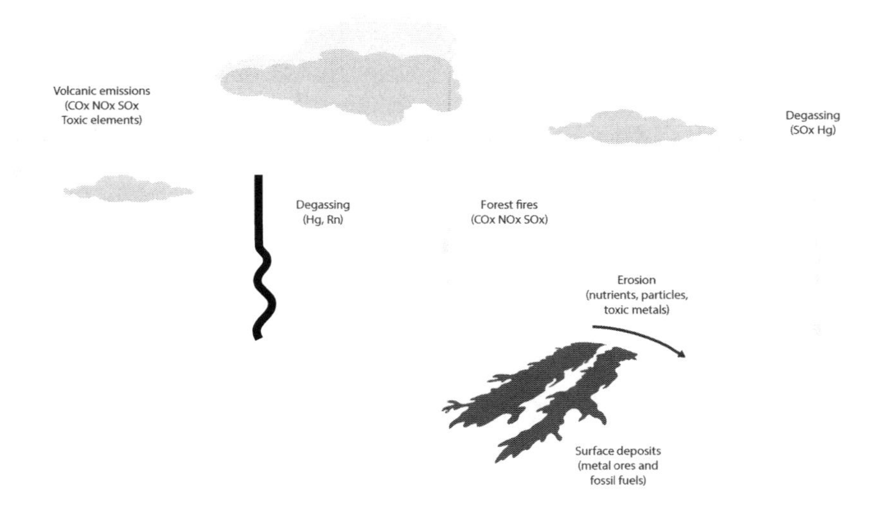

Figure 1.1 Examples of natural sources of pollution. This diagram is simplified for clarity and the element chapters should be consulted for full details of natural processes. For example, several sulfur-containing gases (not shown) are emitted into the atmosphere from the oceans, but these ultimately oxidise to SOx, which is shown on the diagram (right).

'pollution', from an early period of Earth history, is that of the 'great oxygenation event' (GOE) of some 2.5 billion years ago, when oxygen first accumulated in the atmosphere as a result of the evolution of photosynthesising cyanobacteria in the oceans. Oxygen is toxic to anaerobic organisms, which must have declined significantly at this time; they persist today in oxygen-free environments such as lake-bottom sediments and animal intestines. The GOE also illustrates that some substances that we might not think of as toxic can be very harmful to some organisms.

Anthropogenic pollution: past, present and future

The earliest forms of anthropogenic pollution are likely to date back far into prehistory, occurring in settlements where sewage and drinking water were not adequately separated. Pollution of the air probably has a long history too, arguably dating back to the earliest use of fire. A much later milestone is likely to have been the development of metal working in ancient times, when primitive extraction and smelting techniques could conceivably have caused localised problems of poor air quality and possibly water pollution. The earliest laws against pollution were most likely passed for the protection of water quality, in Roman times for example, and in medieval times a ban was placed on the burning of poor quality coal in London. In the 18th century, polluting activities escalated dramatically at the onset of the Industrial Revolution and, in the current era, gross pollution from heavy industry continues in many parts of the world, most especially in the 'newly industrialised nations' such as China and India. In comparison, many of the former industrial powerhouses of Europe, America and elsewhere are now 'post-industrial' economies with less heavy industry. Despite this, pollution problems persist in such countries, both from the legacy of past industrial activities (especially in contaminated soils and sediments) and from present-day activities such as food production, power generation, waste management, transportation and remaining industries.

Looking ahead, pollution problems are likely to persist and potentially increase in magnitude in the future, despite progress on regulatory, technologically and economically-driven pollution control measures. A rapidly increasing world population and, more particularly, global per-capita increases in resource exploitation and material consumption mean that environmental pollution will remain a significant concern for the foreseeable future, with continuing risks to human health and vulnerable ecosystems.

Sources and receptors

Pollution arises chiefly from the human disturbance of natural elemental cycles and the transfer of elements and their compounds to parts of the physical environment where they would normally be found at lower concentrations or, in some cases, not at all. In most cases this entails the release of elements from the Earth's crust by extraction of minerals and fossil fuels (and their processing and/or combustion), followed by the utilisation of these raw materials in the manufacture of material goods or the provision of power. However, there are exceptions; for example, in the manufacture of nitrogenous fertilisers, N is transferred from the atmosphere to the Earth's surface {→N} and Br (and some Cl) is derived from the sea {→Br →Cl} (Fig. 1.2). Deliberate, accidental or negligent releases of process wastes to the environment during processing

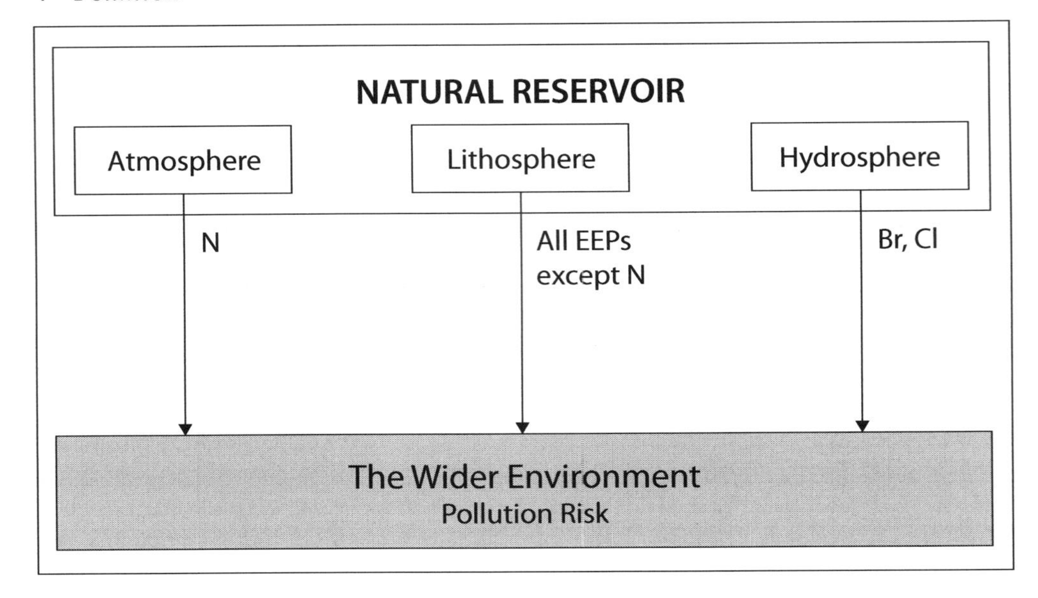

Figure 1.2 Most of the 'elements of environmental pollution' (EEPs) are extracted from the Earth's crust (the lithosphere) and transferred into the air, waters and soils; however, Br, Cl and N are notable exceptions.

and final product use can cause pollution. End-of-life product disposal (e.g. landfill and waste incineration) creates other pollution hazards. Major pollution sources are shown in Table 1.2, along with primary physical 'receptors', i.e. the components of the physical environment (air, soils, waters) that are typically the initial recipients of the different pollution types. Of course, receptors of pollution also include the humans, animals, plants and other organisms that inhabit these environments (see Pathways section, below).

Table 1.2 indicates that most pollution problems can be attributed to just a few broad categories of human activity: agriculture, manufacturing industry, transportation, power generation and waste management. These activities are described briefly below, together with indications of where more detailed discussions appear in the appropriate element chapters.

Agriculture

Modern agriculture is highly dependent on the use of pesticides and fertilisers to minimise crop damage and target maximum yield (Fig. 1.3). Pollution occurs when these chemicals (and manures) are washed off, or through, soils into surface waters and groundwaters. Fertiliser pollution causes eutrophication and other impacts {→N →P}, while pesticides can be directly toxic to humans and wildlife {→C →Cl →P}. Another pollution impact arises from the burning of biomass, including forested areas, in order to clear land for agricultural use; impacts arise mainly from particulate pollution of the air {→C} and the input of eroded soils from the cleared land into watercourses. The large energy demand of modern agriculture, including the manufacture of agricultural chemicals, is a further source of pollution. The scale of agricultural pollution is magnified by the continually increasing demand for meat

Table 1.2 Major pollution sources, primary receptors and pollutants from routine anthropogenic activity (i.e. not including industrial accidents).

Sources	*Primary receptor(s)*	*Key pollutants*[a]
	Agriculture	
Fertilisers	Soil	Nitrate, phosphate, metals
Pesticides	Soil	Cl, P, POPs
Biomass burning	Atmosphere	CO_X, PM
Soil erosion	Rivers and lakes	PM
Organic farm waste	Soil, rivers and lakes	ROM, nitrate, phosphate
	Atmosphere	Ammonia
	Manufacturing industry	
Mining and processing	Soil	Metals
	Surface water	Metals, CN^-
	Groundwater	Metals, CN^-
Metals smelting	Atmosphere	PM, metals, SO_X
Manufactories	Soil	Metals, POPs
	Atmosphere	PM, NO_X, SO_X, CO_X
	Rivers and lakes	Metals, POPs, ROM
	Seawater	Metals, POPs, ROM
	Transportation	
Vehicle emissions	Atmosphere	CO_X, NO_X, SO_X, PM, metals
	Power generation	
Fossil fuel combustion	Atmosphere	CO_X, NO_X, SO_X, PM, metals
Combustion residues	Atmosphere	Metals, PM
	Rivers/lakes (ash ponds)	Metals, PM
	Waste management	
Landfill	Groundwater	Metals
	Atmosphere	Methane
Incineration[b]	Atmosphere	PM, POPs, metals, SO_X, NO_X, CO_X
Unregulated waste dumps (mainly in Asia and Africa)	Groundwater	POPs, metals
	Atmosphere	
	Rivers and seawater	
Sewage/wastewater treatment	Rivers and seawater	ROM, metals, POPs

a For the purposes of this table, 'metals' can also be taken to include metalloids such as arsenic.

b Includes use of old tyres, industrial solvents and other wastes for firing of cement kilns.

CN^- = cyanide; CO_X = oxides of carbon (carbon monoxide, CO and carbon dioxide, CO_2); NO_X = oxides of nitrogen; PM = particulate matter {→C}; POPs = persistent organic pollutants {→Cl}; ROM = respirable organic matter; SO_X = oxides of sulfur.

NB: The second column shows the *primary* receiving media for specific pollutants from specific sources; in many cases pollutants will subsequently be transported to other parts of the environment and may have impacts there (see Pathways section, below).

Figure 1.3 Pesticide spraying, United Kingdom.
Credit: Chafer Machinery (Flickr).

around the world; the conversion of plant feed to meat and dairy produce has much lower efficiencies than the production of arable crops for direct human consumption and requires larger areas of land and greater inputs of energy and agricultural chemicals.

Power generation

Most power, for domestic and industrial use, is generated by the combustion of fossil fuels, which generates atmospheric emissions of the oxides of C, N and S (COx, NOx and SOx), together with fine particulate matter, some of which contains trace amounts of toxic elements. The combustion of fossil fuels, particularly coal, creates residual solid waste in the form of ash and, if it is not suitably managed, this can be a source of pollution by toxic elements like arsenic {→As} and selenium {→Se}. The impacts of COx, NOx and SOx are described in detail in the appropriate chapters {→C →N →S} and the various metal and metalloid chapters include discussion of power generation as an important source of toxic elements to air and, subsequently, to soils, waters and organisms. Nuclear power generation is based on the neutron bombardment of uranium in fuel rods; pollution occurs following leaks of radioactive cooling water or catastrophic failures, as seen at Chernobyl, Fukushima and elsewhere {→U}. For both fossil and nuclear fuels, in addition to the direct emissions arising from combustion and fission, respectively, pollution is also generated from the extraction and processing of the fuels and other associated activities, including the construction of power stations. The main advantage of renewable sources of energy, such as solar panels and wind turbines, is that they do not directly cause pollution when operating; only their manufacture, construction and disposal are sources.

Transportation

The internal combustion engine, based on fossil fuel combustion, is another form of power generation (see above) and, as such, is responsible for many of the same pollutants, particularly CO_X, NO_X and SO_X. Electric vehicles produce the same pollutants (via power stations) unless the electricity is produced using non-fossil fuel sources. Exhaust emissions of petrol and diesel vehicles (Fig. 1.4) also lead to the problem of poor air quality in urban areas, where inhabitants are exposed to potentially injurious air pollutants such as CO, NO_2 (also a precursor of tropospheric ozone), SO_2, toxic metals and fine particulate matter {→C →N →S →metals}.

Figure 1.4 Heavy traffic, Birmingham, United Kingdom. Vehicle emissions are a major source of air pollution, particularly in urban areas.

Credit: Joe CmdrGravy (Flickr).

Waste management and sewage treatment

Traditionally, the approach to managing domestic and some other wastes has been to bury it in landfill sites, although this is decreasing in importance in countries where regulations have been passed to minimise its use. The microbial breakdown of organic wastes in the low-oxygen environment of a landfill site generates methane (a potent greenhouse gas) and other volatiles that can diffuse into the atmosphere; in some cases the methane is flared or tapped and used for power generation. If rainwater percolates through landfilled waste, leachate can form, potentially transporting metals, POPs and other pollutants to groundwaters, particularly in poorly engineered sites. The main alternative waste management option is incineration, which is a source of air pollutants, including CO_2 (from incineration of organic waste), metals (e.g. from batteries) and POPs, such as dioxins (mainly from plastics {→Cl}). However, technological improvements in recent decades mean that pollutant emissions from modern incinerators are lower than in the past and greenhouse gas emissions can be offset to some extent by operation of incinerators as energy-from-waste plants. Properly regulated recycling operations have low risks of pollution but are not completely immune; for example, paper recycling generates wastes of its own from the removal of ink and adhesives from the recycled materials. Unregulated recycling activities often occur on toxic waste dumps in less developed countries, especially those where electronic wastes are dismantled. These can lead to ill health, particularly among the workers at such locations.

Other major sources of waste materials are domestic sewage and industrial wastewater, which are treated, together with contaminants contained in urban runoff, at sewage and wastewater treatment plants. Filtered, digested and dewatered solids (sewage sludges) are mainly spread on agricultural soils as an organic fertiliser with a smaller fraction disposed of, mainly via incineration (Table 1.3). Agricultural use raises concerns about the input to soils of toxic elements that have entered the sewage treatment plant, mainly in industrial wastewater or urban runoff {→metals}. The liquid fraction (filtrate) is emitted to surface waters following microbial treatment in filter beds or activated sludge units and, in some cases, disinfection by ozone, chlorine or ultraviolet light treatment. Sewage effluent is itself a potential source of pollution, particularly in cases of failing treatment systems or in stormy weather, when combined sewer overflows may discharge sewage as well as stormwater into rivers or coastal waters. In addition to metals, example pollutants in sewage effluent include phosphate and other nutrients {→P}, hormones, hormone disruptors and respirable organic matter {→C}.

Table 1.3 Destinations of sewage sludge in the UK, 2010.

Destination	*Tonnes*	*Percentage*
Agriculture	1,118,159	79
Incineration	259,642	18
Landfill	8,787	1
Other (not defined)	26,248	2
Total	1,412,836	100

Source: DEFRA (2012).

Life-cycle analysis: mining, processing and manufacturing

The focus in the sections above has been on direct emissions of pollutants from human activities; however, when considering the polluting potential of such activities it is important to include all stages that might give rise to pollution. This was hinted at in the case of power generation (above), and transportation can be used as another example. In this case, in addition to direct exhaust emissions, a full consideration of pollutant emissions would also include indirect pollution from associated activities such as vehicle manufacture, including the mining and processing of raw materials, transportation infrastructure (e.g. roads and railways) and the extraction and processing of the fuel required to power vehicles.

Consideration of all the direct and indirect environmental impacts, including pollution, that are associated with a manufactured product (or an anthropogenic process) is termed 'life-cycle analysis' (LCA), based on a 'cradle-to-grave' scrutiny (Fig. 1.5). As such, analysis of pollution arising from a manufactured product should include:

(i) the extraction and processing of the raw materials contained in the product;
(ii) the manufacture of the product from the raw or processed materials;
(iii) the transportation involved in the processing and distribution of the product;
(iv) the waste management processes required in the final disposal or recycling of the product;
(v) the fuel used throughout the life cycle of the product, as required for extraction, manufacture and/or transportation.

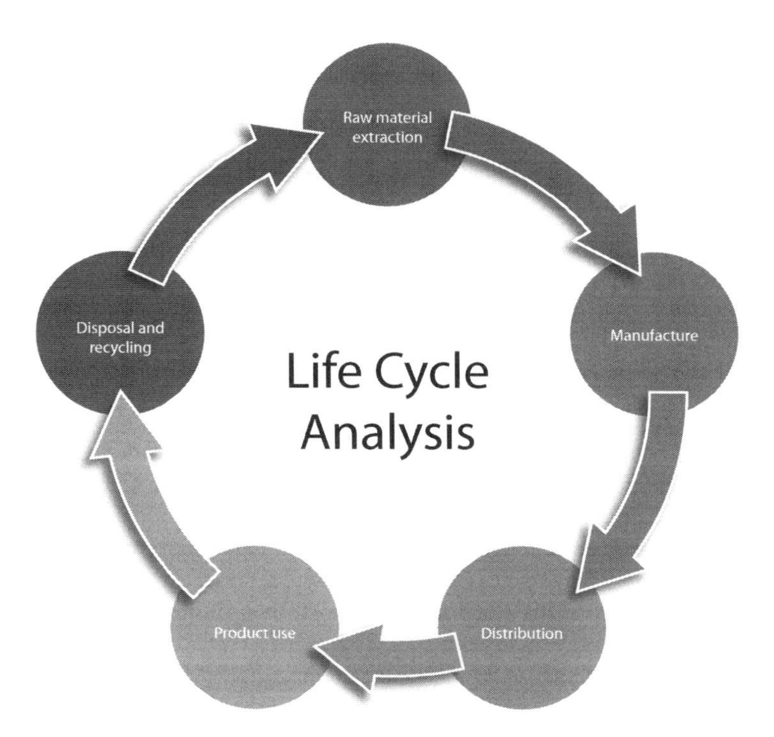

Figure 1.5 Life-cycle analysis.

Points (iii)–(v) have already been considered and now we need to turn our attention to pollution caused by mineral extraction/processing and product manufacture.

The extraction and processing of minerals from the Earth's crust typically involves the crushing of ores to a small grain size to facilitate separation of the desired mineral or element from the host rock. Separation, by techniques like froth flotation, generally entails the use of liquids such as xanthates and dithiophosphates. Crushing and separation raise the possibility of particulate transport to the atmosphere (with subsequent deposition to soils) and to local waterways. The waste materials left behind, in waste heaps and/or tailings ponds, are also potential sources of atmospheric and waterborne particulates and these are likely, especially in metal mining areas, to contain residual toxic elements and cyanides or cyanates. Leaching of rainwater through mine wastes and mine workings leads to the generation of acid mine drainage, contaminating waterways with acidity, metals and ochreous (iron oxyhydroxide) deposits {→S}; this can also occur in coal mining areas because of the presence of metals and sulfides in coal. In metals processing, sulfide ores are smelted to drive off sulfur and concentrate the residual metals; this creates atmospheric metal and metalloid pollution together with SO_2 emissions {→metals →S}. The extraction of copper, nickel and uranium is sometimes accomplished by acid leaching of ore bodies; waste heaps are similarly treated and cyanide is sometimes used to extract gold and silver from ore tailings. Clearly, such techniques pose potential water pollution risks, particularly if sound pollution control measures are not in place and fully adhered to. There has been a large increase in the mining of rare earth elements (REE) in recent years, particularly for use in electronic goods; because the REE ores often contain uranium and thorium, mildly radioactive wastes can be left behind.

Product manufacture, by its very nature, relates to more specific processes than those considered so far; specific pollution problems arise from individual industries. Emissions from such industries, to air and water, are regulated in most countries, but this does not guarantee that pollution will not occur. Table 1.4 lists major product categories and manufacturing processes together with their associated elements, which may be released if pollution occurs. The table refers to *current manufacturing processes* so, for example, lead, which can still contaminate drinking water that flows through old lead pipes, is not shown in this case to be associated with plumbing because the use of lead for this purpose has now been phased out.

Other sources of anthropogenic pollution

It is important to note that some types of anthropogenic pollution do not arise from *routine* pollutant emissions of the kinds described above. Some of the most notorious instances of pollution have resulted from industrial accidents, typically caused by human error or neglect. Particularly notable examples are the industrial disaster at Bhopal, India in 1984,[1] various oil spills in the world's oceans {→C} and major nuclear accidents at Chernobyl, Fukushima and elsewhere {→U}. Furthermore, human activities that are not typically associated with direct pollution risks can sometimes lead to damage; illustrative cases in this regard are agricultural irrigation projects in the USA leading to incidences of selenium toxicity {→Se} and the sinking of drinking water wells in Bangladesh that resulted in widespread arsenic poisoning {→As}. Pollution can also occur as an unintended side effect of warfare and human conflict; examples include the use of organochlorine herbicides by the US military in the

Table 1.4 Manufacturing products and processes and associated elements of environmental pollution.

Product or manufacturing process	*EEPs associated with manufacture*
Alloys	Cd, Cu, Mo, Ni, Pb, Se, V, Zn
Batteries	As, Cd, Hg, Ni, Pb, Zn
Cabling and wiring	Cl, Cu, Pb
Chemical industry	C, Cl, Cr, Fl, Hg, S
Cleaning products	B, C, Mn, N, P
Coinage	Ag, Cu, Ni
Construction materials	C, Cu, Fl, Mo, Pb, S
Cookware	Ag, Fl, Pb
Cosmetics	Cr, Fl, Hg, Ni, Zn
Disinfectants	Br, C, Cl
Drilling fluids	Br, S
Dyeing	Cr
Electronics	Ag, As, Cu, Fl, Hg, P, Tl
Electroplating	Cd, Cr, Ni, Zn
Fertiliser manufacture	Fl, N, P
Flame retardants and fire-fighting	B, Br, Fl, P
Food processing	P, S
Glass and optics manufacture	B, Cr, Pb, Se, Sn, Tl
Jewellery	Ag, Ni
Leather/tanning	Cr
Lighting	Hg
Medical uses, including equipment	As, C, Hg, N, Pb, Tl
Packaging	C, Cl, Fl
Paint	C, Cd, Cr, Cu, Zn
Pesticide manufacture	Ag, As, Br, C, Cl, Cr, Cu, Fl, Hg, P, S, Se, Sn, Tl, Zn
Plastics and stabilisers	C, Cd, Cl, Hg, Pb, Zn
Plumbing	Cl, Cu
Polyurethane foam	Br, Cl
Refrigerants	Cl, Fl, N
Rubber	C, Cl, F, Pb, S, Zn
Solar voltaic panels	Cd, Se
Solvents	C, Cl, Fl
Steel and superalloy manufacture	Cr, Fl, Hg, Mn, Mo, Ni, V, Zn
Textiles	C, N
Weaponry	N, P, U

Note: The table does not include complex, assembled products (e.g. motor vehicles) but the raw materials used to make their constituent parts (e.g. steel, plastics, glass, rubber). Carbon and chlorine are generally listed in relation to industrial organic compounds. Pollutants (especially COx, NOx and SOx) also arise from fossil-fuel usage in virtually all manufacturing processes. EEPs, elements of environmental pollution.

Vietnam War of the 1950s–70s (to clear forested terrain) and the dumping of oil in the Persian Gulf and firing of oil wells by the Iraqi military in the Gulf and Iraq Wars of 1990–91 and 2003, respectively (Fig. 1.6).

Humans can also be exposed to pollutants in the indoor environment. Some of this exposure can be attributed to transfers of exterior pollutants indoors (considered under transport pathways, below), but in some cases the pollutants have an internal origin; for example, the polybrominated diphenyl ethers (PBDEs) present in soft furnishings and electronic devices {→Br}. Other direct indoor pollutants include NOx (gas cookers and domestic heaters), CO (incomplete combustion sources, including domestic heaters) and formaldehyde (plywood, paints and varnishes).

Figure 1.6 Burning oil wells in Kuwait during Operation Desert Storm, 1991.
Credit: US Department of Defense (Wikimedia Commons).

Pathways and environmental fate of pollutants

The previous section described the major sources of pollution and illustrated some of the primary environmental receptors of the various pollutants. A full understanding of pollution science, however, requires consideration of the transport routes or 'exposure pathways' that convey pollutants from their sources to all possible physical and biological receptors (not just primary receptors). This source–pathway–receptor concept is known as 'the pollutant linkage', the pathway providing the link between source and receptor (Fig. 1.7). This is a simple but important concept in pollution *control*, which aims to break the linkage by removing the source, pathway and/or receptor. The main pollutant pathways are shown in Table 1.5.

The pathway between a source and an ultimate receptor may be very short in time and/or distance; for example, the exposure of surface soil organisms to pesticides that have been applied to soils by a farmer. In many cases, however, transport pathways are longer and/or more complicated. For example, many atmospheric pollutants are sufficiently long-lived to be transported around the planet before deposition to a secondary receptor such as a soil or water body; in illustration of this, toxic metals and other persistent pollutants are routinely found by scientists in polar ice cores, deep sea sediments and other remote locations. Despite understandable concerns about such findings, the highest concentrations of pollutants occur near to their sources, before there has been much opportunity for dispersion and dilution to occur in air, water or soil; as such, the most consistently polluted areas tend to be in urban and industrial zones and in regions of intensive agriculture.

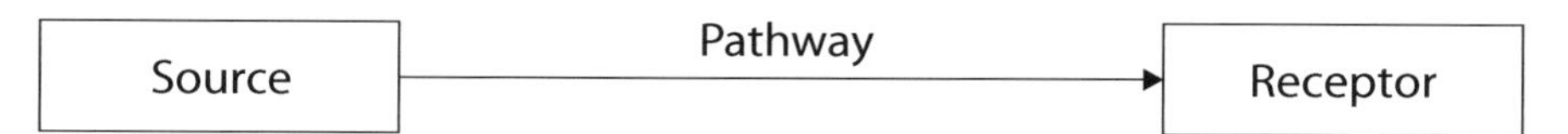

Figure 1.7 The pollutant linkage.

Table 1.5 Pollutant pathways and main driving mechanisms.

Pathway	*Mechanism*
Atmosphere → soils	Wet and dry deposition (most atmospheric pollutants)
Atmosphere → waters	Wet and dry deposition (most atmospheric pollutants)
Soils/solid wastes → waters	Surface runoff (e.g. nutrients, metals, PM); leaching (soluble pollutants)
Waters → soils	Inundation of floodplains and other areas by overflowing, polluted rivers
Freshwaters → seawater	River flow (most aqueous or liquid pollutants and fine PM); deep groundwater seepage (leachates)
Surface (soils and waters) → atmosphere	Degassing (e.g. Hg, NO_x); volatilisation (e.g. organic solvents); wind blow (PM); combustion (various)

Pollutant transport in soils

Soils act as retainers for many pollutants, holding them temporarily or permanently against further environmental transport. Whether pollutants are retained in this way or are released into the air or waters flowing past and through the soil depends on a number of factors. In some cases, pollutants may enter soils in particulate form. Particles on the soil surface are prone to aeolian (wind) and fluvial (runoff) transport unless they move down into the soil column via gravity, perhaps assisted by percolating water. If the particulates are organic in nature they will undergo biodegradation by soil organisms. Otherwise, the fate of particulate pollutants in soils depends in large part on their solubility. Insoluble inorganic pollutants will remain indefinitely as part of the soil matrix. However, many particulates will undergo some degree of **dissolution**,[2] releasing aqueous pollutants into soil porewaters where they will undergo the processes described below.

There are a number of potential pathways in soils for pollutants released from particulate matter or entering the soil in liquid or aqueous form (Fig. 1.8). Volatile liquids such as petroleum and some organic solvents will, to some degree, evaporate from the soil surface or diffuse out of the soil into the atmosphere. Non-volatile pollutants will generally exist in soil porewaters in aqueous form and will be transported from the soil, either in **runoff** or downward percolating water (**leachate**), unless they become temporarily or permanently attached to solid soil particles. The latter may occur via two main processes: **adsorption** (including ion exchange and specific adsorption) and **precipitation.**

Adsorption describes the movement of dissolved ions from soil porewaters to '**ion exchange**' sites on the surfaces of individual soil particles where they are held, at least temporarily, against further leaching. Adsorption occurs via the attraction between the opposite charges of a pollutant ion and an ion exchange surface site. In soils of the world's temperate zones, ion exchange sites are mainly negative, the charges being derived from the structure of the prevalent 2:1 type clays (giving rise to 'permanent charge') and from dissociation of hydroxyl groups in clays and organic matter as soil pH increases (resulting in non-permanent, 'pH-dependent charge'). Therefore, positively charged ions (cations) tend to be preferentially adsorbed and retained in the soil, while anions (repelled by like negative charges on soil particles) are more prone to leaching (see Fig. 1.9). The implications for the fate of pollutants are clear; for example, in temperate soils cationic toxic metals (e.g. Cd^{2+}, Cr^{2+}, Cu^{2+}, Hg^{2+}, Ni^{2+}, Pb^{2+}, Zn^{2+}) and radionuclides (e.g. Cs^{+}, UO_2^{2+}) are far more likely to accumulate and cause 'contaminated land' problems than anionic pollutants like nitrate (NO_3^-), which

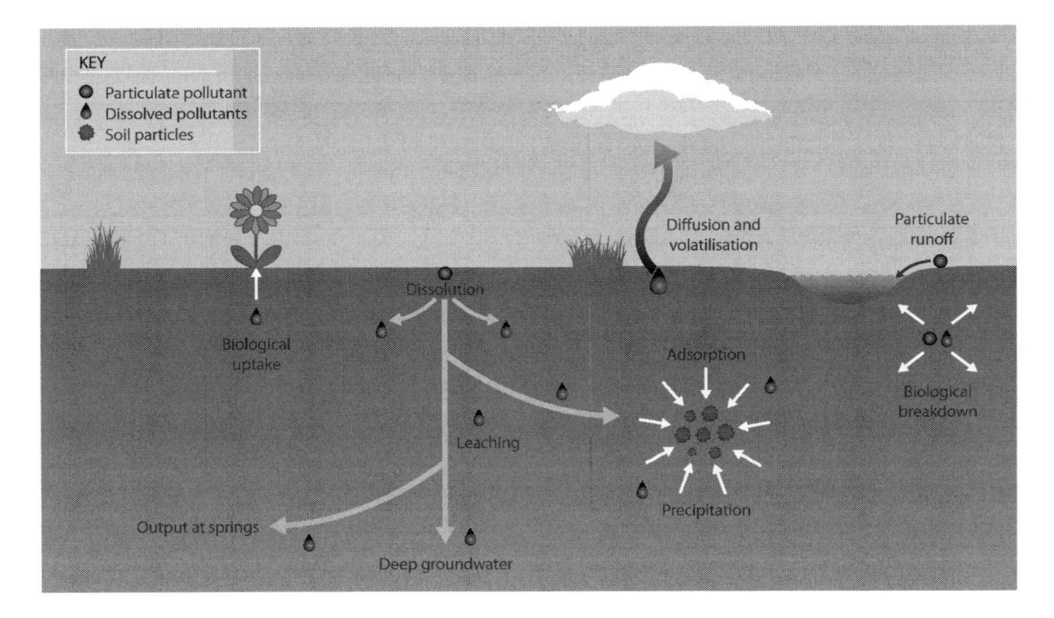

Figure 1.8 Pollutant transport and fate in soils.

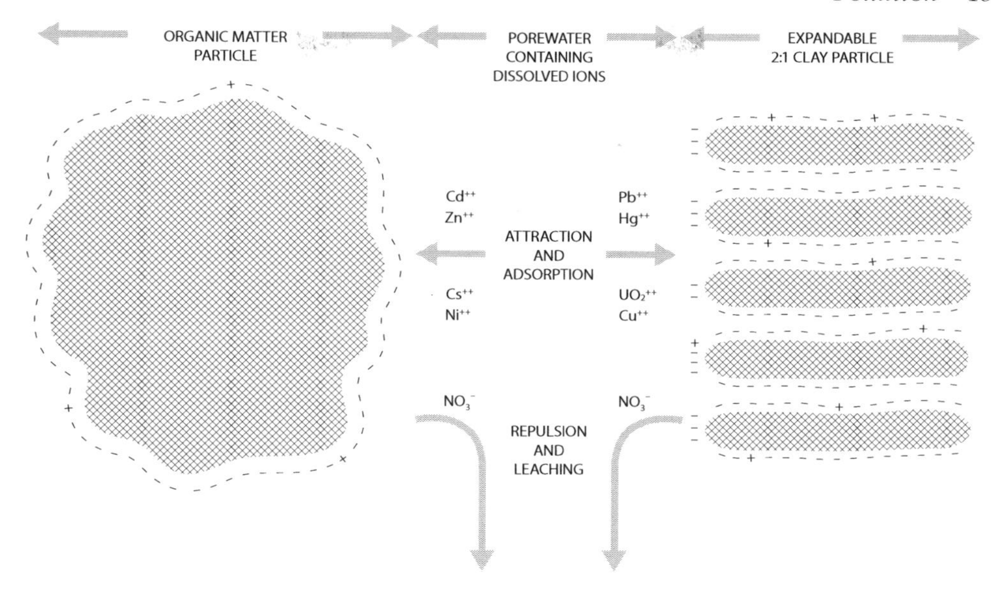

Figure 1.9 The effect of soil cation exchange capacity on pollutant fate. The image is representative of a temperate zone topsoil containing organic matter and 2:1 type clay minerals. Many tropical soils have higher numbers of anion exchange sites (positive charge). Positively charged cations are attracted to the cation exchange sites and are more likely to be retained in the soil compared with negatively charged anions, particularly nitrate, NO_3^-, which is prone to leaching out of the soil and contaminating surface and underground freshwaters. Note the large interior surface area (and therefore increased cation exchange complex) of the clay particle, which is expandable in moist conditions as water enters between the individual layers.

leach readily into groundwaters and surface waters, leading to water quality concerns {→N}.

The capacity of a soil to adsorb cations is known as its **cation exchange capacity** (CEC); soils also have an anion exchange capacity but this typically only becomes important in the most heavily weathered tropical soils. Acidic soils generally have a lower *effective* CEC because their adsorption sites are preferentially occupied by acidic (H^+) ions; therefore, there is usually an inverse relationship between soil pH and the mobility of cationic pollutants. Exceptions to this rule are pollutant elements that are typically present in soils in anionic form (e.g. arsenic and selenium). Acidic soils tend to have higher concentrations of cationic pollutants that are bioavailable; e.g. available to plants via root uptake of soil solution. Pollutant ions may also undergo **specific adsorption,** where the ion is unhydrated and, not being separated from the soil surface by water molecules, may be more permanently incorporated into the soil particle; this is referred to as **inner sphere adsorption,** as opposed to the **outer sphere adsorption** described above.

The fate of an aqueous soil pollutant, i.e. its tendency to accumulate or leach, is also influenced by its propensity to undergo **precipitation** out of solution as a solid; this is the reverse of the dissolution discussed previously and, in the same way, is governed partly by the pollutant's solubility. Other influences on precipitation include the soil's pH and Eh, i.e. its redox potential. For example, in oxic soils (high Eh), aqueous ferrous iron (Fe^{2+}), a common element in most soils, is likely to precipitate as solid ferric (Fe^{3+}) iron oxyhydroxide crusts on the surfaces of soil particles; soil pollutants such as arsenic and toxic metals can undergo 'co-precipitation' within these oxidation crusts and be effectively retained in the soil against leaching, at least unless soil conditions subsequently become anoxic and the crusts dissolve again.

The distribution of soil pollutants between the solid and aqueous phases, governed by adsorption, precipitation or both, is expressed as the soil-water partition coefficient, K_d, which is simply the ratio of the pollutant concentration in the soil's solid matrix to that in the porewater. Soil half-lives, or residence times, of some pollutants are very long because of specific adsorption and/or the insolubility of the typical forms present. Some aerially deposited pollutants are retained mainly in the organic surface horizons of soils; this is partly because of their uptake by plants that subsequently decompose in the topsoil and partly because the pollutants (e.g. copper and lead) are strongly held in the soil's organic fraction by adsorption.

Pollutant transport in the hydrosphere

Water is constantly cycling around the hydrosphere but in the context of pollution it is perhaps useful to start with rainfall. Air pollutants, both gaseous and particulate, can become incorporated into raindrops, and other forms of precipitation like snow and hail, in two ways. First, raindrops, snowflakes and hail crystals form in clouds around tiny particles called 'cloud condensation nuclei' that have entered the atmosphere from both natural and anthropogenic sources. In this way, airborne particulate pollutants are deposited to the Earth's surface upon precipitation. Second, as raindrops fall to the Earth's surface they absorb soluble gaseous and particulate pollutants and, as before, transfer them to the Earth's surface. These two processes are classified as '**wet deposition**', to distinguish them from the '**dry deposition**' of particles and gas molecules, i.e. direct deposition with no aid from precipitation (Fig. 1.10). In rainy conditions, pollutants that are deposited in terrestrial catchment areas, including by dry deposition before rainfall, will either run off the ground surface into streams and rivers (this surface drainage is itself called 'runoff') or leach through soils and permeable bedrocks unless they are first retained in soils by adsorption or, in the case of less soluble pollutants, precipitation (Fig. 1.8). Vertically transported leachate will carry remaining pollutants into aquifers, while lateral flow, particularly in catchments with impermeable bedrocks (i.e. those with few vertical fissures), will transfer them to surface waters via springs.

In catchments, the amount of pollution transported in surface and groundwaters depends in large part on the prevalence of pollutant sources in the area and here a distinction can be made between **point** and **diffuse sources**. The former refers to a specific discharge point; for example, an effluent pipe discharging pollution from a factory into a river. Diffuse pollution is more difficult to control, because it derives from pollutants that have been spread across a wide area of land (e.g. agrochemicals and mine waste) followed by their transport into watercourses via runoff and/or leachate. In a similar way, urban areas are sources of diffuse pollution, because street runoff, including road salt, tyre debris and deposition from vehicle emissions, is transported into storm drains and local watercourses.

How far pollutants are transported within catchments (Fig. 1.11) depends on several factors, including the nature of the pollutant and the characteristics of the

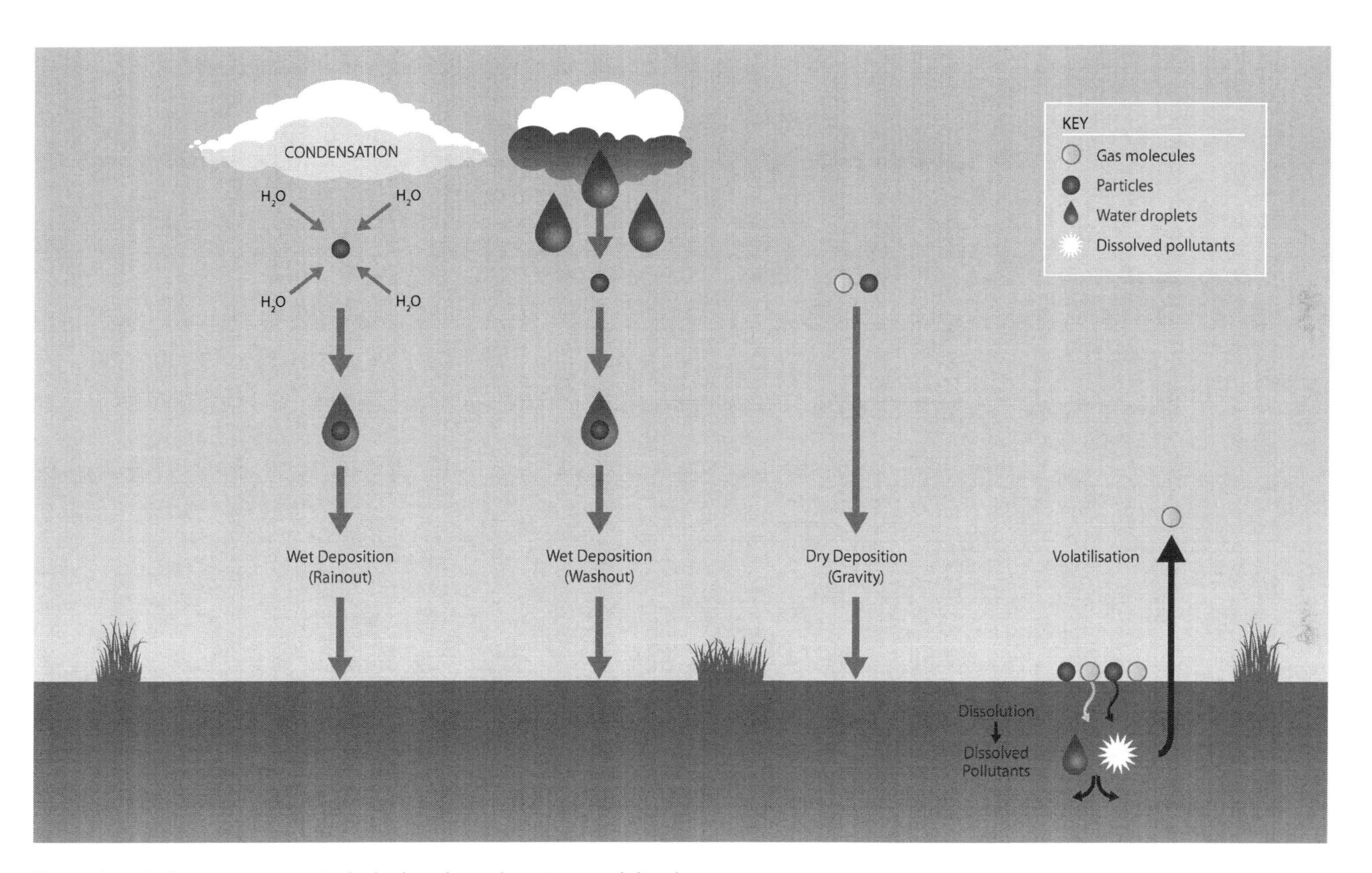

Figure 1.10 Pollutant transport in the hydrosphere: deposition and dissolution.

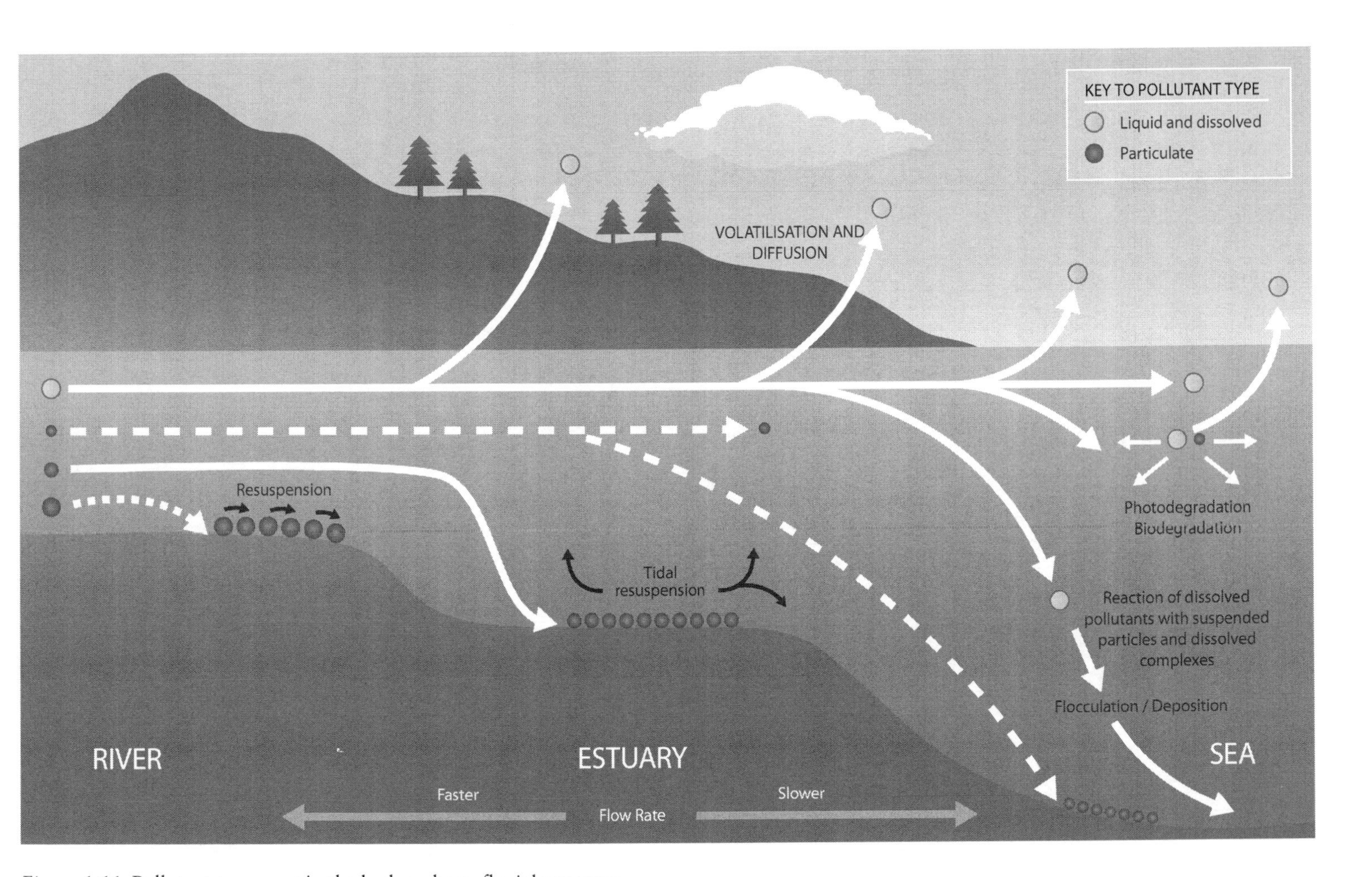

Figure 1.11 Pollutant transport in the hydrosphere: fluvial processes.

catchment. Volatile pollutants may gradually be removed from a river, not by flow but by diffusion into the atmosphere. Dissolved pollutants in rivers are most likely to be transported out to sea, unless they first react with particles or aqueous complexes (see below). Particulate pollutants are either deposited into the bottom sediments or remain suspended for some time, depending on the particle size and density and the strength of stream flow. Flow rates typically fall in estuaries, where topographical relief decreases and where water volume may be spread out over a wider area than further upstream; therefore settling of many of the pollutant particles that have remained in the stream flow takes place here. Estuarine bottom sediments are, however, prone to ongoing resuspension by the deep turbulence created by incoming tides. Sediments are disturbed further by anthropogenic activities like dredging.

Sediment particles, both suspended and settled, are important in the context of their reactivity with aqueous pollutants. The same processes of adsorption/desorption (ion exchange), specific adsorption and precipitation/dissolution as were noted for pollutants in soils apply to sediments; non-pollutant sediment particles, particularly of clay and organic matter, can act as pollutant carriers. Iron and manganese oxides are also important for the distribution and re-distribution of trace elements between the aqueous and dissolved phases, because they precipitate (usually as coatings on particles) and dissolve depending on prevailing conditions, thus 'scavenging' (co-precipitating) and releasing aqueous pollutants as they do so. Resuspension of settled particles into the water column effectively increases the chances of sediment–pollutant interactions.

Non-particulate components of river and estuarine waters are similarly important in pollutant geochemistry. For example, dissolved organic matter (DOM) can form complexes with toxic metal ions and other aqueous pollutants, preventing their reaction with solid particles and therefore keeping them in solution, but also away from reaction with other aqueous components, at least temporarily. Another relevant process in this context is flocculation, which aggregates a number of dissolved complexes, like those just described, into small particles. Inorganic complexes are also important; for example, in saline waters chloride ions will also complex cations (like Cd^{2+}), again keeping them in the dissolved phase. Hard water, which contains elevated levels of calcium (Ca^{2+}) and magnesium (Mg^{2+}) ions, is thought to mitigate the toxicity of toxic cations; for example, these nutrient ions can occupy cation exchange sites on fish gills and therefore decrease the likelihood of toxic metal absorption.

As in soils, the processes described above are influenced, for particular pollutants, by the prevailing pH and Eh (redox) status and also, in estuarine and coastal waters, by salinity. For example, in tidal estuaries pH and salinity increase as seawater mixes with the freshwater; Eh decreases with lowered surface water turbulence and organic matter (OM) enrichment because, on the one hand, less oxygen dissolves into the water and, on the other, OM biodegradation (microbial respiration) increases demand on the water's dissolved oxygen {→C}. Increases in pH will generally promote the adsorption, to particles, of aqueous cations, including toxic metals like Cd^{2+} and Pb^{2+}, but not of anionic pollutants such as nitrate (NO_3^-) and arsenate (AsO_4^{3-}) (see 'Pollutant transport in soils'). Redox status is also important; for example, in organic-rich, anoxic and fine bottom sediments, potentially toxic metal cations react with hydrogen sulfide, a common product of such sediments, to form insoluble metal sulfides.[3] These are just a few simple examples; in reality the geochemistry of the estuarine environment is complex. However, in summary we can see that estuaries and coastal waters are significant zones of pollutant mixing, reaction and distribution between solid and dissolved phases, with implications for pollutant transport (when considering particles and DOM as vectors of pollution), mobility and bioavailability. The distribution of pollutants between the dissolved and particulate phases will determine whether they are more likely to be carried out to sea or retained in bottom sediments, respectively.

The residence time of water bodies (Table 1.6) is important in understanding the likely fate of pollutants in the hydrosphere. For example, fast-flowing rivers can flush dissolved pollutants away relatively easily (although many pollutants can persist in sediments). Where pollutants are transported into water bodies with longer residence times, such as aquifers (groundwater reservoirs) and lakes, such rapid dilution and dispersion mechanisms do not operate. The leaching of persistent pollutants like metals and POPs from landfill sites, for example, may contaminate groundwaters for extended periods of time. Pollutants in aquifers generally move slowly under a hydraulic gradient as water moves from areas of high to low hydraulic head (pressure). Flow speed is also dependent on hydraulic conductivity, based on the porosity and permeability of the aquifer, and the rate of pollutant movement is further controlled by the pollutant's solubility and propensity for adsorption to solid surfaces in the aquifer.

The retention of pollution in lakes with long residence times is exacerbated by the thermal stratification of many lakes (Fig. 1.12) into a relatively warm upper layer (the epilimnion, 'top lake'), a colder, denser, and often hypoxic and nutrient-rich, bottom layer (the hypolimnion, 'bottom lake') and a dividing layer (the metalimnion, 'middle lake'); the stratification leads to limited mixing of the lake waters and therefore less chance for dilution and dispersion of pollutants and hypoxic water. Temperature decreases with depth in the metalimnion giving a distinct temperature gradient (the thermocline). The epilimnion can be well mixed by wind-generated surface currents but little mixing may occur between the other layers except in temperate zones; in such areas, seasonal lake turnover can occur as surface water cools to 4°C (its temperature of maximum density) and sinks, allowing hypolimnetic waters to rise and the lake water to become more mixed. Pollution studies of the Great Lakes of North America give a good illustration of the problems that can occur in large lakes (e.g. →Hg). In the oceans the thermocline (zone of greatest temperature decrease) is generally found between approximately 100 and 1000 m depth; above and below this zone, temperature change is less pronounced.

Aqueous and complexed pollutants, and those bound to the finest particulates, are transported via rivers to the oceans. Dissolved pollutants may also reach the oceans via direct input from deep groundwaters, but this is likely to be a small amount in

Table 1.6 Residence times of water in selected reservoir types; representative figures.

Water body	*Typical residence time (years)*
Atmosphere (air moisture)	3×10^{-2}
River	5×10^{-2} to 5×10^{-1}
Large lake	1×10^{2}
Sea	4×10^{3}
Deep groundwater (aquifer)	1×10^{4}

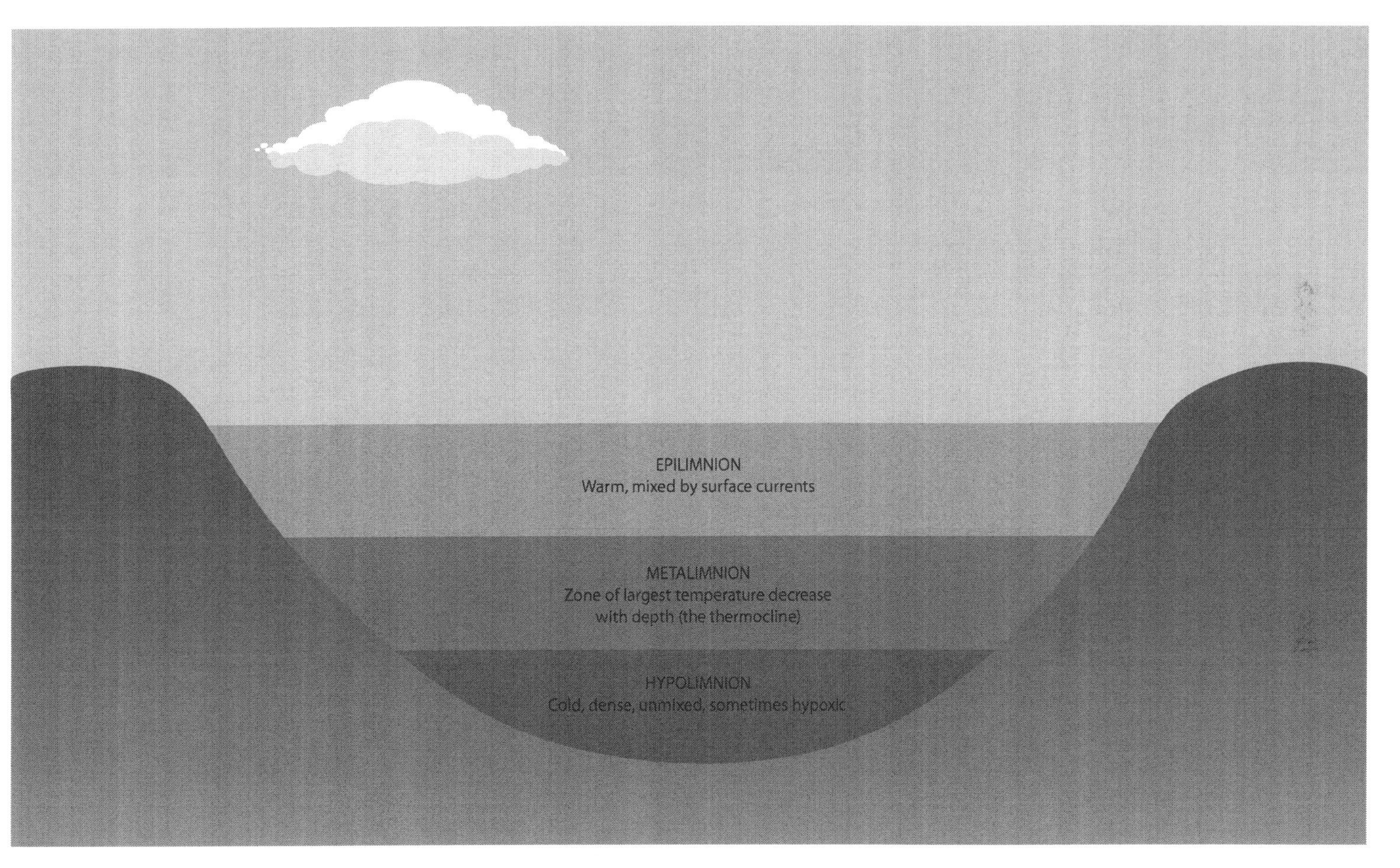

Figure 1.12 Thermal stratification of a lake. This stratification breaks down, allowing greater vertical mixing, if the epilimnion cools sufficiently (see text).

comparison to rivers. Riverine inputs to the oceans add to direct pollutant deposition from the atmosphere and from localised sources such as oil leaks and tank washing. Oil slicks at sea will gradually be removed from the water column by a mixture of volatilisation, biodegradation and sinking of residual material to the seabed. Many pollutants in seawater will undergo sinking as distinct particles, if they have sufficient density to settle to the seabed, or by attachment to such particles. Bottom sediments in coastal areas have notably higher concentrations of pollutants than deep-sea sediments but there are detectable anthropogenic signals in the latter. Pollutants in the dissolved phase, or those associated with less dense particles, will, if not biodegraded or photodegraded, persist in the water column, including the uppermost surface waters that are richest in marine life. These pollutants include toxins, like metals, and nutrients, such as nitrate. Despite the vast size of the oceans and the relatively large scope for dilution and dispersal of pollutants (away from coastal regions at least), the serious nature and persistence of some pollutants in seawater is illustrated by the large accumulations of marine litter (mainly small particles of plastic) in oceanic gyres, observed in the North Pacific and elsewhere.

Pollutant transport in the atmosphere

Atmospheric pollutants are in either gaseous or particulate form and, for both, there are point and diffuse sources. Point sources include chimney and exhaust emissions (Fig. 1.13a); for example, CO_X, NO_X, SO_X and particulate matter from power stations, metals from smelters and POPs from waste incinerators. Diffuse sources include diffusion of ammonia from farm wastes spread across fields, aeolian movement of mine waste particles to adjacent areas and methane emissions from landfill sites and livestock (Fig. 1.13b). Atmospheric transport can expose plants, animals and humans to harmful pollutants; in the case of humans, the pollutants may be transported into homes through open windows (although an additional, non-atmospheric route into homes is the 'tracking in' of particulate pollutants on footwear, clothing and pets).

The period of time that a pollutant remains in the atmosphere, and therefore the distance it travels, is dependent on: (i) the nature of the pollutant and (ii) meteorology. Gases are the pollutants most likely to be held aloft for longest, prior to wet or dry deposition (Fig. 1.10). Particles, which have much larger masses than gas molecules, are far more prone to deposition and this tendency increases with particle mass. Many pollutant particles, particularly those formed in high-temperature combustion processes, are classified as fine particulate matter (<10 μm mass median diameter, known as **PM_{10}**) or ultra-fine particulate matter (<2.5 μm MMD, known as **$PM_{2.5}$**) and such small particles can have atmospheric lifetimes of weeks or months {→C}. The most important meteorological factors influencing pollutant transport are wind speed and direction (indicating **advection**, the lateral movement of air), **convection** (the vertical movement of air) and precipitation, which increases the likelihood of wet deposition (Fig. 1.10). Advection carries pollutants away from the source, dispersing and diluting them, but potentially transporting them regionally or even globally and, in some cases, to remote locations like the Polar or mid-ocean regions. Convection is influenced by ground surface temperature and regional pressure systems. A warm surface conducts absorbed solar energy to the air at ground level, which then rises (convects). Conversely, cold surfaces can produce **temperature inversions**, where the air is colder at ground level than

Figure 1.13 (a) Smoke stack in Leicestershire, UK, a point source of air pollution. (b) Grazing livestock, a diffuse source of air pollution. Credits: (a) Matt Neale; (b) David H-C (both Flickr).

at higher altitude and does not convect, cold air being denser than warm air. Inversions can also prevent the upwards movement of individual parcels of polluted air *within* the atmosphere; understanding this requires an explanation of **lapse rates** (below).

A parcel of air (polluted or otherwise) will rise as long as it is warmer than the surrounding atmosphere. As it does so, it will expand (because of the atmosphere's decreasing density with altitude) and cool as a result of doing 'work' to expand; this cooling may be more easily understood when considering what happens in the opposite case of a gas being compressed by having work done on it and warming in consequence. The drop in temperature of a rising, dry air parcel is known as the dry (or unsaturated) **adiabatic lapse rate**[4] (ALR) and is constant at 9.8°C km^{-1} (Fig. 1.14a). In a *moist* air parcel, water vapour condenses into liquid droplets as the parcel rises and cools. The condensing moisture releases 'latent heat', which warms the air parcel; its temperature decrease with altitude (its ALR) is therefore lower than if no condensation had occurred. This gives a smaller 'wet' (or 'moist', or 'saturated') ALR for moist air, which is variable depending on the moisture content; a typical value is 5°C km^{-1} (Fig. 1.14a).

A completely separate type of lapse rate is the **environmental lapse rate** (EL; Fig. 1.14b), which is simply a measure of the temperature change of the atmosphere with increasing distance from the ground surface; note that the ELR relates to the wider atmosphere at a given location, *not* to an individual air parcel rising within that atmosphere. The upwards movement of a parcel of polluted air emitted at ground level depends on the ELR. The parcel will rise if it remains warmer than the surrounding atmosphere (Fig. 1.14b), but not if it is colder (Fig. 1.14c). If the ground becomes cold (e.g. on a clear night or on a cold winter's day), a temperature inversion may form. Pollution emitted at ground level may begin to rise in some cases, but will ultimately be trapped under the inversion (Fig. 1.14d).

In cold weather conditions (in winter or in perennially cold regions), trapping of polluted air at ground level is most likely to happen when high-pressure conditions occur; clear skies with no cloud blanket, particularly at night, allow daytime ground heat to radiate upwards, subsequently cooling the ground and the air in contact with it. Compared with low-pressure conditions, high pressure is also characterised by relatively low wind speed and precipitation. Therefore, such high pressure conditions increase the likelihood of ground-level pollution accumulating because pollutants are less likely to be removed by adiabatic processes, convection, advection or wet deposition. The presence of a high-pressure system increases the risk of a **'winter smog'** forming; this is characterised particularly by high levels of PM_{10} and SO_2, which are emitted by the burning of fossil fuels for heating and other purposes. The most famous case of a winter smog or 'London smog' (sometimes also called 'classic smog') occurred in London, UK in December 1952. A stationary high-pressure system covered southern England at this time, giving rise to cold, still and foggy air for several days, during which the smoke from thousands of domestic chimneys accumulated and mixed with the fog droplets to give a 'smog' (smoke + fog). Thousands of Londoners died over the following weeks from the deep inhalation of fine particulates and gaseous sulfur dioxide (SO_2) that had also dissolved into the fog droplets giving a weak sulfuric acid {→S}. Over the following decade Clean Air Acts were passed in the UK and London's air improved significantly, but 'pollution episodes' still occur in the UK and elsewhere when weather conditions conspire; particularly serious winter smogs have developed in China in recent years {→C} (Fig. 1.15).

In warmer climates and seasons, stable air masses associated with high-pressure systems can give rise to a different kind of pollution, called '**summer smog**' or 'photochemical smog'. In this case the air stability is not caused by cold conditions (a radiation inversion), but by other types of inversion. A high-pressure system is caused by an air mass pushing downwards towards the Earth's surface as a consequence of global atmospheric circulation patterns. When air rises (in a low-pressure system for example) it expands and cools; the opposite happens when air descends in a high-pressure system – it gets compressed and warmed, just like the air being pumped into the confined volume of a bicycle tyre after being drawn into the pump from a much larger volume of atmospheric air. Descending air can cause a 'subsidence inversion', where the sinking air becomes warmer than the air at ground level, even though the latter may not be particularly cold. This can give rise to problems in sunny, traffic-polluted locations (Fig. 1.16). Such smogs were first widely noted in Los Angeles and are sometimes described as 'Los Angeles type smogs'. In colder areas, subsidence inversions often combine with radiation inversions when high-pressure conditions prevail, creating particularly strong temperature inversions.

The pollutants in a summer smog are often derived from vehicle emissions, although other sources can be important too, depending on the location {→N}. Many vehicles emit **primary pollutants**, particularly NO_2 and unburnt hydrocarbons, directly from their exhaust pipes and these undergo photochemical (i.e. sunlight-driven) reactions to form **secondary pollutants** including tropospheric ozone; this process is discussed in detail in the nitrogen chapter, but it is worth noting here that secondary pollutants take time to form and peak concentrations can occur many miles from the original primary pollution source (tropospheric ozone is indeed classed as a regional pollutant, often affecting rural areas).

We have seen that high-pressure conditions can hinder the transport of air pollutants away from the source area and this can be exacerbated further by topography; for example, some of the world's smoggiest cities (including Los Angeles) lie in geographical basins, meaning they are surrounded on most or all sides by high ground, which further hinders the dispersion of pollutants in still and stable air conditions. Other factors can also be important in the formation of temperature inversions; for example, 'marine inversions' occur where sea breezes bring in a ground layer of cold air and 'mountain slope inversions' involve the flow of air down a slope, followed by its compression and warming above relatively cool air in the valley below.

The foregoing discussion of smogs has focused on the inhibition of pollutant transport and dispersal and this is important because it explains why some cities and regions regularly suffer episodes of gross air pollution. In most circumstances however, pollutants are dispersed and carried away by regionally or globally circulating air masses, driven by the pressure gradient force, which moves air masses from high- to low-pressure areas in a large-scale, curved or circular motion because of the Earth's rotation (the Coriolis effect). Gaseous pollutants that remain aloft and are sufficiently resistant to breakdown accumulate in the global atmosphere and may have effects there over long timescales, the long-term warming effect of CO_2 being an obvious example. Long-range transport also gives rise to transboundary air pollution, where a country and its inhabitants are exposed to pollutants emitted in a different country; examples include regionally transported tropospheric ozone {→N} and acidifying substances {e.g. →S}. Long-range transport also gives rise to deposition of terrestrial pollutants to the sea; these include

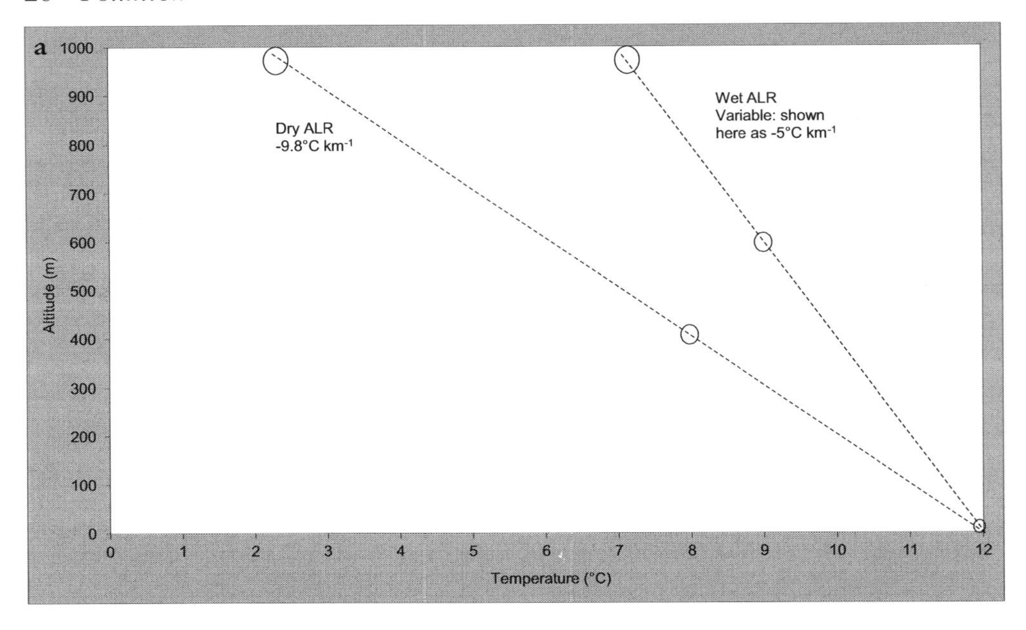

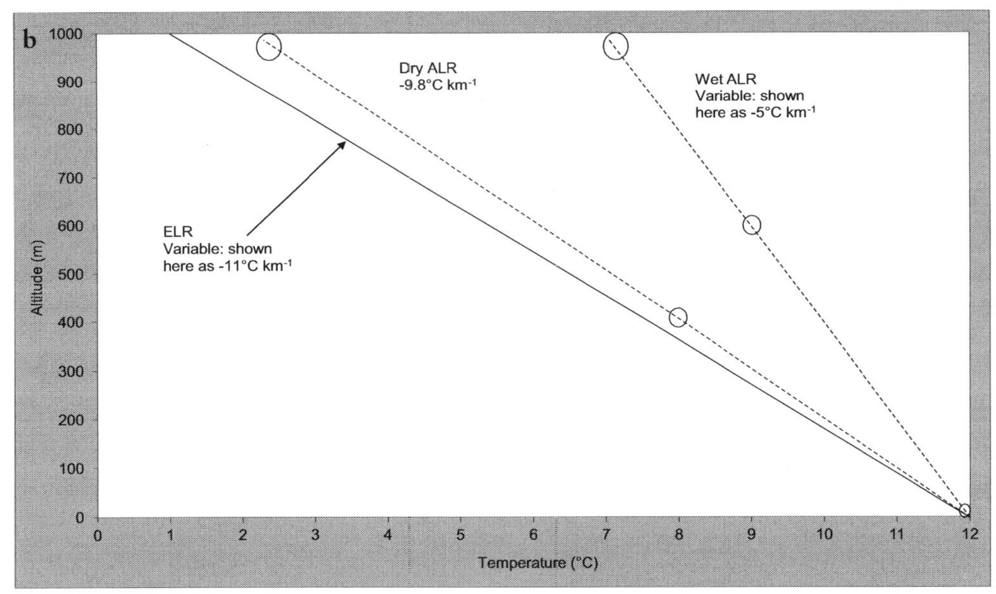

Figure 1.14 Lapse rates. The ground-level temperatures shown for all lapse rates (dotted and solid lines) are arbitrary and have been chosen for clarity of presentation. (a) Dry and wet adiabatic lapse rates. The circles represent polluted air parcels that rise and expand at the DALR or WALR, depending on their moisture content (see text). (b) An environmental lapse rate (solid line) with an illustrative value of –11°C km^{-1}; in this example, a parcel of polluted air with a temperature of 12°C, emitted at ground level, and following the track of either of the ALRs, will continue to rise because it will remain warmer than the surrounding atmosphere (shown by the ELR) at all altitudes. (c) A smaller ELR of –3°C km^{-1}; in this case the same polluted air parcel will not be able to rise because it is colder and denser than the surrounding atmosphere at all

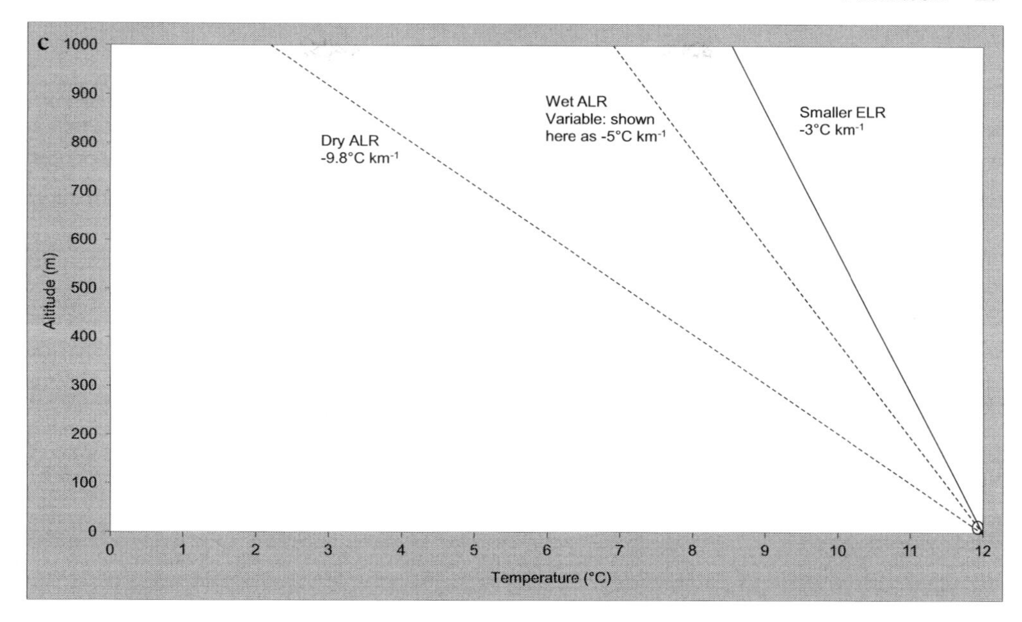

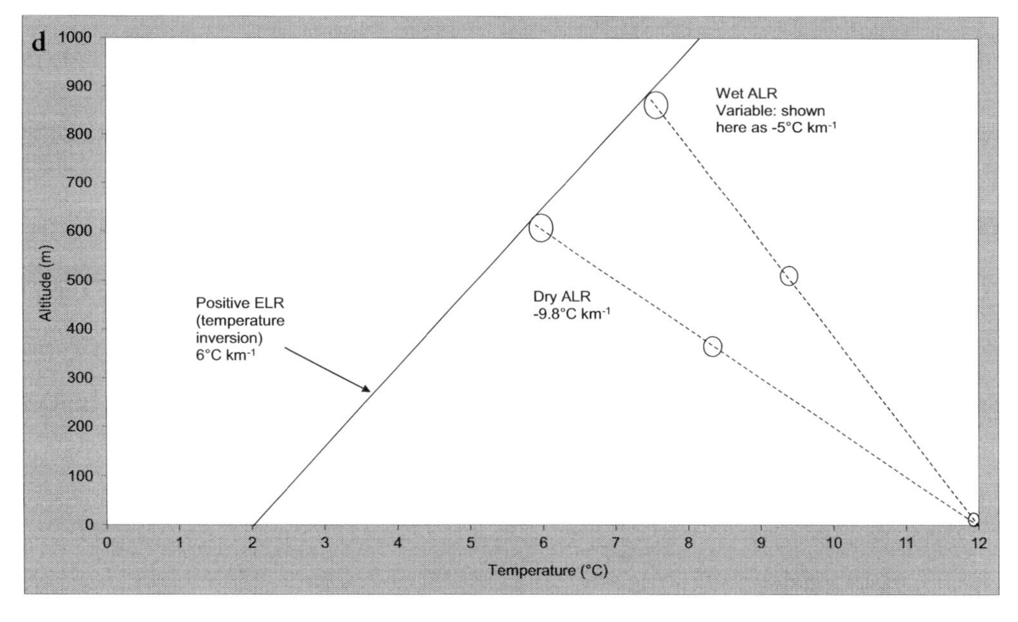

altitudes. It will be effectively trapped at ground level, even though there is no temperature inversion. (d) A positive ELR of 6°C km^{-1} giving a temperature inversion; in this diagram the temperature of the atmosphere at ground level is shown as being colder than previously, representative of a cold, clear night or a winter's day. The polluted air parcel, being in this example initially warmer than the surrounding atmosphere, is able to rise at the dry or wet ALR until it has cooled down adiabatically (because of expansion) to the ambient temperature of the atmosphere, at which point it will stop rising; it is said to be 'trapped' beneath the temperature inversion. Realistically, the rising air parcel is likely to also cool to some degree via heat conduction to the cooler air surrounding it, but this would be in addition to the adiabatic temperature falls shown.

Figure 1.15 Winter smog in Beijing, China.
Credit: Brian Jeffery Beggerly (Flickr).

Figure 1.16 Photochemical smog in Dubai.
Credit: Mahaodeh (Wikimedia Commons).

fine particles, bearing metals and other toxic materials, and soluble gases, including CO_2 which can cause problems of ocean acidification {→C}.

Biological receptors: exposure pathways

The previous sections have considered the physical components of the environment, i.e. atmosphere, soils and waters, as receptors of pollution. However, the ultimate focus of concern is normally the exposure to pollutants of the biological receptors that inhabit these environments (with the exception of some pollutant types, such as greenhouse gases, where the environmental impact is physical in nature, rather than directly biological). The main exposure routes for living organisms are shown in Table 1.7 and the potential impacts of exposure are discussed in detail in the next section.

Table 1.7 Biological exposure routes.

Exposure route	*Most vulnerable groups*
Inhalation of polluted air	Urban dwellers; industrial and urban workers with inadequate protection
Ingestion of contaminated soil and dust	Young children; grazing animals
Ingestion of contaminated foodstuffs	Vegetable gardeners; grazing animals; predators in the food chain
Ingestion of contaminated water	Wildlife, and humans with untreated water supplies, in polluted catchment areas
Skin absorption	Industrial and agricultural workers with inadequate protection; aquatic organisms
Leaf intake	Crops and wild plants growing in polluted areas
Root uptake	Crops and wild plants growing in polluted soils
Microbial uptake	Bacteria, fungi and other micro-organisms in polluted soils and waters

The impacts of environmental pollution

Detailed discussions of the effects of individual pollutants and pollutant types are contained in each of the element chapters. Therefore, what follows in this section is not an exhaustive list but a broad overview of some of the general environmental impacts that are likely to occur in soils, waters and the atmosphere, followed by an introduction to toxicity and ecotoxicity, where the focus is on biological receptors.

The impacts of soil pollution

The deliberate application of agrochemicals to soils helps to provide the food we depend on, but it can have unintended consequences. Fertiliser applications mostly raise concerns about water quality, but excess nutrients in the soil may have effects on plant biodiversity as fast-growing plants become dominant over less competitive species, particularly in nutrient-poor soils {→N}. Nitrogenous fertilisers can also acidify the soil, leading in some cases to the increased bioavailability of toxic elements and the loss (via leaching) of micronutrients. Sewage sludge is often applied to agricultural soils as a fertiliser but can contain high levels of toxic elements {→metals}, as can rock phosphate {→Cd}. Pesticides, the other main type of agrochemical, are typically non-selective and beneficial organisms, including soil micro-organisms, may be depleted in addition to the intended targets {→C →Cl →P}. In urban and industrial areas, soils are subject to pollutant leaks and spills, as well as deposition of pollutants from the atmosphere; the pollutants introduced via these routes, including metals and POPs, can have direct toxic impacts on soil macrofauna and micro-organisms {→metals →Cl}. The retention in soils of radionuclides, deposited following nuclear accidents for example, can render soils unsuitable for use for extended periods of time {→U}.

The impacts of freshwater pollution

Industrial effluents add pollutants to rivers, lakes and other freshwater bodies, and these can have toxicological impacts on aquatic organisms {e.g. →metals →Cl} or physical effects (especially particulate matter {→C}), including increased turbidity (reduced light for photosynthesising organisms) and blanketing of feeding and spawning sediments. Respirable organic matter from specific industries (e.g. paper mills) and sewage treatment plants can increase the biochemical oxygen demand (BOD) in freshwaters, depleting the water of oxygen with consequent impacts on aerobic organisms {→C}. Oxygen depletion is also caused by excess nutrient inputs from fertilised farmland, which, via the process of eutrophication, cause algal blooms that are subsequently biodegraded in the same way as other organic wastes {→N →P}. Pesticide applications to soil can enter water bodies via runoff and leaching, and have unintended toxic effects on aquatic organisms {→Cl →P}. In urban areas, diffuse runoff transports toxic trace elements from various sources to local watercourses, potentially affecting aquatic organisms {→metals}. Urban waterways can also become contaminated by excess salinity (mainly excess chloride ions) from the use of road salt in winter; populations of sensitive aquatic organisms can be reduced as a result {→Cl}. Another major form of water pollution is acidification, either by the aerial deposition of acidifying substances {→N →S} or the acidic drainage of metal and coal mines {→S}; biological effects are caused by the release of toxic metals, particularly aluminium, from soils, sediments and bedrocks within an acidified catchment or stream.

The impacts of seawater pollution

Some of the most well-known and visible forms of marine pollution are oil spills from ocean tankers. Spilled oil has physical and toxic effects on marine wildlife; for example, the matting of seabirds' feathers reduces their buoyancy and preening causes ingestion of the oil {→C}. Gross oil pollution generally arises from accidents, but ships routinely undertake tank cleaning at sea, which can cause toxic impacts depending on the nature of the tank residues. The anti-fouling paints used on ships' hulls can cause imposex in aquatic organisms and fatalities at higher doses {→Cu →Sn}. Marine litter (Fig. 1.17), including plastic material and netting, can strangle and injure marine mammals and other animals. Inputs of riverine pollutants (see above) add to these oceanic sources of pollution. Coastal waters are particularly affected by riverine inputs and other sources such as sewage outflows; the resulting pollution of bathing waters can have direct

Figure 1.17 Plastic debris on the shore of Laysan Island, Hawaii, located near the centre of the North Pacific gyre, where marine litter concentrates.

Credit: US Fish and Wildlife Service (Flickr).

impacts on humans as well as wildlife. Leaks of cooling water from nuclear power plants can expose marine life to radionuclides with their associated hazards, including mutagenicity {→U}. Deposition of atmospheric pollutants into the seas also causes environmental impacts; an illustration of this is the acidification of sea water by increased CO_2 absorption {→C}.

The impacts of air pollution

The most well-known impact of air pollution is undoubtedly global warming caused by anthropogenic emissions of CO_2 and other greenhouse gases {→C →F →N}. In the 1980s and 90s, stratospheric ozone depletion, caused by releases of ozone-depleting chemicals and still occurring today, became equally familiar, being known popularly as the 'ozone hole' {→Cl}. In addition to these global atmospheric pollution concerns is the prevalence of poor air quality, in cities in particular, and the consequent impacts on human and ecological health. There are many local air pollutants, but perhaps of most concern are SO_2, NO_2, O_3 and fine PM (often bearing toxic metals), which directly and indirectly cause human health problems, particularly respiratory and cardiovascular diseases {→C →N →S →metals}. Today the most serious air pollution problems are found in 'newly industrialised' countries, but post-industrial countries continue to experience widespread problems of poor air quality, from vehicle emissions and other sources, and this has deleterious effects on human health, crop plants and ecological receptors.

Principles of toxicity

Some of the most serious impacts caused by human disturbance of elemental cycles and reservoirs are physical in nature, the anthropogenic greenhouse effect being a clear example. However, the main concern in many cases is the direct toxicity of pollutants, often trace elements, that humans and other organisms are exposed to. In this context, it is important to acknowledge that some of the elements considered to be toxic pollutants when present in excess are also essential to organisms at lower concentrations; this is clearly the case for essential (but also potentially toxic) elements like Cu, Se and Zn. Natural deficiencies of such elements in rocks and soils can cause ill health and death, just as excesses can (Fig. 1.18); in areas where essential elements are naturally deficient, their introduction, from what would normally be considered pollution sources, may actually have positive effects. Nonetheless, it should be remembered that, in such cases, essential elements may be accompanied by other, non-essential, toxic substances that may cause problems. Although deficiency is an important problem for many elements (e.g. Se deficiency is more problematic in general than Se toxicity), and is mentioned in some of the element chapters {→As →Cr →Cu →Se →Zn}, the primary focus of this book is pollution and the problems that occur when anthropogenic activities cause elements to be present in *excess*.

When discussing toxicity it is necessary to define some key terms. The presence of an excess amount of a toxic pollutant in the environment is a **hazard**, meaning, in this

Figure 1.18 Chlorotic spots on leaf of *Brassica rapa* indicating copper deficiency.

Credit: Forest and Kim Starr (Flickr).

context, a substance that can cause death, injury or damage to health. Such a pollution hazard poses a **risk** to the environment and its inhabitants, i.e. a possibility of harm occurring as a result of exposure to the hazard. The likelihood of harm occurring on exposure depends on the amount of the pollutant that is taken into an organism, i.e. the pollutant **dose**. A useful phrase here is 'the dose makes the poison', a shortened version of a quote from the Renaissance scientist, Paracelsus and, in this context, the earlier discussion of essential trace elements like Cu and Zn is relevant. Other elements illustrate the importance of the dose; for example, fluorine (as fluoride) can improve dental health at low doses, but causes a skeletal disease called fluorosis if the dose is too high {→F}.

Important concepts associated with the dose are those of **solubility** and **bioavailability**. The aqueous solubility of a pollutant influences its concentrations in drinking waters (accessed by humans and other animals), in soil porewater (accessed by plants, soil micro-organisms and soil invertebrates) and in fresh and marine waters. Bioavailability relates to the proportion of a pollutant taken into an organism that is absorbed by it (the remainder being excreted). This will depend on the solubility of the pollutant within the body of the organism; this includes its solubility in gastric juices (known as its **bioaccessibility**) and in fatty tissues (**lipophilicity**). The bioaccessibility of most toxic metals increases in the acidic conditions of the mammalian stomach, although the metalloid arsenic is more bioaccessible in the small intestine, which has a higher pH. Lipophilicity, which influences how much of a pollutant is absorbed by body tissues, is typically expressed in laboratory studies as the octanol–water partition coefficient, K_{OW}, which is the ratio of the molar concentrations of a chemical dissolved in: (i) octanol, which, like fat, is a non-polar substance and (ii) water, a polar substance. A high K_{OW} for a pollutant therefore indicates lipophilicity and possible bioaccumulation (see below). In aquatic systems (including soil porewaters) a further influence on bioavailability (and toxicity) is the chemical form of dissolved pollutants; those present as free ions (Cd^{2+}, Pb^{2+}, etc.) are generally more reactive within an organism than those contained in dissolved complexes (organic or inorganic).

Higher than desirable doses of pollutants may result from **acute** or **chronic** exposure. Acute exposure refers to a single, large dose of a pollutant (or several exposures over a short period of time) and is typically characterised by symptoms that develop quite quickly. Chronic exposure relates to a series of lower level but continuous doses over a longer period of time, where health problems may not become immediately apparent. In some cases the effects of acute exposure may also be delayed; an example is the development of a tumour resulting from an earlier acute exposure to a carcinogen. It is often difficult to link a subject's exposure to a toxin to a specific health outcome; proving cause and effect in outcomes like carcinogenicity is particularly difficult because of **confounding factors** that may also be associated with the health effect, such as smoking, diet, occupation and exposure to other environmental pollutants or risk factors. Epidemiologists, who study the distribution of diseases and their possible causal factors, seek to account for such confounding factors but are not normally able to eliminate all risk factors with total confidence.

Pollutants can be **synergistic**, where exposure to two or more at the same time causes deleterious effects that are greater than the sum of the effects that would be caused by each pollutant individually. Pollutants and chemical species can also be **antagonistic**, the presence of one ameliorating the effects that the other might be expected to cause when present in isolation. An example of synergism is the respiratory damage caused by the inhalation of fine PM together with SO_2 {→S} and antagonism can be exemplified by the reduction in metal toxicity to fish if Ca^{2+} ions are also present (see above, 'Pollutant transport in the hydrosphere').

A number of terms that are commonly used in ecotoxicological studies also need to be defined. **Bioaccumulation** is the net retention of a pollutant by an organism, while **bioconcentration** (expressed as the bioconcentration factor, BCF) is the ratio between a pollutant concentration present in the organism and that present in its surrounding environment. Another term, **biomagnification**, relates to the increase in pollutant concentration in increasingly higher trophic levels in an ecosystem. Certain species, or individuals within a population, often exhibit **resistance** or **tolerance** to a pollutant. There are various mechanisms of tolerance; for example, the binding, or **sequestration**, of toxins in phytochelatins and metallothioneins (proteins within organisms that form complexes with metals), cellular granules and plant vacuoles. Such tolerance mechanisms allow some plant species to **hyperaccumulate** toxins in such structures, away from parts of the plant where they might do harm; other plants exhibit resistance via **exclusion** mechanisms, where transport from the root to the shoot and leaves is actively resisted. Many organisms are capable of **biotransformation** of toxic substances; for example, the microbial biodegradation of toxic organic chemicals into less harmful metabolites.

The ecotoxicity of pollutants is often expressed in terms of $\mathbf{LD_{50}}$ or $\mathbf{LC_{50}}$ (LD = lethal dose; LC = lethal concentration), meaning the concentration of a pollutant at which 50% of the organisms in a population exposed to it are killed. The effects of particular pollutants can also be expressed in the terms of a **LOAEL** (lowest observed adverse effect level) or a **NOAEL** (no observed adverse effect level), each of which are based on the findings of a number of separate observations or studies. The concentration corresponding to a LOAEL or NOAEL for a particular toxin is typically divided by a safety factor of 10 or more to give standards and guidelines for safe pollutant concentrations. Much of the ecotoxicological literature using such terms is based on laboratory experiments, partly because field studies are difficult to control, particularly in terms of measuring exposure to the toxin(s) of interest; however, many laboratory experiments have characteristics that limit extrapolation to real world situations (Table 1.8).

Some groups of humans and other organisms are particularly vulnerable to the toxic impacts of pollutants. For example, young infants are at heightened risk of exposure to toxic elements in contaminated soils and dusts because of their specific

Table 1.8 Limitations of laboratory-based ecotoxicological studies.

Limitation
Exposure of organisms to single elements or compounds that are not present in such isolation in the natural environment
The use of soluble/bioavailable forms of toxins in many cases, whereas mineral or chemical forms in the environment may have limited solubility
The use of short-term exposures rather than long-term (and perhaps low-level) exposures, which are likely to be more realistic
Inter- and intra-species differences in the extent of uptake and effects of toxins (limiting the relevance of animal experiments to human health, for example)
The ethics of using live animals in experiments

behavioural patterns, particularly crawling on the ground and 'mouthing' dirty objects and fingers. Grazing animals (both wild and domestic) are at risk in polluted areas, partly because they are known to ingest soil particles as they graze. Physiological differences can also be important; as an example, the higher stomach pH of infants puts them at greater risk of a condition known as methaemoglobinaemia, which can result from water pollution by nitrates {→N}. Groups at particular health risk from inhaling air pollutants include the very young, the very old and those with respiratory ailments, including those living in the most polluted areas.

The toxic effects of pollutants on humans are many and varied, as will be seen in the individual elements chapters, from mild symptoms that do not cause lasting damage to the more serious outcomes that are summarised in Table 1.9. Ecotoxicological impacts are summarised in Table 1.10; this lists effects at the individual organism level, but it is important to acknowledge that pollution is also recognised as one of the threats (along with habitat loss, etc.) to whole ecosystems and therefore to the 'ecosystem services' they provide, such as crop pollination, waste decomposition and the provision of clean water.

Responses to environmental pollution

Pollution management is not the primary focus of this book, but this section gives a broad overview of the actions that can be taken in response to environmental pollution. Examples are also included in the individual elements chapters. The approach to pollution management and control primarily involves:

1 Recognition of an impact or problem, usually via monitoring and assessment.
2 Consideration of an appropriate level of response, which typically will be dependent on the level of concern of political leaders, as influenced by their electorates.
3 A response, if required, usually utilising one of, or a mixture of: legislation, technology, remediation techniques, economic tools and, in some cases, initiatives to promote desired changes in behaviour.

Monitoring and assessment

Pollution is often clearly visible, a plume of dark smoke from a chimney stack for instance; however, the existence, and certainly quantification, of potentially harmful pollutants in a particular environment are usually revealed by monitoring. Proactive monitoring of environmental pollutants and naturally occurring compounds can be invaluable in signalling potential problems; the discovery of the 'ozone hole' in the 1980s is a prime example. Most monitoring is reactive, however, or undertaken at locations where pollution is known to occur or might be expected, such as busy city centres and industrial zones. In Europe, North America and other parts of the world, authorities routinely monitor air and water quality and, in some cases, soil quality. Measured concentrations are normally compared with guideline levels (e.g. Tables A1–A3, Appendix 2) to gauge the likelihood of environmental or human health impacts and to ascertain whether pollution control measures need to be implemented, or in some cases strengthened.

Table 1.9 Categorisation of some of the main human health impacts of exposure to environmental pollutants.

Organ or process affected by pollutant exposure	*Pollutants (or associated elements) causing the stated effect*
Bone: deformation and brittleness	Cd, F
Cancer formation (**carcinogenicity**): various organs	As, C[1] (e.g. 1,3-butadiene), [Cd], Cl[2], [Cr], [F], [Ni], [P[3]], [Pb], [U[4]]
Cardiovascular system: heart and blood	As, CN^-, F, N[5], [V]
Cell death (**apoptosis**)	[U[4]]
Endocrine system: hormone disruption	Br[2], C[1], Cl[2]
Foetus: birth defects (**teratogenicity**)	C[1] (e.g. PAHs[6]), Hg, [U[4]]
Gastrointestinal tract	As, Cu, F, Hg, Sn[7]
Genetic material damage (**mutagenicity**)	C (e.g. PAHs[6]), [Cr], [U[4]]
Hair loss	Se, Tl
Immune system (**immunotoxicity**)	Cl[2], [V], [Zn]
Kidney: reduced function (**nephrotoxicity**)	As, [Br[8]], Cd, F, Sn[7]
Liver: reduced function (**hepatoxicity**)	As, Cl[2], Cu, [Se], Sn[7]
Metabolism (leading to various disorders; e.g. weakness, anaemia, weight loss)	As, Cd, Cu, F, Hg, P[3], Pb, [Zn]
Nervous system (**neurotoxicity**): including behavioural effects	As, [Br[8]], C[1] (e.g. toluene, xylene), Cl[2], [CN^-], [Cu], F, Hg, [Mn], [P[3]], Pb, Se, Sn[7], Tl, [Zn]
Reproduction: reduced function	Cl[2]
Respiratory system: reduced function	[Br[8]], CN^-, Cu, [Ni], N[9], O_3, S[10], PM[11], Sn[7]
Skin: lesions, keratosis, dermatitis	As, Cl[2], Cr, Hg, Ni, Sn[7]

Notes: Pollutants shown in square brackets are only *suspected* of causing the stated effects on environmental exposure (i.e. the effects have not been noted in the environment, but have been observed: (i) in occupational settings; (ii) following accidental exposure; or (iii) in epidemiological studies where confounding variables may not always have been adequately accounted for).

CN^- = cyanide; Mn = manganese; Tl = thallium; V = vanadium (see chapter 20).

1 In the form of non-halogenated organic compounds.
2 In the form of organohalogen compounds.
3 In the form of organophosphates.
4 And associated radionuclides.
5 In the form of nitrate, NO_3^-.
6 Polycyclic aromatic hydrocarbons.
7 In the form of tributyltin and other organotins.
8 In the form of methyl bromide, CH_3Br.
9 In the form of nitrogen dioxide, NO_2.
10 In the form of sulfur dioxide, SO_2.
11 Particulate matter.

Table 1.10 Ecotoxicological effects of pollutants.

Organ or process affected by pollutant exposure	*Pollutants (or associated elements) causing the stated effect*
Animal growth and development	[Cr], Cu, Hg, Sn[1]
Biodiversity changes (species intolerance)	Al, Cl, Cr, Cu, U[2], Zn
Bone	Cd, F, Se
Cancer formation	Cl[3]
Embryos and larvae (teratogenicity)	Cu, Hg, P[4], Se, [U[2]]
Endocrine system	Br[3], C[5], Cl[3], Cu, Mn, Sn[1]
Gastrointestinal tract	Mn, Tl
Genetic material (**genotoxicity**)	F, [U[2]]
Gills: respiration, osmoregulation	Al, Cr, Cu, Mn, Pb, Se, Zn
Immune system	C[5]
Kidney	Cd, Pb
Leaves: chlorosis, discoloration, necrosis	Br, Cd, Cr, Cu, F, Mn, Ni, O_3, S[6], Se, [V], Zn
Liver	Cu, F[7], Hg, Se, Zn
Metabolism	Al, As, Cd, Cl[3], Cr, F, Mn, Mo, Pb
Microbial processes (e.g. mineralisation)	Ag, Cd, Cu, Ni, Pb, [V], Zn
Nervous system (neurotoxicity)	C[5], [Cr], Pb, Hg, N[8], P[4],Tl
N-fixation by algae, including lichens	S[6]
Olfactory system	Cu
Photosynthesis	Cd, Cr, F, Ni, O_3, S[6]
Plant growth and development	As, B, Cd, Cl[3], Cr, Mn, Ni, O_3, S[6], Zn
Plant respiration	Cd, F, S[6]
Plant roots	Al, Cr
Plant wilting	Cr
Reproduction	[Ag], Al, C[5], Cl[3], Cu, F, Hg, Mn, Pb, Sn[1]
Transpiration	Cd, Ni

Notes: Pollutants shown in square brackets are *suspected* of causing the stated effects on environmental exposure (e.g. the effects have been noted in laboratory studies on plants or in field studies where the co-presence of other pollutants may have been confounding factors). In a few cases, the listed effects and symptoms may overlap; for example, hormonal changes (endocrine system) caused by organotin, leading to reduced reproductive success.

Ag = silver; Al = aluminium; B = boron; Mn = manganese; Mo = molybdenum; Tl = thallium; V = vanadium (see chapter 20). O_3 = ozone (see N chapter).

1 In the form of tributyltin and other organotins.
2 And associated radionuclides.
3 In the form of organohalogen compounds.
4 In the form of organophosphates.
5 In the form of non-halogenated organic compounds.
6 In the form of sulfur dioxide, SO_2.
7 In the form of perfluorooctane sulfonate.
8 In the form of nitrate, NO_3^-.

Level of response

Having recognised the existence of a particular pollution source or impact, the level of response to it will depend in large part on public and political enthusiasm for environmental protection. We may consider environmental sustainability to be a desirable goal, but our individual and collective actions often work against it. Economic growth depends on consumption of natural resources and this brings with it gross disturbances of elemental cycles and reservoirs, creating pollution and other environmental impacts. We might consider pollution as being caused by someone or something else – for example, an intensive farm, a factory or a coal-fired power station – but the farm, factory and power station are producing goods and services that are demanded by individuals. We all contribute to pollution, although some of us more than others (e.g. see Table 2.1, in the carbon chapter). Our food (especially meat and dairy products), clothes, buildings, transportation and material goods all involve the generation of pollution to some degree, not least because of the energy inputs required in their provision.

The degree of pollution control undertaken in a society tends to reflect a balance between concern about environmental damage and the desire to continue 'business as usual'. The level of governmental response to pollution problems is in large part dependent on public attitudes to the environment and on the pressure that electorates apply (or not) to their politicians. Politicians have to be re-elected (in democracies at least) and are therefore likely to be short-termist in their outlook; not an ideal situation for many of the long-term pollution problems that face us. It may be argued that elected politicians should simply act in what they perceive as the wider public (and long-term) interest when it comes to controlling pollution; however, it is unlikely that a particular political action will be taken if it is liable to be unpopular with voters. If this seems counter-intuitive, consider the likely repercussions at the ballot box if a governing party acts to aggressively curb the use of private vehicles, the main sources of pollution in many modern cities.

In this context, it is clear that individual attitudes and actions are important, as well as collective responses. A problem here is that one person's idea of being 'environmentally friendly' may be very different to another's and may depend on the level of inconvenience caused to the individual. For example, it might consist of reusing a plastic carrier bag at the local shop but choosing to drive the short distance there in a car instead of walking for 10 minutes. Furthermore, many consumers may not be willing to buy, or able to afford, ethical (e.g. less polluting) alternatives. Another potential problem is that the opinions (and levels of concern) of individuals are often shaped by broadcasters, the press and other media outlets that can sometimes prioritise eye-catching stories over responsible science reporting, raising the prospect of confusion among the voting public about the scale and impacts of pollution; for example, arguments about the 'reality' of anthropogenic global warming continue to be aired, in spite of the broad consensus among climate scientists.

An added factor influencing the scale of individual and community actions is the concept known as 'nimbyism', based on the 'not in my back yard' acronym, where a person's level of concern over pollution may depend on their individual exposure to it. Nimbyism is arguably global in dimension, as well as local; in recent decades there has been a growing 'export of pollution' (and toxic wastes in some cases) from increasingly post-industrial countries to 'emerging economies' in Asia, South America and elsewhere. Thus, consumption continues unabated in a cleaner 'developed world' at the expense of pollution elsewhere. Nimbyism can also extend to non-polluting but visually intrusive developments such as wind turbines and solar farms and this can sometimes negate potential reductions in pollution. For example, there is a general acceptance that renewable energy generation is better for the environment than fossil fuel combustion, but a massive obstacle to its wider adoption in many areas is public opposition on visual impact grounds.

Despite these factors, there are actions that many concerned individuals take to reduce their environmental 'footprint' (Table 1.11). More importantly perhaps, such actions, if taken by a discernible percentage of a community's population, are more likely to prompt its political leaders to adopt more stringent measures to curb pollution. Such measures may include anti-pollution legislation and technological and economic solutions (see below) and state-sponsored initiatives to encourage the wider uptake of 'environmentally friendly' practices such as energy efficiency and waste minimisation. They may also include the development of large-scale infrastructures that are best delivered by central and/or local governments. One example of the latter is an integrated transport policy that facilitates efficient mass transportation and, ideally, encourages individuals out of private vehicles. A second example is the widespread provision of waste management tools, services and facilities based on the 'waste hierarchy' concept, where waste prevention and minimisation are the ultimate goals, followed by reuse, recycling and recovery and where disposal is seen as a last resort.

Table 1.11 Examples of individual actions that are likely to directly or indirectly reduce pollution.

Actions to reduce pollution	
Choosing cycling, walking or public transport over private car travel	Undertaking energy efficiency measures at home and work
Minimising waste production and reusing and recycling waste materials	Making ethical shopping and investment choices
Reducing meat and dairy consumption or considering vegetarian or vegan diets	Joining or supporting anti-pollution pressure groups and campaigns

Responses: legislation

Legislation is necessary for the long-term management of pollution from the main sources: industry, power generation, agriculture, transport and waste disposal. Pollution prevention may be the ideal, but pollution control is often the focus of legislation because generation of some pollution is inevitable in the life-cycle of most products and processes. Pollution control laws are enforced by regulatory bodies, often via specific *regulations*. For example, in the member states of the European Union (EU), national pollution control regulations are largely derived from EU legislation in the form of Directives. Legislation is also required to establish the appropriate regulatory authorities or agencies to respond to individual pollution incidents.

Legislation is implemented to control pollutants with known environmental and health effects; if the effects are unknown or unclear, the 'precautionary principle' should normally be applied, whereby emissions or use of a substance are restricted or banned until (or unless) it is shown to be safe. Another important concept is 'integrated pollution prevention and control' (IPPC), which seeks to protect the whole environment; this is important because pollution control applied to a single receptor (e.g. a river) may simply result in the transfer of process wastes to another media (e.g. the atmosphere).

Transboundary and/or global-scale pollution are not effectively controlled by domestic laws in individual countries. In such cases, nation states may formulate international agreements that commit signatories to appropriate action (Table 1.12). The Montreal Protocol is often put forward as evidence of the potential efficacy of such agreements because it led to the phasing-out of stratospheric ozone-depleting compounds. The chemicals industry had marketable replacements, however, and a large behavioural shift by the world's consumers was not required to solve the problem. In contrast, the required shift away from fossil fuels is proving far more difficult for the world's economies and its populations and, consequently, the Kyoto Protocol on greenhouse gas emissions has been beset by difficulties, including the withdrawal in 2012 of Canada, one of the world's largest emitters of carbon dioxide.

Table 1.12 Examples of international protocols and conventions relating to transboundary pollutants.

Protocol or Convention	*Year agreed*	*Pollutants covered*
Convention on Long-range Transboundary Air Pollution	1979	Most air pollutants
Montreal Protocol	1987	Ozone-depleting substances
Kyoto Protocol	1997	Greenhouse gases
Stockholm Convention	2001	Persistent organic pollutants

Responses: technology and novel techniques

Technological innovations are instrumental in reducing pollution and progress has been made in many areas, reducing emissions from energy production, industry, transport and waste management. Technologically driven reductions have been countered, at least in part, by extra pollution from a growing world population and increased levels of consumption, but continuing improvements in technology are vital as we look to the future. The adoption of 'geoengineering' solutions to address global warming have been proposed; some of these can seem far-fetched, but may nevertheless be required if the difficult behavioural changes that are required on a global scale are not forthcoming.

Technologies are already available to reduce and replace the use of fossil fuels in power production and industrial processes. These include renewable technologies (such as solar, wind, wave, tidal and geothermal), energy efficiency measures (e.g. insulation materials and smart meters), nuclear power and the use of biomass and biofuels. Any shift from fossil fuels will not be straightforward, however. For example, proponents of shale gas, which is extracted by the hydraulic fracturing ('fracking') technique, argue that, because its carbon content is lower than that of coal, its use will reduce greenhouse gas emissions until viable energy alternatives are developed. Opponents counter that fracking continues to transfer carbon from the lithosphere to the atmosphere and that viable renewable technologies are already available, despite their use often being opposed for various reasons, including visual intrusion (see above). Second, debate continues about the future role of nuclear power. Many scientists point to its low overall risk compared with the health and environmental impacts of burning fossil fuel, but it is already being phased out in countries like Germany, which, in the wake of the Fukushima incident in 2011 {→U}, announced it would close down its 17 nuclear power plants by 2022. Biofuel and biomass crops are further alternatives to fossil fuels, but have various drawbacks; the most serious is perhaps their encroachment onto land that was previously pristine natural habitat (e.g. rainforest) or food-producing, agricultural land.

There may be difficulties relating to the take up of new technologies, as we have seen, but technology has led to pollution reductions in many sectors (see Table 1.13 and element chapters) and has much potential for further progress. For example, in transport, technologies that have helped to reduce emissions from individual vehicles have included improved fuel efficiency, low sulfur and unleaded fuels, diesel particulate filters, hybrid and electric vehicles and catalytic converters, which have been adopted by most countries and have reduced NOx emissions from petrol-powered vehicles by more than 90%. Fuel cells remain in the future for private cars but are already used on a small scale in commercial vehicles.

Proposed technologies to counter global warming include orbiting solar reflectors and carbon capture and storage (CCS), which involves the extraction of CO_2 from the atmosphere or pollution sources, followed by its burial underground. Caution is required for such proposals because they are reactive responses to pollution impacts rather than proactive pollution controls and they have important limitations and risks. For example, the long-term provision of solar reflection would carry ongoing financial costs, would not reduce ocean acidification {→C} and, if ever terminated, would most likely lead to rapid global warming and climate change in a still heavily carbonised atmosphere. Other techniques proposed to reduce atmospheric CO_2 concentrations include carbon sequestration by plants. For example, one proposed approach is to produce 'biochar', where plant matter is converted to charcoal by heating it with a limited supply of oxygen, followed by incorporation of the biochar into soils, where it remains stable and is able to sequester carbon for centuries.

Table 1.13 Examples of technological approaches to pollution control. Further details of these and other technologies are given in the element chapters.

Technology	*Sector*	*Chapter*
High-temperature incineration	Waste management	{→Cl}
Anaerobic digestion	Organic waste treatment	{→N}
Flue-gas desulfurisation	Power generation	{→S}
Low NOx burners	Power generation	{→N}
Acid mine drainage treatment	Mining	{→S}
Membrane technology	Chemical industry (chlor-alkali process)	{→Hg}

Responses: techniques for contaminated land remediation

Air and water are both fluid media and their pollutants tend to have relatively short residence times. Consequently, responses to air and water pollution largely rely on preventing further release of pollutants via control technologies such as particle filters or chemical treatments. Soil pollution, more commonly referred to as contaminated land, is different because land can remain contaminated long after it first received an input of pollution, particularly in the case of toxic metals, which tend to bind strongly to soil particles. Therefore, when monitoring and assessment reveals the existence of contaminated land it may be retrospectively cleaned, or 'remediated', using various physical, chemical and biological techniques. Similarly, sediments in water bodies can remain contaminated for long periods and in such cases remediation may also be required; for example see Box 9.1 {→Cl}.

Contaminated land remediation may involve the removal of polluted topsoil to secure landfill, to be replaced by imported clean topsoil. Other physical techniques include solidification and cementation but, as with removal, such methods do not clean the soil and therefore waste a natural resource. Soils may be treated chemically; for example, metals and organic pollutants are sometimes removed by soil-washing using water and chemicals such as detergents or acids. Alternatively, the approach to metals may be to reduce their bioavailability and mobility, usually by soil liming to increase soil pH and/or by adding organic matter. Biological techniques include the use of micro-organisms, to break down organic pollutants and immobilise toxic metals for example, or phytoremediation, which involves growing metal-tolerant plants on contaminated land to cover and stabilise otherwise unvegetated soils that can be a source of air and water pollution. Phytomining has also been proposed, whereby hyperaccumulating plants (see above) remove toxic metals from the soil, followed by recovery of the metals from the harvested and ashed plants.

Responses: economic measures

Economic tools constitute the other main response to pollution. An important principle in economics, and in pollution management, is cost benefit analysis, where the financial (and possibly other) costs of an action taken are 'weighed' against its financial, social, environmental and other benefits. For example, industrial pollution is typically controlled using so-called 'best available techniques not entailing excessive cost' (BATNEEC). What constitutes an excessive cost in each case is effectively decided on a cost benefit analysis based on the cost of the pollution control options and their likely benefits to human health and the environment. The 'NEEC' clause is applied because the costs of pollution control have to be paid by someone – the taxpayer, the consumer or the polluter – and there would likely be unwanted consequences in each case. The use of taxpayers' money means other essential services (healthcare, education, etc.) have to be curtailed. Passing the cost to the consumer may mean goods and services become prohibitively expensive, at least for the poorest in society. Alternatively, if excessive costs are not subsidised by taxpayers or passed on to the consumer in higher prices, the polluter may either: (i) go bankrupt, meaning its goods or services are no longer provided to the consumers needing or wanting them or (ii) relocate the operation to a country with less stringent pollution controls and costs, which simply shifts the pollution elsewhere.

Another key concept in the economics of pollution is the 'polluter pays principle' (PPP), which seeks to shift the costs of pollution, social as well as financial, away from society (i.e. the taxpayer) and onto the polluter (i.e. the producer of goods or services); in effect, this is an 'internalisation' of external costs, the latter being defined as costs that are not borne by the polluter. This will normally increase product costs (not desirable in a competitive market) and, in theory, PPP therefore encourages the polluter to increase process efficiency and minimise waste, resulting in less pollution without incurring excessive costs. Applying PPP to a widespread global pollutant like CO_2 is complex but has been attempted by the implementation of carbon taxes, which are levied on the carbon content of fuels. An alternative economic approach is carbon emissions trading, whereby a government sets a cap on total emissions allowed in the country (established in some cases by international commitments under the Kyoto Protocol) and sells the rights to emit this pollution in the form of permits. Polluters can trade the permits depending on their success, or otherwise, in reducing pollution.

Summary

Pollution occurs mainly when human activities disrupt natural elemental cycles or reservoirs. The main sources of anthropogenic pollution can be broadly categorised as energy production, industry, agriculture, transportation and waste management. Pollution can be explained using the source–pathway–receptor concept, known as the 'pollutant linkage'. Pollution can have environmental, ecotoxicological and human health impacts. The main responses to pollution are based on legislation, technology and economic measures.

Notes

1 In Bhopal, thousands of people died when methyl isocyanate (C_2H_3NO) leaked from a factory producing carbaryl pesticide.
2 Words that appear in bold type in the remainder of this chapter denote important terms and concepts in pollution science.
3 In certain geological circumstances, such sediments may subsequently be lithified into a type of rock called black shale, which typically has some of the highest toxic metal contents of any rock; for similar reasons fossil fuels also contain higher concentrations of potentially toxic metals than some other geological materials {→metals}.
4 'Lapse rate' refers to the rate of temperature change; adiabatic means without gain of heat from, or loss of heat to, an outside body via conduction.

Reference

DEFRA (Department for Environment, Food and Rural Affairs). 2012. *Waste Water Treatment in the United Kingdom: 2012*. DEFRA, London.

2 Carbon

Environmental reservoirs and chemical forms

Lithosphere

The Earth's crust is the largest environmental reservoir of carbon. Carbon (C) is particularly concentrated in fossil fuels (coal, oil and gas), with an estimated crustal mass of 4×10^6 Mt C and extractable reserves estimated at $0.6–1.6 \times 10^6$ Mt C. Most crustal C (6.5×10^{10} Mt C; Fig. 2.1) occurs in other sedimentary rocks, mainly as carbonates and kerogen. Calcium carbonates are common, particularly aragonite and calcite, which both occur in limestones but with different crystal structures. Kerogen occurs in rocks formed from organic-rich sediments, such as black shales and mudstones. Elemental C is found in graphite and, far more rarely, diamond. Carbon is present in soils in organic matter (OM) and sometimes in mineral fractions. Soils have varying OM contents depending on environmental conditions (particularly climate); in peat, partially decayed OM comprises virtually all of the dry mass. Unconsolidated sediments in water bodies contain organic matter as well. Some marine sediments and permafrost soils also hold carbon in the form of methane hydrates (referred to also as methane clathrates); there is little agreement on the precise quantity, but it is likely to be of a similar order of magnitude to fossil fuel resources, with estimates ranging from 1.5 to 7×10^6 Mt C.

Atmosphere

The main form of C in the atmosphere is carbon dioxide (CO_2), the most naturally abundant 'greenhouse gas' in dry air (see Box 2.1). The generally accepted pre-industrial (*c.* 1750) CO_2 concentration, estimated from data-gathering techniques such as ice core analysis, is ~280 ppm (i.e. 280 molecules of CO_2 per million molecules of dry air). Geological and other evidence indicate that, prior to this, CO_2 concentrations fluctuated naturally over time, linked to geological processes and astronomical ('Milankovitch') cycles. Concentrations over the last 8×10^5 years appear to have ranged between ~180 ppm and ~300 ppm and, going back further in geological time, CO_2 levels have been much higher, particularly in times of heightened tectonic activity (e.g. >1000 ppm, approximately 50 mn years ago). The global atmospheric CO_2 concentration also oscillates annually because of hemispherical and seasonal variations in biological production/absorption (see below).

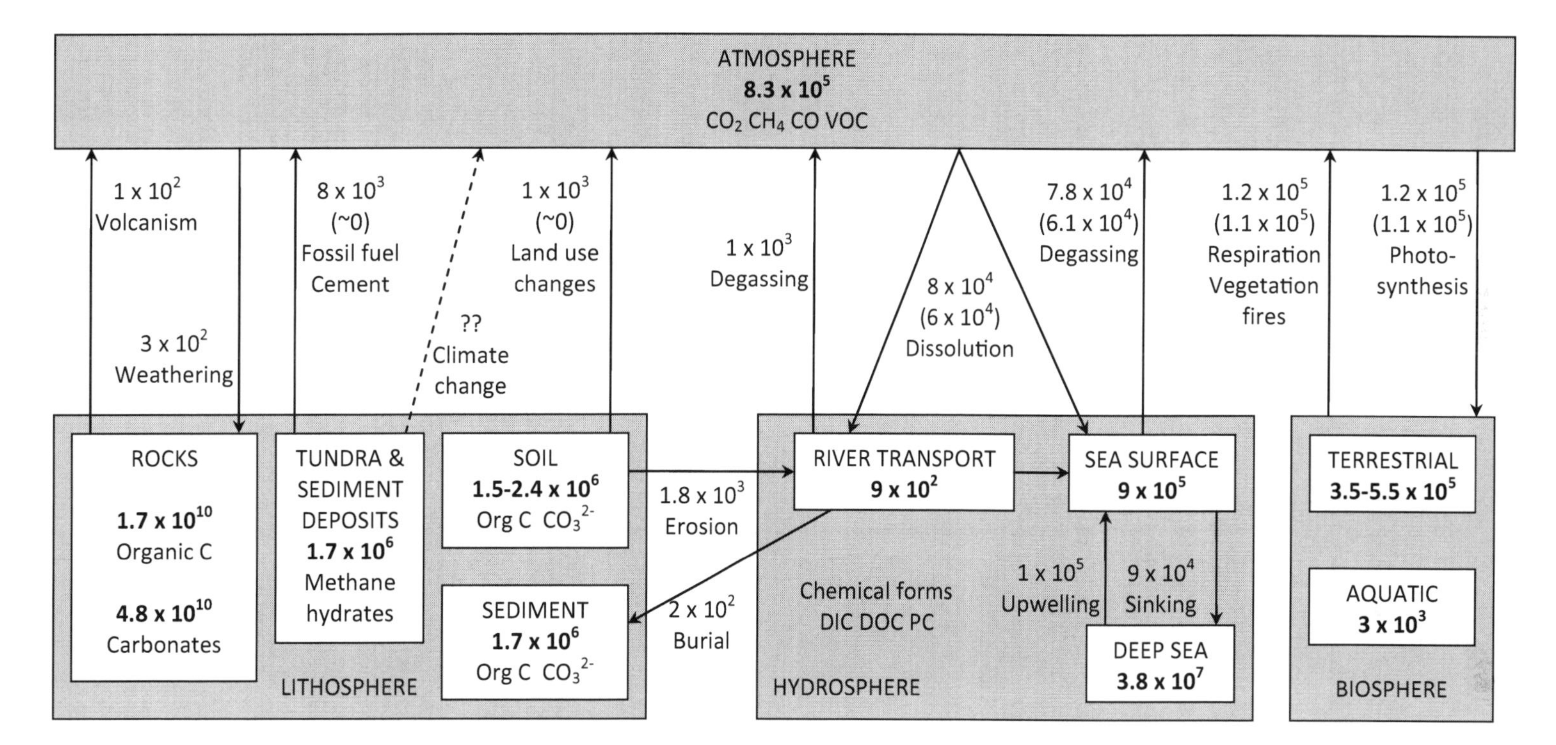

Figure 2.1 The carbon cycle. Units for estimated reservoirs (in boxes) and fluxes (adjacent to arrows) are Mt C and Mt C a^{-1}, respectively. Figures in brackets show pre-industrial fluxes, where estimated. For clarity, the biosphere is shown separately; in reality the processes of photosynthesis and respiration occur within waters and within and upon soils. CO_2 used and produced by these processes within water bodies, chiefly in the oceans, enters and leaves the oceans via dissolution and degassing, respectively, as shown. CO_3^{2-} = carbonates; DIC = dissolved inorganic carbon; DOC = dissolved organic carbon; PC = particulate carbon (see text). The dotted arrow indicates the possibility of methane release from permafrost and other deposits with continued increases in global temperatures.

Source: Diagram based mainly on data in IPPC (2013).

Box 2.1 The natural greenhouse effect.

Short-wave solar radiation reaching the Earth is either reflected back into space by clouds and reflective particles, absorbed by gases (and some particles) in the atmosphere or absorbed as heat by the Earth's surface. Surface heat is ultimately conducted back into the atmosphere as long-wave, infra-red radiation. A portion of this 'thermal' radiation is absorbed by naturally occurring 'greenhouse gases' such as CO_2, CH_4 and H_2O, warming surface temperatures to a global average of ~15°C. With no greenhouse effect, temperatures would be an estimated 33°C lower (–18°C). Each greenhouse gas absorbs infra-red radiation of specific wavelengths (corresponding to specific molecular vibrations) so, taken together, they retain an appreciable amount of energy; those wavelengths that are not absorbed by any gases radiate into space through an 'atmospheric window'.

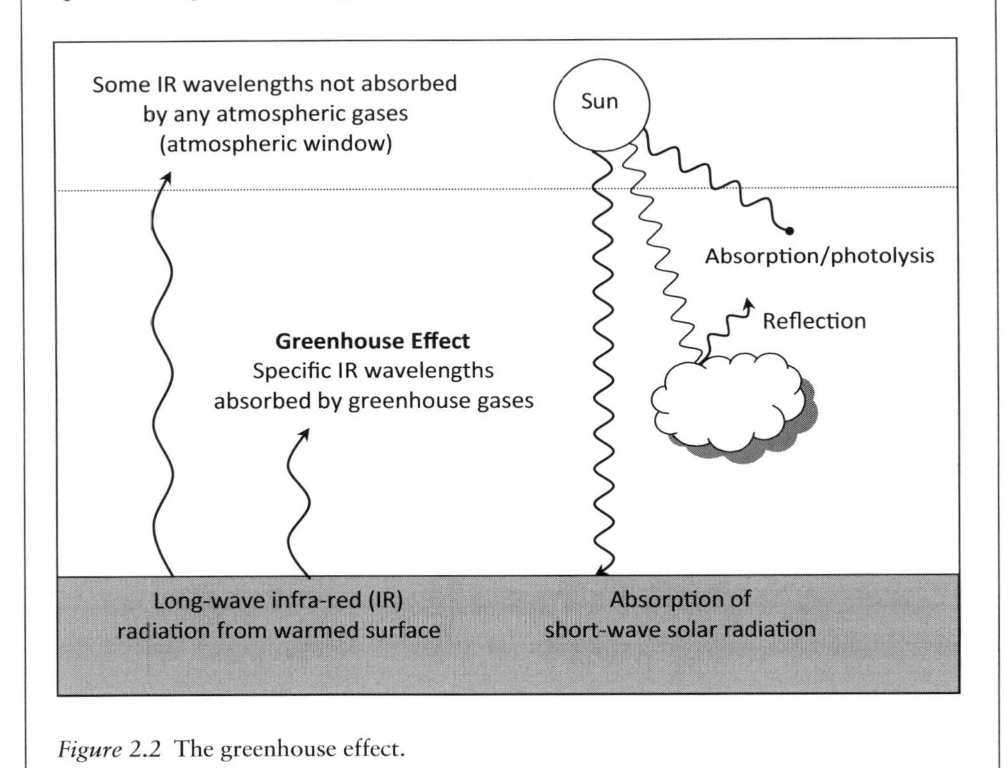

Figure 2.2 The greenhouse effect.

The other main C gases present in the atmosphere are methane (CH_4), which is also a greenhouse gas, and carbon monoxide (CO), which has a short atmospheric lifetime, being rapidly oxidised to CO_2; the estimated pre-industrial concentrations of CH_4 and CO are 0.7 ppm and 0.1 ppm, respectively.

Hydrosphere

Carbon occurs in freshwaters and seawaters mainly as dissolved inorganic carbon (DIC) and, to a lesser extent, as dissolved organic carbon (DOC: e.g. humic and fulvic acids) and carbonaceous particulates, such as fragments of eroded carbonate rock and undecomposed organic solids. The main forms of DIC are dissolved CO_2 and aqueous carbonate (CO_3^{2-}) and bicarbonate (HCO_3^-) ions, at an approximate ratio of 1:10:100.

Biosphere

Carbon is the key component of organic life forms. The building blocks of organisms – carbohydrates, proteins and lipids – are carbon-based. Life is central to the global cycling of C via the processes of photosynthesis and respiration (see below). Terrestrial vegetation contains ~5×10^5 Mt C and marine biota 3×10^3 Mt C.

The natural carbon cycle

The natural carbon cycle is characterised mainly by fluxes to and from the atmosphere from the lithosphere, hydrosphere and, in particular, the biosphere. This is mainly because carbon compounds are used extensively by living organisms in the fundamental processes of photosynthesis and respiration. In photosynthesis, primary producers (algae, cyanobacteria and plants) build carbohydrates from atmospheric CO_2, water and sunlight, providing the basic food and energy requirements of the food chain and emitting oxygen into the atmosphere. In the process of cellular respiration, organic matter is broken down by plants, animals and micro-organisms, providing energy for the respiring organism and releasing CO_2 (and water). The movement of carbon through the biosphere is sometimes called the 'fast carbon cycle'. By contrast, the 'slow carbon cycle' refers to longer term processes of C-cycling involving the lithosphere (outputs from weathering and inputs from rock formation).

Inputs to the atmosphere

The largest input of C to the atmosphere derives from the microbial biodegradation of organic matter via aerobic respiration (see equation 2.1, using glucose, $C_6H_{12}O_6$, as a simple example).

$$C_6H_{12}O_6 + 6O_2 \rightarrow 6CO_2 + 6H_2O \text{ (+ energy)} \qquad [2.1]$$

Aerobic respiration of organic matter involves the use of oxygen as the oxidant but, in anoxic conditions, other chemical species are used by micro-organisms as the terminal electron acceptors (e.g. equations 2.2 and 2.3, showing the use of nitrate and sulfate, respectively, to oxidise organic matter, which is signified here as CH_2O, equivalent to $C_6H_{12}O_6$).

$$5CH_2O + 4NO_3^- + 4H^+ \rightarrow 2N_2 + 5CO_2 + 7H_2O \qquad [2.2]$$

$$2CH_2O + SO_4^{2-} + 2H^+ \rightarrow H_2S + 2CO_2 + 2H_2O \qquad [2.3]$$

A separate type of anaerobic respiration, methanogenesis, produces CH_4. This occurs particularly in wetlands, where organic matter is microbially biodegraded in anoxic marshes and swamps, for example. Additional contributions arise from the breakdown of organic matter by gut bacteria in the gastrointestinal tracts of insects (especially termites) and higher animals (especially ruminants).

Globally, most respiration occurs in the northern hemisphere's winter, where the larger terrestrial biomass of the temperate regions (compared with the southern hemisphere) results in more decomposition globally at that time as plants die off and deciduous trees shed their leaves. The combination of atmospheric CO_2 inputs from respiration and outputs from photosynthesis (mostly in the northern summer, see below) gives an oscillation of ~6 ppm in the global atmospheric CO_2 concentration, with a peak and trough at the end of the northern hemisphere winter and summer, respectively (Fig. 2.3). Respiration occurring within soils and water bodies leads to degassing of C into the atmosphere from these reservoirs. Carbon is also transferred from the biosphere to the atmosphere via natural vegetation fires; this may be in the form of particulate matter ('soot' or 'black C'), CO_2 (complete combustion in air) or CO and CH_4 (incomplete combustion). The Earth's crust is an additional source of atmospheric C; on average, some 100 Mt C a^{-1} are emitted from volcanoes.

Outputs from the atmosphere

A molecule of CO_2 has an average atmospheric lifetime of approximately 5 years before it is absorbed from the atmosphere by photosynthesis, oceanic dissolution or weathering of rocks. Because of the constant biotic recycling of CO_2, its *absolute* atmospheric lifetime is not easy to calculate but is thought to be at least several

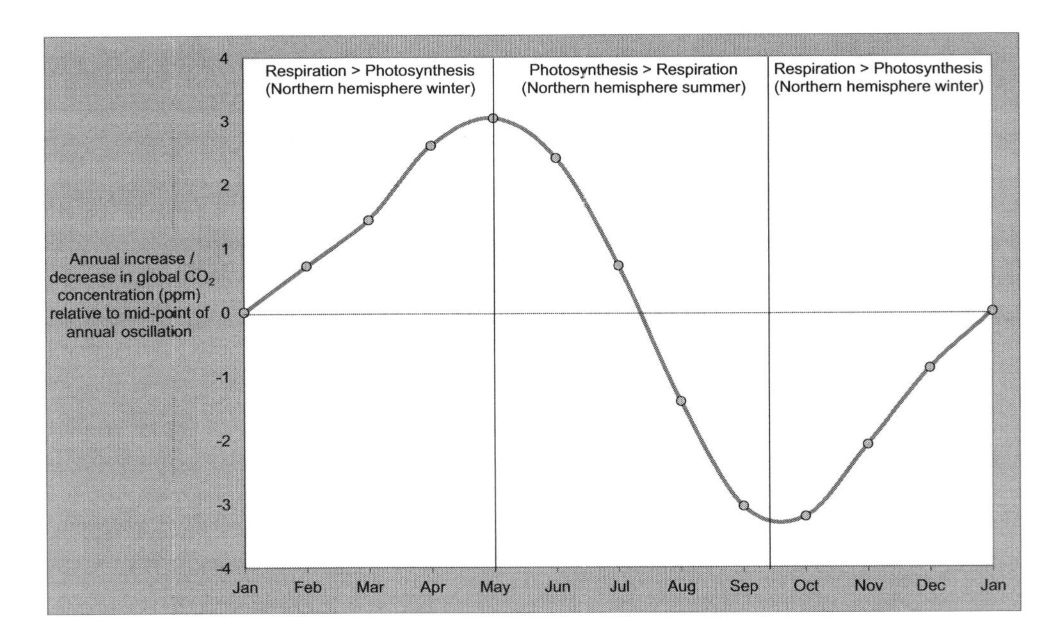

Figure 2.3 Annual oscillation in global CO_2 concentration caused by variation in respiration and photosynthesis rates (see text).

decades. Photosynthesis, which drives net primary production (equation 2.4), accounts for the absorption of 1.2×10^5 Mt C a^{-1} by terrestrial plants. Marine photosynthesis, by phytoplankton in the surface ocean, adds to the overall output of CO_2 from the atmosphere.

$$CO_2 + H_2O + light \rightarrow CH_2O + O_2 \quad [2.4]$$

Carbon dioxide is readily water soluble and its dissolution in rainfall and surface water is an important output of atmospheric C. The naturally acidic pH of rainfall (pH 5.6) is caused mainly by the dissolution of CO_2 in cloud and rain droplets, forming carbonic acid, H_2CO_3 (equation 2.5). Globally, most CO_2 dissolution occurs at the interface between the atmosphere and the ocean surface, particularly in colder waters (CO_2 dissolution being temperature dependent); this adds significantly to the amount of dissolved C in the oceans.

$$H_2O + CO_2 \rightarrow H_2CO_3 \leftrightarrow H^+ + HCO_3^- \quad [2.5]$$

When carbonic acid dissociates to H^+ and HCO_3^- ions (equation 2.5), the H^+ can go on to react with a carbonate ion, CO_3^{2-}, to form a second HCO_3^- ion; the dissolution of CO_2 in a water body therefore reduces its overall number of aqueous CO_3^{2-} ions, which in turn reduces the buffering capacity of the water, increasing its acidity.

Each year, an estimated 300 Mt of atmospheric CO_2 is used up in the weathering of rocks such as carbonates and silicates (e.g. calcium silicate, $CaSiO_3$, equation 2.6; note that half of the CO_2 is consumed in this reaction). Such weathering is enhanced by the relatively elevated CO_2 concentrations existing in the pore space of soil (because of respiration there), the soil being an important location for the weathering of mineral fragments.

$$CaSiO_3 + 2CO_2 + 2H_2O \rightarrow CaCO_3 + SiO_2 + CO_2 + 2H_2O \quad [2.6]$$

The main removal mechanism of atmospheric CH_4 is reaction with hydroxyl ($^{\cdot}OH$) radicals; a much smaller proportion is absorbed by methanotrophic (methane-metabolising) bacteria in soils. As with CO_2, there is an annual oscillation in global CH_4 concentration because removal occurs mainly in the northern summer, while production is steady over the year. The atmospheric lifetime of CH_4 is ~9 years.

Lithosphere, hydrosphere and biosphere cycling

Annually, rock weathering and soil erosion add $\sim 1.8 \times 10^3$ Mt C to rivers and approximately half of this is transported to the oceans; the remainder enters sediments or volatilises from biodegrading soil OM particles. In rivers, C is transported in solution as DOC and DIC (mainly carbonate, CO_3^{2-}) and in particulate form as residues of inorganic C erosion and organic C respiration (humus). Once C has reached the oceans it is transferred between surface and deep waters by upwelling and sinking.

Dissolved calcium and carbonate ions derived from weathering are used by many marine organisms, including plankton (e.g. coccolithophores and foraminifera), shelled animals (such as molluscs) and coral organisms, to build protective calcium carbonate structures, a process called calcification. These are either in the form of

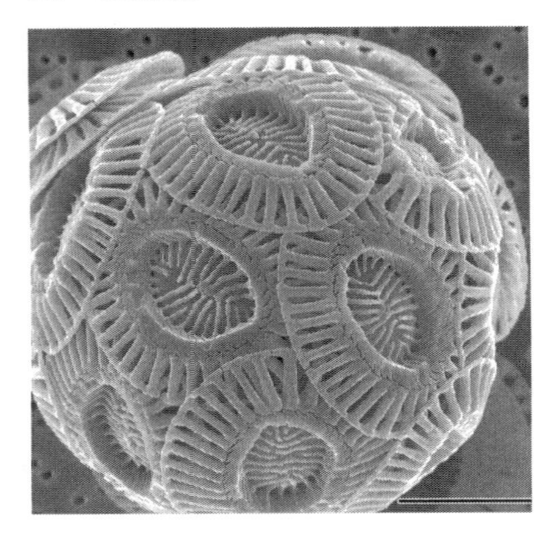

Figure 2.4 Emiliania huxleyi, a single-celled marine phytoplankton (coccolithophore) that produces calcified scales (coccoliths).

Credit: Alison R. Taylor (Wikimedia Commons).

calcite (e.g. coccolithophores – Fig. 2.4, foraminifera and crustaceans) or the more soluble aragonite (corals and some molluscs) and magnesium calcite (some coralline algae).

Each year, approximately 200 Mt C are deposited in the sediments of the seafloor (as organic and inorganic particles). A portion of this oceanic C is subsequently lithified during diagenesis (sedimentary rock formation). Organic C forms fossil fuels and kerogen-rich shales; inorganic carbon precipitates (including calcareous structures) ultimately form carbonate deposits of chalk and other limestones.

Anthropogenic disturbance of the carbon cycle

Atmospheric pollution: greenhouse gas emissions

The natural C cycle has been altered significantly by human activity. Carbon is in the fossil fuels we burn for our energy needs, the organic compounds we use in manufacturing, the wood we use for construction and the food we eat. The earliest forms of environmental pollution affecting humans are likely to have involved the cycling of C, particularly smoke inhalation from fires and exposure to human waste in growing settlements. In modern times the main anthropogenic disturbance of the natural C cycle is the extraction and combustion of fossil fuels.

According to comprehensive reviews of the scientific literature by the Intergovernmental Panel on Climate Change (IPCC, 2013), anthropogenic activities since the beginning of the industrial era (~1750) have emitted 5.5×10^5 Mt of C into the atmosphere, mainly as CO_2. By burning fossil fuel for power and heating, the human race continues to transfer larger amounts of C from the Earth's crust to the atmosphere every year (Fig. 2.5). Total fossil fuel C emissions, together with those from cement manufacture are now $\sim 9 \times 10^3$ Mt C a^{-1} (3.3×10^4 Mt CO_2 a^{-1}), which equates to an average of ~1.3 t C a^{-1} for every person on the planet; however, the amounts of fossil fuel used per capita (i.e. per person) vary dramatically between the nations of the world, with inhabitants of the richest countries accounting for significantly more C emissions than those living in poorer nations (Table 2.1).

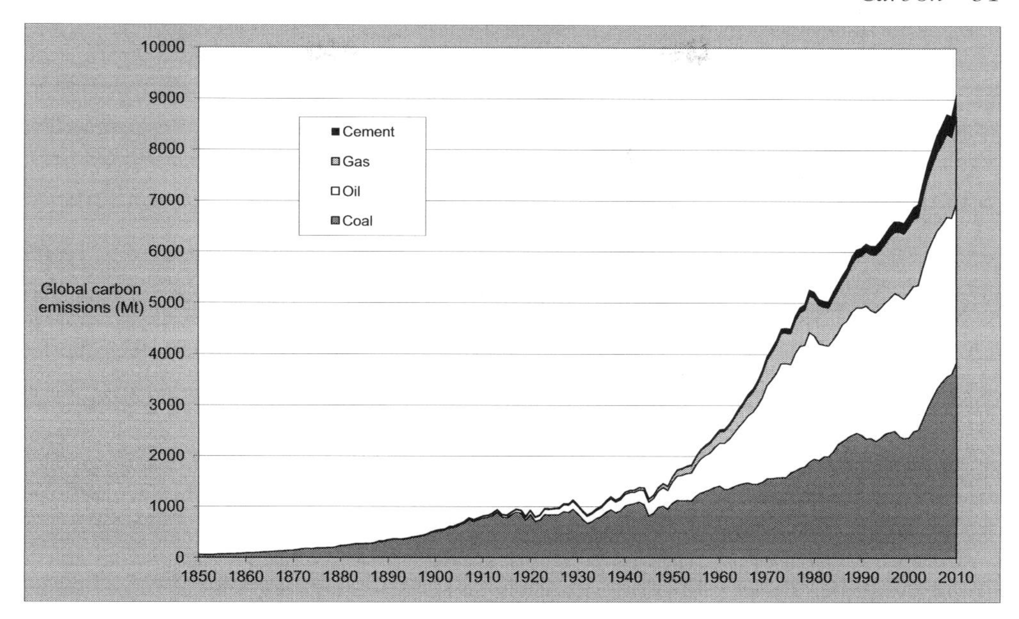

Figure 2.5 Global carbon emissions, 1850–2010 (carbon emitted as CO_2). The main source is fossil fuel combustion (coal, oil and gas) with a smaller contribution from cement manufacture.
Source: Data from Boden et al. (2014).

Table 2.1 Per capita carbon emissions in selected countries.

Country	*Estimated carbon emissions per capita in 2013 (t C a^{-1})*
Qatar	10.94
Trinidad and Tobago	10.30
Brunei	6.26
United States of America	4.71
Australia	4.57
Canada	4.00
Russian Federation	3.32
Netherlands	2.99
South Africa	2.50
United Kingdom	2.16
New Zealand	1.97
China (mainland)	1.68
Mexico	1.07
Brazil	0.59
India	0.45
Pakistan	0.25
Bangladesh	0.10
Haiti	0.06
Uganda	0.03
Chad	0.01

Note: To convert from C (atomic mass: 12) to CO_2 (molecular mass: 44), multiply by 3.67 (i.e. 44/12).
Source: Data from Carbon Dioxide Information Analysis Center, www.cdiac.ornl.gov.

An additional CO_2 source is deliberate biomass burning, usually to clear naturally vegetated land for agriculture and, in some cases, urbanisation; globally this is redressed to some extent by forest regrowth in other areas, but there is an estimated net input to the atmosphere from this source of ~1×10^3 Mt C a^{-1} (IPCC, 2013). Atmospheric CO_2 is also generated in cement production, by the heating of calcium carbonate to produce lime, CaO, a key ingredient of cement:

$$CaCO_3 + heat \rightarrow CaO + CO_2 \qquad [2.7]$$

Most of the anthropogenic sources of CH_4 are associated with the biodegradation of OM in anoxic or hypoxic conditions (Table 2.2). Methane is also produced by incomplete combustion of fuels and this is the usual source of CO pollution. In addition to C *gases*, fuel and wood combustion releases *particulate* C, which is the major source of seasonal 'haze' problems (particularly in Asia) and 'winter-type' smogs in cold regions (see below). The main sources of particulate C are industry, diesel-powered vehicles, domestic heating, vegetation burning for agricultural clearance and large-scale power stations, particularly those that are coal-fired.

An estimated 57% of the anthropogenic CO_2 produced since 1750 has been absorbed by the oceans (30%) and terrestrial ecosystems (27%), leaving some 2.4×10^5 Mt of C (the remaining 43%) in the atmosphere (IPCC, 2013). This excess has increased the global atmospheric CO_2 concentration from ~280 ppm in pre-industrial times to >400 ppm in 2014 (Fig. 2.6). Rock weathering operates too slowly to absorb much of the excess CO_2, so it continues to accumulate in the atmosphere. Furthermore, photosynthesis does not result in a significant net absorption because it is part of a balanced cycle with respiration, which releases roughly the same amount of CO_2 each year; however, the 'fertilisation effect' of additional atmospheric CO_2 means that a small proportion of the extra C is absorbed by increased forest growth.

Methane, another greenhouse gas, is also rising above pre-industrial levels and at a greater rate than the increase in CO_2. Anthropogenic activities have added 3×10^3 Mt CH_4 to the estimated pre-industrial total of 2×10^3 Mt and 17 Mt are being added to the atmosphere annually (IPCC, 2013); this has increased the pre-industrial CH_4 concentration by >150% to approximately 1.8 ppm.

Table 2.2 Anthropogenic sources of methane.

Source	*Reason*
Livestock	Organic matter digestion in stomachs of ruminants
Landfill sites	Biodegradation of organic matter in compacted waste
Sewage treatment plants	Anaerobic digestion processes
Rice cultivation	Anoxic conditions in flooded paddy fields
Artificial dams	Biodegradation of organic matter in bottom sediments
Fossil fuel use	Mainly leakage from gas pipelines
Biomass burning (land clearance)	Incomplete combustion

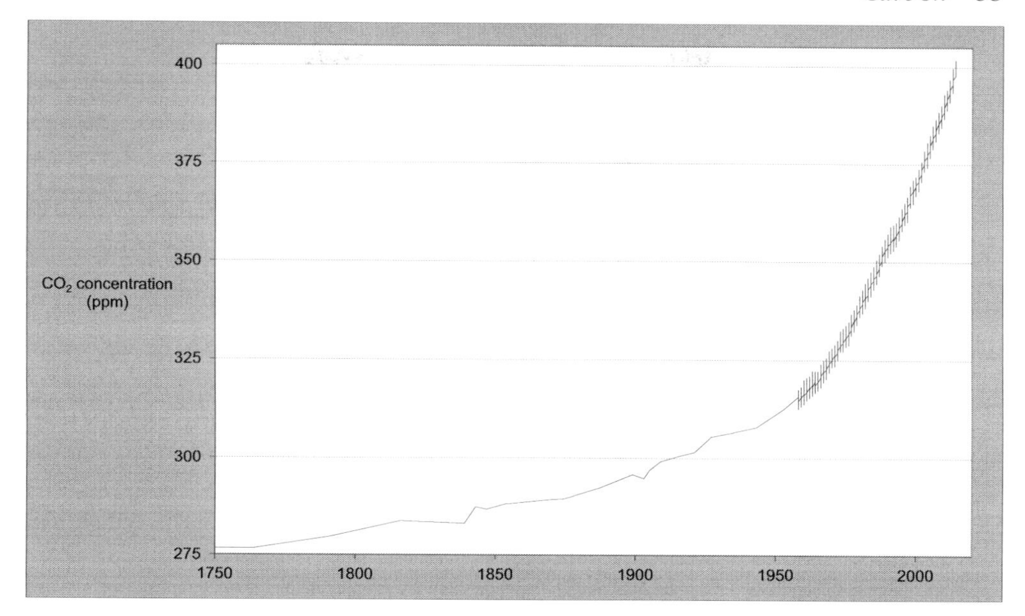

Figure 2.6 Atmospheric CO_2 concentrations from 1750 to 2014.

Data sources: 1750–1953 ice core data – CDIAC (2014) and monthly measurements at Mauna Loa, Hawaii, March 1958 to May 2014 – NOAA/ESRL (2014).

Water pollution: organic wastes and fuels

Watercourses can be polluted by respirable organic matter (ROM) contained in effluents from sewage treatment plants, farms, food processing and industrial sources (e.g. paper mills); other sources include soil erosion and urban runoff. Fossil fuel use can also pollute waters; at sea, oil leakages from ocean-going tankers and pipelines are a hazard, while on land, oil and gasoline can leak into groundwaters from underground tanks. The ash residues of fossil fuel combustion are often disposed of in artificial dumping areas called ash ponds and failures of such structures are a further hazard.

Industrial organic compounds: manufacture and use

The manufacture and use of industrial organic compounds (IOCs) can pollute water, air and soil. Table 2.3 shows some key categories of IOCs and their main uses. Separate categories of organic compounds that cause concern in relation to environmental pollution are the organohalogens {→Br →Cl →F}, organophosphates {→P} and organometals {→metal(loid)s}.

Atmospheric pollution by IOCs is chiefly related to volatile organic compounds (VOCs) and unburnt hydrocarbons (HCs). The VOCs include organic solvents like benzene, toluene and xylenes (Fig. 2.7), which volatilise from the products containing them, such as paints, adhesives and petrol (the latter during distribution, refuelling and combustion). These solvents have high vapour pressures and are therefore volatile at ambient temperatures. Some other toxic VOCs, such as phthalates and formaldehyde, are considered mainly as indoor air pollutants and are present only at relatively low

Table 2.3 Industrial organic compounds and their uses. Structural (rather than molecular) formulas are shown for some of the simpler compounds.

Category	*Example compounds*	*Uses*
Organic solvents[a]	Ethylene, CH_2CH_2 Hexane, $CH_3(CH_2)_4CH_3$ Benzene, C_6H_6 Toluene, $C_6H_5(CH_3)$ Xylene, $C_6H_4(CH_3)_2$ Methanol, CH_3OH Ethyl acetate, $CH_3COOCH_2CH_3$ Acetone, $(CH_3)_2CO$	Industrial degreasing; cleaning products; additives – e.g. in paints, adhesives and varnishes
Pharmaceuticals	Numerous compounds (e.g. antibiotics, painkillers, anticonvulsants, mood stabilisers)	Human and veterinary medicine
Synthetic hormones	Ethinylestradiol, $C_{20}H_{24}O_2$	Birth control pill
Carbamates	Alidcarb, $C_7H_{14}N_2O_2S$ Carbaryl, $C_{12}H_{11}NO_2$ Carbofuran, $C_{12}H_{15}NO_3$	Insecticides
Synthetic pyrethroids	Deltamethrin, $C_{22}H_{19}Br_2NO_3$ Permethrin, $C_{21}H_{20}Cl_2O_3$	Insecticides
Neonicotinoids	Imidacloprid, $C_9H_{10}ClN_5O_2$ Clothianidin, $C_6H_8ClN_5O_2S$	Insecticides
Aldehydes	Formaldehyde, HCHO	Resin and disinfectant production
Alkylphenol ethoxylates	Nonylphenol ethoxylate, $C_{19}H_{32}O_3$	Additives in detergents, paints, pesticides, plastics and rubbers
Phthalate esters ('phthalates')	High molecular weight phthalates – e.g. $C_{24}H_{38}O_4$ Di(2-ethylhexyl) phthalate	Plasticisers in PVC manufacture (e.g. in flooring, tiles, food containers, medical equipment)
	Low molecular weight phthalates – e.g. Diethyl phthalate, $C_{12}H_{14}O_4$	Binders in body-care products, solvents for organic molecules, detergents, varnishes
Bisphenols	Bisphenol A, $C_{15}H_{16}O_2$	Plastics and resins in food and drink packaging

a Some of these compounds are also used in applications not specifically related to their solvent properties; e.g. benzene in the formulation of plastics and synthetic fabrics and toluene diisocyanate in polyurethane manufacture.

concentrations in the wider atmosphere. Both have sources in household objects and materials; particularly PVC products in the case of phthalates and pressed wood products (e.g. plywood), paint and varnish for formaldehyde. Styrene, used in the production of plastics (mainly polystyrene), synthetic rubbers and resins, is another VOC that may be recorded in indoor air and in polluted urban areas. Other important VOCs include chlorinated compounds such as chloroethene {→Cl}. Incomplete combustion of fuel in vehicles, industry and power generation (and also in wood fires) leads to the production of unburnt HCs such as polycyclic aromatic hydrocarbons (PAHs) and 1,3-butadiene (Fig. 2.7).

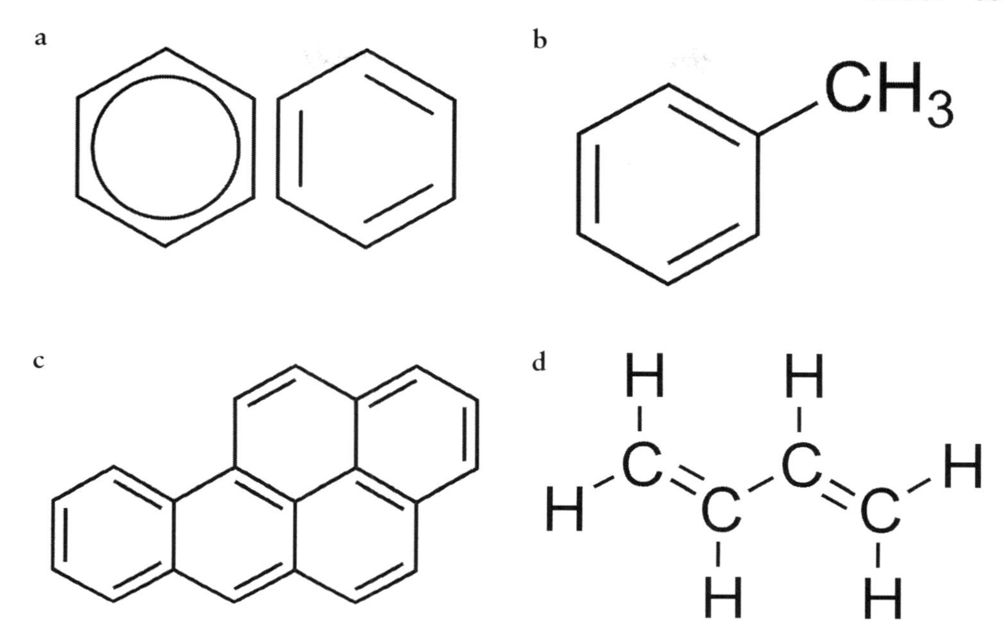

Figure 2.7 Molecular structures of some important industrial organic pollutants: (a) benzene, shown in both of the traditional formats used for the benzene ring (C atoms, each bonded to a H atom, occur at the interstices of the hexagon); (b) toluene, where one of the H atoms of the benzene ring has been replaced by a methyl group, CH_3; (c) benzo(a)pyrene, an important polycyclic aromatic hydrocarbon; (d) 1,3-butadiene.

Credits: (a) Byran Derksen, (b) Emeldir, (c) Calvero and (d) Abpong (all Wikimedia Commons).

Surface waters may be contaminated by IOCs from domestic and industrial effluents; for example, sewage outflows can contaminate rivers with traces of pharmaceutical compounds that have not been fully metabolised by the human body, along with other compounds in domestic wastewater, such as nonylphenol and phthalates. Bisphenol A, which is used extensively in plastics and packaging materials, is also discharged into surface waters via sewers and degradation of waterborne plastic litter. Carbamate, pyrethroid and neonicotinoid insecticides may enter surface waters and groundwaters via runoff and leaching. Typical sources of groundwater contamination by organic compounds also include chemical spills, leaking pipelines, underground storage tanks (solvents and petroleum hydrocarbons) and waste disposal sites (domestic and chemical). For example, benzene, toluene and xylene are often detected in groundwaters that have been contaminated by spills or leaks associated with fuel storage or petroleum refining. These three compounds (and petrol itself) are categorised as 'light non-aqueous phase liquids' (LNAPLs) because they are less dense than water and, being non-aqueous, do not infiltrate the groundwater itself when a leak or spill occurs. Instead they remain at the surface of the groundwater (i.e. at the water table) and are thus easier to remediate than DNAPLs ('dense non-aqueous phase liquids'), which, being denser than water, penetrate deep into the aquifer. DNAPLs include coal tar, creosote and some organochlorine compounds {→Cl}.

Impacts of carbon-based pollutants

Impacts of air pollution: global warming

The natural greenhouse effect (Box 2.1) is enhanced by anthropogenic emissions of greenhouse gases, leading to global warming as more of the Earth's outgoing thermal radiation is retained in the atmosphere. Most of the enhanced greenhouse effect is caused by CO_2, but CH_4 is also of concern, despite its much lower concentration, because it has 20-year and 100-year global warming potentials (GWPs) of 84 and 28, respectively; i.e. a given mass of CH_4 'traps' 84 and 28 times more thermal radiation over these timescales than an equivalent mass of CO_2. Emissions of black carbon (soot) particles also have a role in global warming because they absorb sunlight, further enhancing the greenhouse effect; this is in contrast to some other particle types (e.g. sulfates), which reflect solar radiation and contribute to the phenomenon of 'global dimming' {→S}. As well as CO_2 and CH_4, the United Nations Framework Convention on Climate Change (UNFCCC) lists nitrous oxide {→N}, sulfur hexafluoride {→F}, hydrofluorocarbons {→F} and perfluorocarbons {→F} as anthropogenic greenhouse gases.

In 2013–14, the Intergovernmental Panel on Climate Change (IPCC) released its *Fifth Assessment Report* on climate change, based on extensive research of the scientific literature published over many years. Much of the information in this section is based on the peer-reviewed evidence detailed in the comprehensive report and is stated with 'high' or 'very high' levels of confidence by the IPCC. The summary of the *Physical Science Basis* part of the report (IPCC, 2013) includes the following statements:

> Warming of the climate system is unequivocal
>
> It is extremely likely that human influence has been the dominant cause of the observed warming[1]
>
> Continued emissions of greenhouse gases will cause further warming and changes in ... the climate system.

The IPCC also concludes that substantial and sustained reductions in anthropogenic emissions will be required and that climate change impacts will continue for centuries, even if emissions cease.

The research reviewed by the IPCC indicates that the increases in the atmospheric concentrations of CO_2 and other greenhouse gases have raised the globally averaged, combined land–ocean surface temperature by 0.85°C between 1880 and 2012 (one of the main temperature records is shown in Fig. 2.8). Taken together, the various datasets indicate that the last three decades have each been progressively warmer than any previous decade going back to 1850 and each of the 13 years since 2001 has been among the warmest 15 years on record; however, the *rate of temperature increase* since 1998 has been lower (0.05°C per decade) than the overall rate since 1951 (0.12°C per decade). Climatologists attribute this to temporal variability in atmosphere–ocean interactions leading to greater absorption of excess heat in deep ocean waters in recent years. Furthermore, the extent of sea-level rise over recent years (see below) indicates

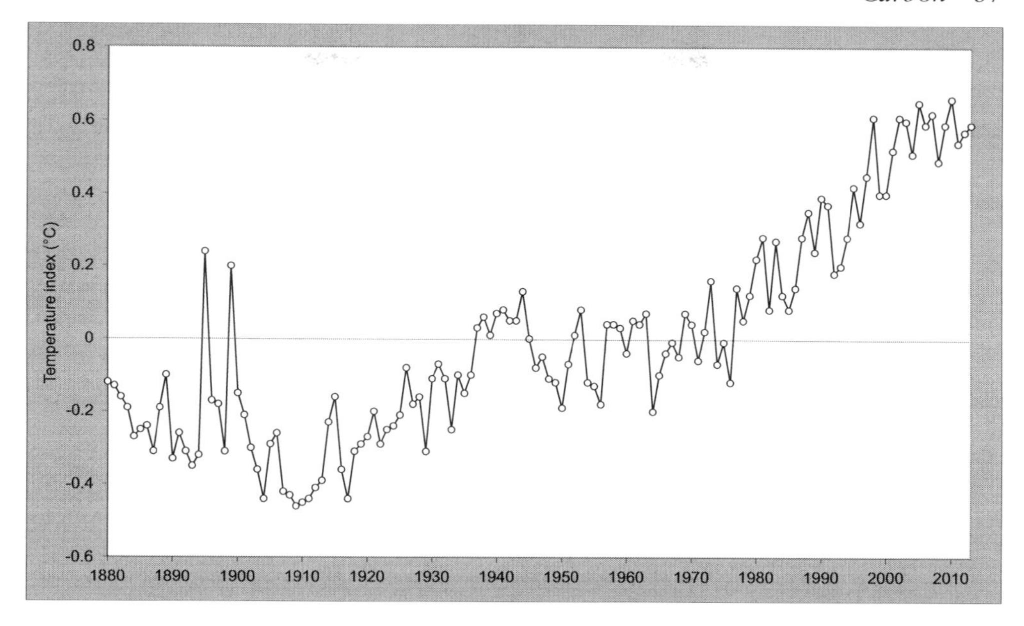

Figure 2.8 The Land Ocean Temperature Index (LOTI), 1880–2014. Each data point shows the annual mean temperature anomaly compared with the mean 1951–80 temperature. Based on a combination of: (i) land-surface air temperatures and (ii) sea-surface water temperatures.

Source: Data from NASA/GISS (2014).

a continuation of previous warming because rising sea levels are caused almost entirely by temperature-related factors: melting land ice and thermal expansion of seawater.

A direct consequence of the global warming already observed is the ongoing melting of polar and glacial ice around the world. The Greenland and Antarctic ice sheets were each losing approximately 3×10^4 Mt of ice per year in the 1990s. During the following decade this had increased to $>2 \times 10^5$ Mt a^{-1} and 1.5×10^5 Mt a^{-1}, respectively. In addition, 2.75×10^5 Mt of glacial ice is melting each year on a global scale. The Arctic has warmed substantially in recent decades and, since 1980, there has been a decrease in annual mean Arctic sea ice cover of up to 4% per decade; the decline in average *summer* ice cover has been $\sim 1 \times 10^6$ km^2 per decade (Fig. 2.9). Over the same period, the area of sea ice around the melting Antarctic ice sheet has increased by up to 1.8% per decade, an area of up to 0.2 million square kilometres. Globally, snow cover and permafrost thickness and extent have also decreased over recent decades.

Melting of land ice, along with thermal expansion of warmer ocean water, has led to a total global sea-level rise of 0.19 m over the last century (1901–2010), equating to a globally averaged, annual rise of 1.7 mm a^{-1} (Fig. 2.9). The rate of sea-level rise has increased in recent decades to an estimated 3.2 mm a^{-1} since 1993. More than a third of this is attributable to thermal expansion, the remainder mostly being derived from melting land ice (a smaller net contribution comes from anthropogenic land-use changes, particularly the pumping of deep groundwater to the surface and the removal of wetlands and forests, which store water). Sea-level rises are not only a hazard for low-lying states like the Maldives, Bangladesh and the Netherlands, but also for

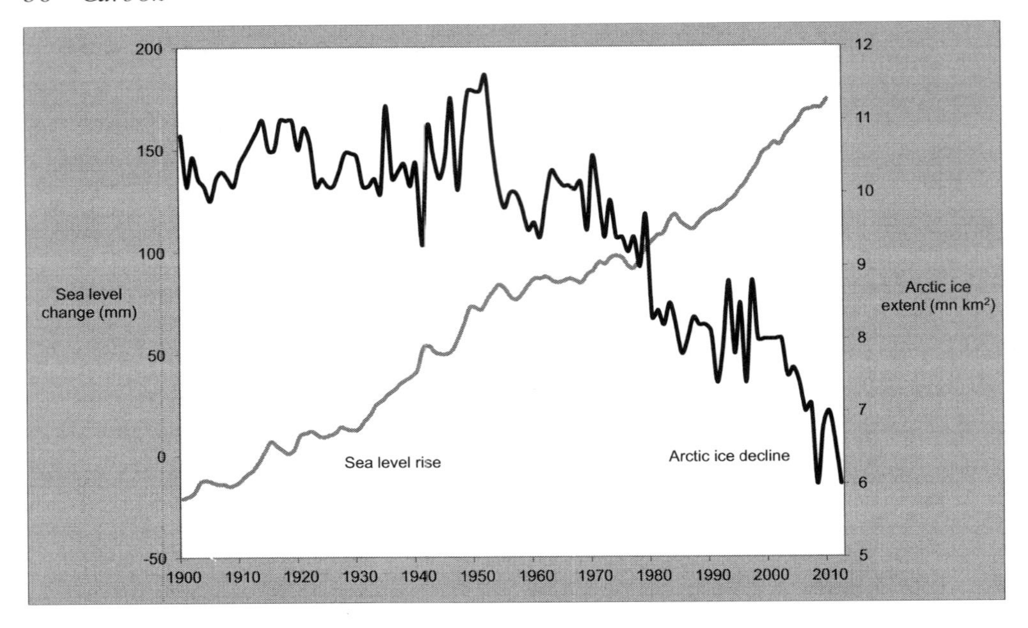

Figure 2.9 Average Arctic ice cover in summer (July–September), 1900–2012 and global mean sea level, 1900–2010, relative to the 1900–05 mean.

Source: Adapted from IPCC (2013).

coastal cities and communities around the world. According to the United Nations Environment Programme, approximately half of the world's population lives within a few tens of kilometres of the coast, mostly in coastal cities.

The threat of climate changes caused by a warmer atmosphere is also of concern. Climate varies naturally, driven by phenomena such as the El Nino-Southern Oscillation (ENSO) and the North Atlantic Oscillation (NAO), but there is an overall trend of rising temperatures and a warmer atmosphere increases the likelihood of climate extremes. The IPCC states, with varying degrees of statistical certainty, that since 1950: (i) the numbers of cold days and nights have decreased globally; (ii) the numbers of warm days and nights have increased globally; (iii) the frequency of heavy rainfall has increased in more areas than it has decreased, with more intense rainfall in Europe and North America; and (iv) the frequency of heatwaves has increased in Australia, Asia and Europe, with the probability of heatwave occurrences doubling in some areas. Increases in drought conditions in some areas (Mediterranean and West Africa) have been balanced by decreases in other areas (North America and Australia). In recent years, devastating floods, tropical cyclones, droughts and heatwaves that have resulted in fatalities, injuries and disease have sometimes been linked by commentators to global warming; climate scientists make clear that no individual weather event can be attributed to a warmer climate, but state that increased frequency of such events is predicted by climate models.

In the *Impacts, Adaptation and Vulnerability* part of the *Fifth Assessment Report*, the IPCC (2014) details both the observed effects and the risks of climate change on ecosystems and human systems. Table 2.4 summarises the stated *observed* effects and Table 2.5 lists key *risks* that span all regions and environmental sectors. In both cases the tables are limited to statements that the IPCC makes with high or very high confidence based on peer-reviewed scientific research (many other observed effects and risks are stated in the report with medium confidence).

In relation to how climate may change in the coming decades, climate scientists have made predictions of future global warming, or 'radiative forcing', based on four scenarios (representative concentration pathways, 'RCPs') of future greenhouse gas emissions (see Box 2.2). All scenarios except RCP 2.6 (decreasing emissions from 2020) are likely to lead to global warming in excess of 2°C (above pre-industrial levels) by 2100 (IPCC, 2013). This raises particular concern because parties to the UNFCCC in 2010 agreed that a *maximum* increase of 2°C would be required to allow a reasonable chance of avoiding the worst impacts of climate change. The IPCC says that a >66% probability of achieving this aim will require the total amount of

Table 2.4 Observed impacts, vulnerability and exposure caused by climate change, stated with high or very high confidence.

Significant vulnerability of ecosystems and many human systems to climate-related extremes such as flooding, cyclones, wildfires and drought causing alteration of ecosystems, disruption of food production and water supply and consequences for human health and well-being
Shifts in geographical ranges and seasonal activities of terrestrial animal and plant species in many regions
Coral bleaching and changes in species ranges due to ocean water warming (combined with effects of ocean acidification; see next section of this chapter)
Shifts in the abundance, distribution and migration of marine species
Increase in number and size of marine 'dead zones' (hypoxic areas) due to higher sea temperatures (in combination with effects from other human activities {→N})
Negative impacts on crop yields have been more common than positive impacts
Climate change-related hazards are an additional burden to people living in poverty

Source: IPCC (2014).

Table 2.5 Key risks of climate change spanning all regions and sectors, stated with high confidence.

Death, injury and disruption to livelihoods, food supplies and drinking water
Food insecurity and breakdown of food systems linked to warming, drought and precipitation variability, particularly in regions with poorer populations
Severe harm due to inland flooding and limited coping and adaptive capacities of large urban populations
Loss of rural livelihoods and income of rural residents due to insufficient access to drinking and irrigation water and reduced agricultural productivity and food insecurity
Hazards affecting infrastructure networks in combination with a high dependency of people on critical services (e.g. electricity, water supply, and health and emergency services) that may break down during extreme events
Loss of marine and terrestrial ecosystems and the services they provide for livelihoods
Mortality, morbidity and other types of harm during periods of extreme heat, particularly for urban populations of the elderly, infants, people with chronic ill-health and expectant mothers

Source: IPCC (2014).

Box 2.2 Radiative forcing and representative concentration pathways.

Radiative forcing (RF) is the change in the Earth's net energy budget (i.e. incoming solar energy minus outgoing thermal energy) resulting from changes in the atmospheric concentrations of greenhouse gases and aerosols since 1750, the pre-industrial period. The current anthropogenic RF value is 2.3 W m^{-2}; i.e. the atmosphere now holds 2.3 watts of extra energy (per square metre of the tropopause) than in 1750 because of anthropogenic greenhouse gas (GG) emissions. Climate scientists have predicted future RF values under four 'representative concentration pathways' (RCPs), based on future GG concentration scenarios (Table 2.6). The RCP number relates to the globally averaged RF that is predicted to result from each scenario by 2100.

Table 2.6 Explanation of representative concentration pathways scenarios.

	RCP 2.6	*RCP 4.5*	*RCP 6*	*RCP 8.5*
Decline in GG emissions from:	2020 'Mitigation'	2040	2080	No decline 'Business as usual'
C emission in 2100 (Mt a^{-1})[a]	-4.2×10^{2}[b]	4.2×10^{3}	1.4×10^{4}	2.9×10^{4}
CO_2 conc. in 2100 (ppm)[c]	421	538	670	936
Predicted RF in 2100 (W m^{-2})	2.6	4.5	6.0	8.5

a Currently ~9×10^{3} Mt a^{-1}, but rising by ~32% per decade (see Fig. 2.5).
b Negative figure because of very low or zero emissions, together with C sequestration.
c Currently 400 ppm.

Data from RCP Database (2014).

anthropogenic CO_2 ever emitted to the atmosphere to be less than 7.85×10^{5} Mt C; by 2011 the cumulative total had reached 5.15×10^{5} Mt.

Under all four RCP scenarios, extremes of high temperature are forecast to increase this century, with higher frequency and duration of heatwaves and intense rainfall; the latter in tropical and mid-latitude areas. Predictions for 2100 (IPCC, 2013), using RCP 2.6 and RCP 8.5 as the minima and maxima (or 'best' and 'worst case scenarios'), include: the warming of surface seawater by 0.6°C to 2°C; the reduction of Arctic ice in September by 43% to 94%; and the reduction of glaciers (excluding on Antarctica) by respective ranges of 15–55% to 35–85%. Projected surface air temperature and sea-level changes are shown in Table 2.7. Sea-level rise is predicted to continue beyond 2100 in any scenario, as the oceans will continue to warm because of the very long-term transfers of surface-to-depth heat transfer. The IPCC (2013) predicts, with high confidence, that predicted global warming over this century and beyond increases the risk of extinction of many terrestrial and freshwater species and that millions of people, especially in south and east Asia will be affected or displaced by coastal flooding.

Table 2.7 Projected temperature and sea-level changes under the four representative concentration pathways scenarios, relative to a 1986–2005 reference period.

		2046–2065		2081–2100	
	Scenario	Mean	Likely range	Mean	Likely range
Global mean surface air temperature change (°C)	RCP 2.6	1.0	0.4 to 1.6	1.0	0.3 to 1.7
	RCP 4.5	1.4	0.9 to 2.0	1.8	1.1 to 2.6
	RCP 6.0	1.3	0.8 to 1.8	2.2	1.4 to 3.1
	RCP 8.5	2.0	1.4 to 2.6	3.7	2.6 to 4.8
Global mean sea-level rise (m)[a]	RCP 2.6	0.24	0.17 to 0.32	0.40	0.26 to 0.55
	RCP 4.5	0.26	0.19 to 0.33	0.47	0.32 to 0.63
	RCP 6.0	0.25	0.18 to 0.32	0.48	0.33 to 0.63
	RCP 8.5	0.30	0.22 to 0.38	0.63	0.45 to 0.82

a Not including possible collapse of marine Antarctic ice sheets, which would substantially increase sea-level rise above the likely ranges shown.

Source: Data from IPCC (2013).

Scenarios and predictions such as these are necessary, but there are many uncertainties to acknowledge when contemplating the future outcomes of global warming, particularly the likelihood of positive and negative feedbacks (see Box 2.3). Furthermore, the Earth system is complex and not fully understood, particularly the interaction of the atmosphere with the oceans, which absorb excess heat. Other uncertainties also require serious consideration; for example, the 'global dimming' that currently reduces the extent of global warming (see →S) may be replaced by 'global brightening' because of actions to reduce particulate pollution in our cities. Linear projections allow policy-makers and electorates to understand some of the possible consequences of 'business as usual', but the transition to a warmer planet is likely to be somewhat unpredictable and non-linear in nature.

Impacts of air pollution: ocean acidification

Ocean acidification (OA) is sometimes said to be the forgotten impact of anthropogenic CO_2 emissions, the 'other CO_2 problem'; however, it is taken very seriously by scientists. Some 30% of the anthropogenic CO_2 emitted into the atmosphere in the industrial era has been absorbed by the oceans. Currently the absorption rate is estimated at approximately 25%, equating to some 2.4×10^7 t of CO_2 per day. Because of continuing increases in global anthropogenic emissions, more CO_2 is dissolved into seawater each year and this increases its acidity by the production of carbonic acid, H_2CO_3 (equation 2.8).

$$CO_2 + H_2O \leftrightarrow H_2CO_3 \leftrightarrow HCO_3^- + H^+ \quad [2.8]$$

As a result, seawater pH has decreased from ~8.2 to ~8.1 since pre-industrial times and the oceans are expected to become more acidic in future (Fig. 2.10). It should be remembered that the pH scale is logarithmic and this seemingly small difference

Box 2.3 Positive and negative feedbacks.

Greenhouse gas emissions and global warming have initial consequences (e.g. melting of ice) that can act as feedback 'triggers', altering natural environmental processes and cycles in ways that can amplify (positive feedback) or dampen (negative feedback) future greenhouse gas concentrations and/or the rate and extent of global warming (Table 2.8). The feedbacks add to our uncertainty and concern about the risks of a warming planet.

Table 2.8 Examples of global warming feedbacks.

Initial trigger	*Feedback*	*Feedback mechanism: effect of initial trigger*
		Positive feedbacks
Melting ice	Albedo	More solar energy absorbed by relatively dark land and sea surfaces, compared with high reflectivity ('albedo') of snow and ice
Melting of permafrost soils	Methane release	5–25 × 10^4 Mt C (as CH_4, a potent greenhouse gas) could be released by 2100 (IPCC RCP 8.5 scenario, low confidence). Release of methane hydrates from seafloor deposits is less certain.
Warmer atmosphere	Water vapour	Increased atmospheric capacity for water vapour, a greenhouse gas
		Negative feedbacks
Increased surface heating	Blackbody (or Planck) radiation	Thermal radiation of a surface increases to the fourth power of its absolute temperature, so loses a greater proportion of its surface energy as it warms. This is included in global climate models.
Weaker atmospheric lapse rate	Lapse rate	Lapse rate (decrease in air temperature with altitude) enhances RF as less infra-red radiation is lost to space from relatively cold air at high altitude, compared with that transferred to Earth's surface from relatively warm air at ground level. Weakening of lapse rate reduces this effect.
		Feedbacks with greater uncertainties
Increased cloud cover (because of larger water vapour content of atmosphere)	Cloud cover	Clouds retain more of Earth's outgoing infra-red radiation (positive feedback) but also reflect more sunlight (negative). Latter feedback lessened by modelled poleward movement of clouds in a warmer climate away from tropics (thus increasing surface heating) to latitudes that receive less sunlight and will therefore reflect less solar radiation than tropical cloud. Likely net feedback positive but low confidence, especially low-altitude clouds (IPCC, 2013).
More CO_2 and heat in atmosphere	CO_2 fertilisation	Increased terrestrial and oceanic C uptake (e.g. forests extend into higher latitudes) leading to increased CO_2 absorption by fixation into woody tissues. But future climate change predicted to decrease biological C uptake.

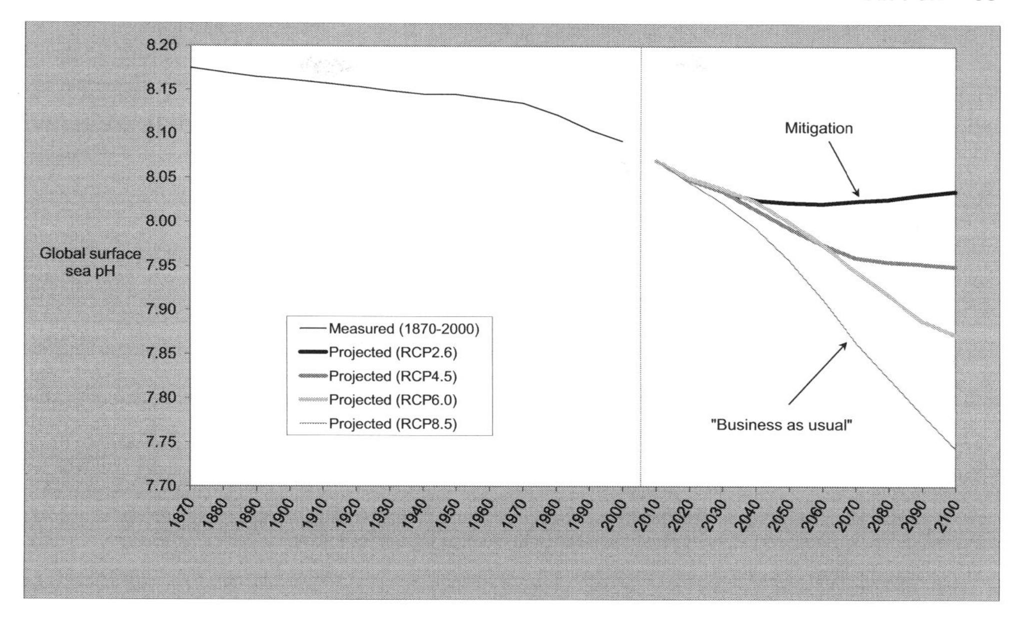

Figure 2.10 Measured change in surface sea pH (1870–2000) and projected changes in the 21st century (RCP scenarios). See Box 2.2 for an explanation of RCPs.

Source: Adapted from Bopp et al. (2013).

equates to a 26% increase in acidity. The oceans are acidifying much more rapidly than at any time in at least tens of millions of years and some areas have higher acidification rates than others. In general, the coldest waters (i.e. in the Polar Regions) are acidifying most quickly because they have a greater capacity for CO_2 absorption.

Ocean acidification is primarily of concern because it reduces the ability of marine organisms to build calcium carbonate structures like shell and coral. Surface seawater is generally saturated with dissolved calcium and carbonate ions, derived from weathering of limestones and other rocks;[2] however, increasing acidity reduces the seawater's $CaCO_3$ saturation state because excess H^+ ions react with CO_3^{2-} ions to form HCO_3^- (bicarbonate) ions. This overall effect of CO_2 absorption by seawater can be summarised as:

$$H_2O + CO_2 + CO_3^{2-} \rightarrow 2HCO_3^- \qquad [2.9]$$

Natural processes of carbonate supply (weathering) are far slower than the current rate of OA, leading to the risk of carbonate undersaturation, which can cause thinning and deformation of shell and coral materials as calcification becomes more difficult and dissolution of organisms' existing structures occurs. There is particular concern for those organisms using the aragonite and magnesium calcite forms of $CaCO_3$ (e.g. corals and some molluscs) because these forms are approximately 50% more soluble than calcite. Reduced calcification in coral is in addition to the threat from coral bleaching, which is also caused by anthropogenic CO_2 (but via global warming; see above). Corals are important and highly valued habitats (Fig. 2.11) and these threats are of particular concern. Reduced calcification and growth has also been reported in molluscs, including mussels and oysters, and dissolution of the shells of pteropods (a small snail) has been observed in the cold waters of the Southern Ocean. Non-calcifying organisms may also be affected by ocean acidification; for example, growth of some fish larvae may be inhibited. Adult fish do not appear to be directly affected, but there may be food chain effects. Some marine organisms (e.g. seagrasses and some algal species) are tolerant of increased CO_2 dissolution or appear unaffected by it and may, on an individual level, benefit from the changes OA brings; however, the changes are predominantly adverse (Fig. 2.12).

Figure 2.11 Coral reef in Biscayne Bay, Florida, USA. Coral reefs are highly valued habitats, among the most biodiverse on the planet.

Credit: US National Parks Service (Wikimedia Commons).

Impacts of air pollution: poor air quality

The previous two sections have detailed the global-scale impacts of anthropogenic CO_2 (and CH_4) emissions. Human disturbance of the C cycle causes significant air pollution on a local scale as well, mainly in relation to carbon monoxide (CO), volatile organic compounds (VOCs), unburnt hydrocarbons (HCs) and particulate carbon. Carbon monoxide is often monitored and regulated in urban areas because of the risk of hypoxia (oxygen deficiency), particularly in poorly aerated environments like road tunnels; however, most deaths from CO inhalation are caused by faulty domestic appliances in homes. Volatile organic compounds are problematic because of their toxicity when inhaled, ingested or absorbed; for example, benzene is known from occupational studies and other research to be carcinogenic and styrene is classed as a possible carcinogen. Unburnt HCs such as PAHs are absorbed by the body when inhaled and converted into metabolites that are carcinogenic, teratogenic and

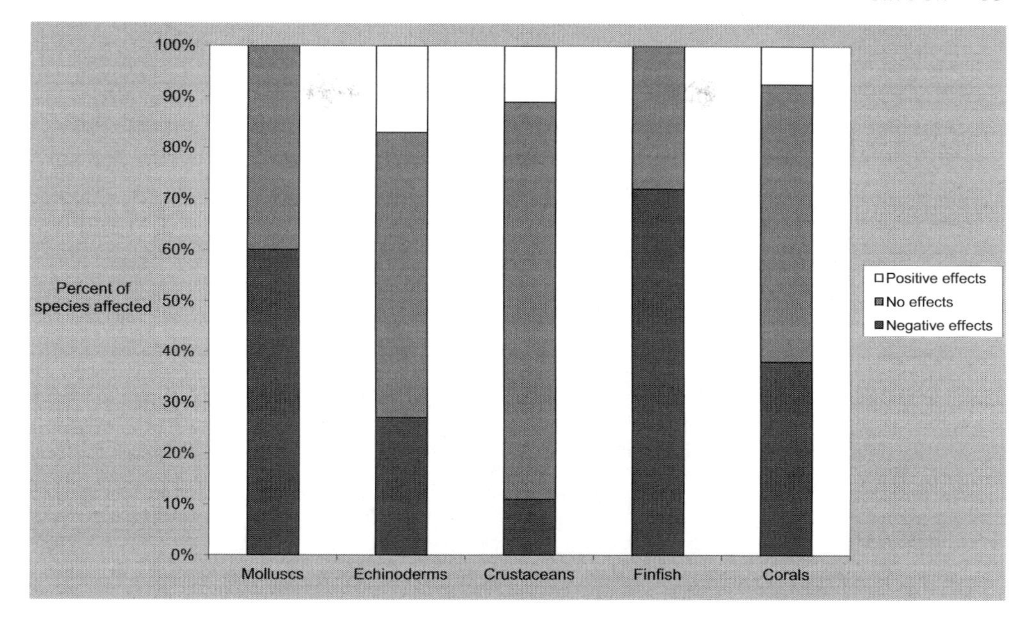

Figure 2.12 Sensitivity of commercially and ecologically important organisms to ocean acidification.

Source: Data from IGBP/IOC/SCOR (2013).

mutagenic. There are thousands of PAH isomers, but of most concern is the carcinogen, benzo(a)pyrene, which is stable and, being lipophilic, is bioaccumulative; it can be adsorbed onto smoke particles and is one of the earliest known toxic pollutants to have had an observed effect on humans, being identified as the cause of scrotal cancer in 19th-century chimney sweeps. Another unburnt HC, 1,3-butadiene, is also carcinogenic and, despite rapid breakdown in the atmosphere, is sometimes recorded at low levels in urban air. Particulate carbon, emitted from fossil fuel and wood combustion is particularly important because particulate matter (PM) affects more people than any other air pollutant; fine PM can damage the respiratory system (see Box 2.4) and is carefully monitored in many urban and industrial areas.

Box 2.4 Particulate matter.

Fine and ultra-fine particles are defined as those with maximum diameters of <10 µm (PM_{10}) and <2.5 µm ($PM_{2.5}$), respectively; far too small to be visible to the naked eye (Fig. 2.13). The PM_{10} fraction is inhaled deep into the lungs where physical damage to the bronchioles can occur. $PM_{2.5}$ is inhaled more deeply still and can also damage the lung alveoli; it can also be absorbed into the blood stream, in some cases carrying metals and other toxins. PM pollution is associated with increased rates of asthma, pulmonary disease and increased mortality and the World Health Organization estimates that each year PM inhalation causes >2 mn excess deaths around the world, many from indoor air pollution (WHO, 2011a). Natural sources of fine PM include volcanoes, salt crystals from evaporating sea spray, crustal material (e.g. Saharan dust) and natural vegetation fires, but anthropogenic sources are of particular concern in polluted areas. Most anthropogenic PM is particulate C ('soot') derived from fuel combustion and deliberate biomass burning (e.g. to clear land for agriculture), but other sources and types of PM exist, such as tyre and mechanical wear in vehicles and nitrate particles derived from N gases {→N}. Carbonaceous soot particles typically carry small amounts of other elements, including toxic metals; where sulfurous coal is burned, fine PM is accompanied by SO_2, leading to a possible synergistic effect {→S}.

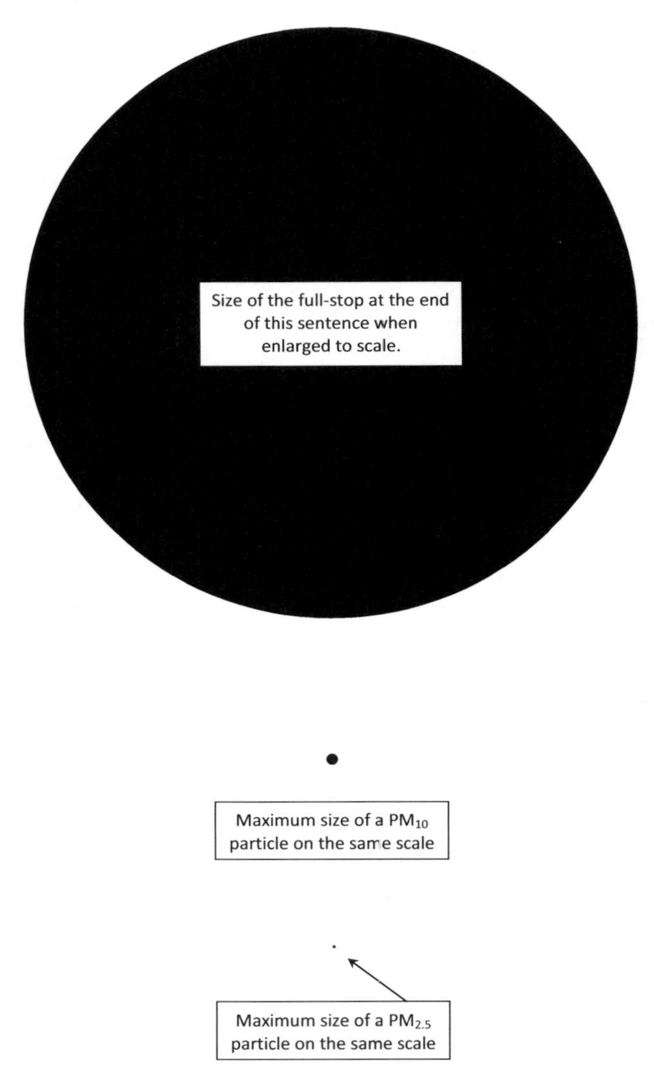

Figure 2.13 PM_{10} and $PM_{2.5}$ are so small that they are invisible to the naked eye. This image compares their sizes to that of a full stop, or 'period'.

In many post-industrial countries, the 'winter-type smogs' characterised by sooty smoke (PM pollution) are not as severe as witnessed in the past (e.g. London, 1952 {→S}), but problems still occur. For example, in the San Francisco Bay area of the USA, domestic wood smoke and vehicle emissions contribute to PM pollution that is exacerbated in the cold winter months by temperature inversions associated with the 'Pacific high', a dominant pressure system in the area; in consequence, a 'Spare the Air' season is declared each winter, during which domestic wood burning is banned. In some countries PM pollution can be very severe (Fig. 2.14). In recent years, China has suffered particularly intense pollution episodes during the winter months when large amounts of PM emitted from domestic and municipal heating, vehicles, industries and power plants have been trapped by high-pressure systems and temperature inversions. Concentrations of PM_{10} in Chinese cities have regularly reached several hundred μg m^{-3} during prolonged episodes (the WHO recommended daily maximum is 50 μg m^{-3}). In other parts of the world, plumes of particulate C are sometimes created when large areas of vegetation are cleared for agriculture by burning; problems of this kind have been particularly prevalent in southeast Asia (Box 2.5).

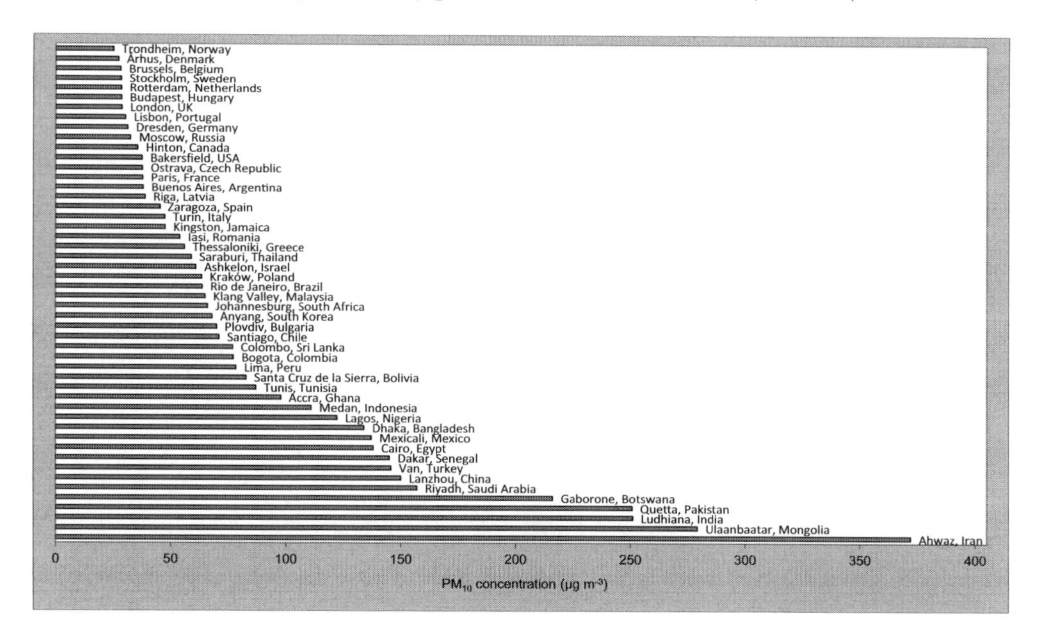

Figure 2.14 The most PM-polluted city in each of 50 selected countries. Concentrations relate to individual years in the 2006–2010 period (majority in 2008 or 2009), except Quetta, Riyadh (2003/04) and Gaborone (2005). Annual averages of PM_{10} concentrations are shown; maximum hourly and daily concentrations are higher. In arid countries PM pollution can be exacerbated by imports of natural dust; for example, the most polluted city, Ahwaz, Iran, has a concentration of heavy industry and traffic, but also suffers from intermittent dust storms.

Source: Data from WHO (2011b).

Box 2.5 Asian Haze and the Asian Brown Cloud.

Forest and peatland fires have frequently caused regional air pollution in southeast Asia since the late 1990s, most recently in 2013 when hundreds of Malaysian schools were shut down and Singapore residents were advised by their Prime Minister to stay indoors. Most of the fires are in Indonesia, where vegetation is cleared for agricultural purposes. The fires generally occur in the dry season between May and October and the smoke generated can be carried across national boundaries (Fig. 2.15) and into densely populated cities, adding to local sources of particulate C (e.g. industry and vehicle emissions). In severe episodes, as well as school closures, flights have been cancelled and people have suffered respiratory problems. Most countries in the region have been affected, in particular Malaysia, Singapore, Brunei, Indonesia and Thailand. With the exception of Indonesia, southeast Asian nations have agreed and ratified the ASEAN Agreement on Transboundary Haze, which aims to take actions to limit the problem. A similar phenomenon to the 'Asian Haze' is the 'Asian Brown Cloud', which covers parts of India and Pakistan each year in the dry season; the source of the problem appears to be particulate C emissions from domestic wood fires, vehicles and industries.

Figure 2.15 Satellite image of pollution over southeast Asia in 1997, one of the worst early incidences of Asian Haze (there have been several since). The white area denotes PM pollution from forest fires in Indonesia, the pollution also covering large parts of Malaysia. The light and dark grey areas denote tropospheric ozone associated with the primary pollution; ozone has a longer atmospheric lifetime than PM and thus travels further afield {→N}.

Credit: NASA (Wikimedia Commons).

Impacts of water pollution: hypoxia and anoxia

The entry of elevated amounts of biodegradable, or respirable, organic matter (ROM) into water bodies can be hazardous for aquatic life. The most serious effect is hypoxia or anoxia caused by the sudden increase in oxygen consumption, or 'biochemical oxygen demand' (BOD), caused by respiration of the ROM by aerobic bacteria. Increased BOD is problematic because oxygen diffuses into water at a very slow rate, only increasing significantly if the water is agitated by turbulent flow, and water does not have a large saturation capacity for dissolved oxygen (DO). For example, lowland freshwaters at 10°C have a saturation DO concentration of ~11 mg L^{-1} and, because DO saturation is temperature dependent, the concentration is lower than this in warmer waters (Table 2.9). The DO requirements of freshwater fish like bass and trout (6.5 mg L^{-1}) and invertebrates such as caddisfly and mayfly larva (4 mg L^{-1}) are not much lower than saturation concentrations and such organisms die if the increased oxygen demand caused by ROM pollution reduces DO below these levels. The reduced DO concentration in warmer waters has implications for thermal pollution as well as warmer climates.

A number of materials can increase the BOD of natural waters (Table 2.10). Food wastes can be particularly problematic; for example, milk leaking from crashed dairy tankers can cause fish kills by entering drains and, ultimately, local rivers. In other cases milk has been dumped on land and into rivers by farmers protesting against quotas; for example, a major fish kill was caused by such a protest in Germany in 2009. A major, but less direct, cause of BOD increase is eutrophication {→N →P}. Pollution of water bodies by ROM may also cause problems of turbidity (blanketing and reduced photosynthesis) and discoloration.

Table 2.9 Saturation concentrations of dissolved oxygen in freshwater at sea level.[a,b]

Temp (°C)	*DO[c] (mg L^{-1})*	*Temp (°C)*	*DO[c] (mg L^{-1})*	*Temp (°C)*	*DO[c] (mg L^{-1})*
0	14.62 (11.2)	10	11.29 (8.8)	20	9.09 (7.2)
1	14.22	11	11.03	21	8.91
2	13.83	12	10.78	22	8.74
3	13.46	13	10.54	23	8.58
4	13.11	14	10.31	24	8.42
5	12.77 (9.9)	15	10.08 (7.9)	25	8.26 (6.6)
6	12.45	16	9.87	26	8.11
7	12.14	17	9.66	27	7.97
8	11.84	18	9.47	28	7.83
9	11.56	19	9.28	29	7.69

a Waters can temporarily have DO levels above those shown because of supersaturation, caused by photosynthesis in aquatic plants and algae.
b The saturation DO level at a given temperature decreases at higher altitudes because of the reduction in atmospheric pressure; example correction factors are 0.94 (500 m); 0.89 (1000 m).
c Saturation DO is also reduced by salinity; figures in brackets show saturation DO concentrations in seawater at selected temperatures.

Table 2.10 Biochemical oxygen demand of various types of respirable organic matter.

Material	*BOD (mg L^{-1})*
Treated sewage	<5
Raw sewage	300–400
Dairy washings	1,000–2,500
Cattle slurry	15,000–20,000
Pig slurry	20,000–30,000
Silage effluent	60,000–70,000
Milk	80,000–120,000
Industrial wastewaters:	
Oil refineries	100–300
Tanneries	1,000–3,000
Bottling plants	200–6,000
Food processing	100–7,000
Paper factories	250–15,000
Distilleries/sugar, molasses factories	600–32,000

Source: Data from SEPA (2014) and, for industrial wastewaters, UNEP-CEP (1998).

Impacts of water pollution: endocrine disrupting chemicals

Some of the organic compounds emitted into waters from domestic and industrial sources can disrupt the hormone (endocrine) systems of humans and animals and are known as hormone disruptors or endocrine disrupting chemicals (EDCs); these include bisphenol-A, phthalates, nonylphenol and ethinylestradiol (the synthetic oestrogen used in contraceptive pills). These compounds disrupt the synthesis and actions of natural hormones and this can affect development and reproduction in exposed organisms and their offspring. The greatest damage appears to occur during the foetal–infant development period and in puberty. The main mechanisms of endocrine disruption are mimicking of hormones and disruption of proteins that control the transport and delivery of hormones in the body. Some EDCs, including phthalates and nonylphenol, mimic natural oestrogen and are termed xenoestrogens. The main health effects that have been linked with EDC exposure are listed in Table 2.11.

Compounds of particular concern, because of their widespread use in commercial products, are bisphenol-A and phthalates. Bisphenol-A has been linked to the disruption of thyroid function, insulin production, immune response and the reproductive system; it has also been linked with breast and prostate cancers. Concerns

Table 2.11 Health effects in humans and animals that have been linked with endocrine disrupting chemicals.

Breast, prostate, testicular, ovarian, thyroid and other cancers
Birth defects
Impaired immune response
Neurological and learning problems
Disorders of the reproductive system, including intersex characteristics
Early onset of puberty

about phthalates relate to reduced spermatogenesis, shortened gestation, early onset of puberty and disruption to thyroid function and insulin production. Pollution of water bodies by the synthetic oestrogen ethinylestradiol (via sewage effluents) is associated with the development of intersex conditions in fish and amphibians, including egg production in males. The European Union has proposed an environmental quality standard (EQS) for this EDC and, if adopted, much effort will be required to meet the requirements (Table 2.12). Some oestrogenic growth promoters, which are used in livestock and enter waters via agricultural waste, have been discounted as potential environmental EDCs because of their rapid breakdown in sunlight; however, recent research indicates that some are able to reform during darkness and may be more persistent than previously believed. Nonylphenol and octylphenol ethoxylates have relatively weak oestrogenic effects but are common in environmental waters because of their widespread use in detergents and other commercial products.

Many of the possible effects of EDCs have been reported from experimental studies using laboratory animals and caution should always be applied in such cases (see chapter 1). Indeed, there is some debate among scientists about the magnitude of endocrine disruption that can safely be attributed to environmental exposure to EDCs in humans and wildlife. A major WHO/UNEP report on EDCs in the environment (Bergman et al., 2012) concludes, however:

> There is a growing concern that maternal, fetal and childhood exposure to EDCs could play a larger role in the causation of many endocrine diseases and disorders than previously believed.

Concern is also expressed in the report that most chemicals in current use have not been tested for endocrine disruption.

Impacts of water pollution: drinking water contamination

Organic solvents such as toluene, benzene and xylenes evaporate from contaminated surface waters; however, if they spill or leak into the ground (e.g. during petroleum refining, storage or transport) they do not bind readily to soils and can rapidly enter

Table 2.12 Proportions of national river lengths in European Union member states that are predicted by modelling techniques to exceed an EU-proposed environmental quality standard for 17α-ethinylestradiol of 0.035 ng L^{-1} (annual average).

Percentage	*European Union member states*
>30%	Belgium, Bosnia, Germany, Macedonia, Netherlands, Poland, Romania, Serbia, Slovakia
25–30%	Czech Republic, England, Hungary
20–25%	Albania, Bulgaria, Denmark, Greece, Portugal
15–20%	Austria, Italy, Switzerland
10–15%	Croatia, Luxembourg, Spain
>10%	Estonia, Finland, France, Ireland, Latvia, Lithuania, Norway, Slovenia Sweden, Wales/Scotland

Source: Data from Johnson et al. (2013).

groundwaters, remaining there for long periods. For this reason, maximum contaminant levels in drinking waters are often established, mainly to protect against neurotoxicity in the case of toluene and xylenes and increased risk of cancer in the case of benzene (Table 2.13). Drinking waters are also monitored for the PAH, benzo(a)pyrene, which can enter waters from atmospheric deposition and urban runoff and, in the case of groundwaters, from water tanks and distribution pipes that have bituminous linings. Long-term exposure to benzo(a)pyrene in drinking water may increase the risk of cancer and cause reproductive disorders. Most of the other organic compounds that are routinely monitored in drinking waters are organochlorines {→Cl}.

Concerns have been raised about the potential for groundwaters and drinking waters to be contaminated by the process of hydraulic fracturing, or 'fracking'. Fracking is a process of natural gas (CH_4) extraction from shales and some other formations including coalbeds; the shale gas is extracted by chemicals that are pumped into the rock at high pressure. The potential for pollution arises from: (i) surface spills and effluents, both of the fracking fluids and the wastewaters generated by the process (see also →Br); (ii) the direct injection of fracking fluids into aquifers and; (iii) the migration of gases into groundwaters, potentially contaminating groundwaters. In 2011, the US Environmental Protection Agency began a detailed study of the potential impacts of fracking on drinking waters and a year later published a first progress report, which did not set out to draw conclusions but which included in its appendix lists of the chemicals used (USEPA, 2012). The lists run to several pages and contain predominantly organic compounds, including many of those discussed in this chapter; a large number of inorganic compounds are also listed. In 2013, an extensive study found dissolved methane, ethane and propane contamination of drinking waters associated with proximity to shale gas wells (Fig. 2.16); 12 households located <1.5 km away had CH_4 concentrations above the threshold for immediate remediation.

The contamination of rivers and drinking water by pharmaceuticals (e.g. painkillers and antibiotics derived from domestic sewage effluents and farm wastes) has been reported more frequently in the scientific literature in recent years, although this may be because of advances in analytical techniques rather than any increase in their disposal from hospitals or other potential sources. Concentrations of such pharmaceuticals are very low and measurable harm to humans or wildlife is currently thought to be unlikely; endocrine-disrupting pharmaceuticals such as ethinylestradiol are of more concern, however (see above).

Table 2.13 Maximum contaminant levels in US drinking waters.

Pollutant	*Maximum contaminant levels (mg L^{-1})*
Toluene	1
Benzene	5×10^{-3}
Xylenes	10
Benzo(a)pyrene	2×10^{-4}

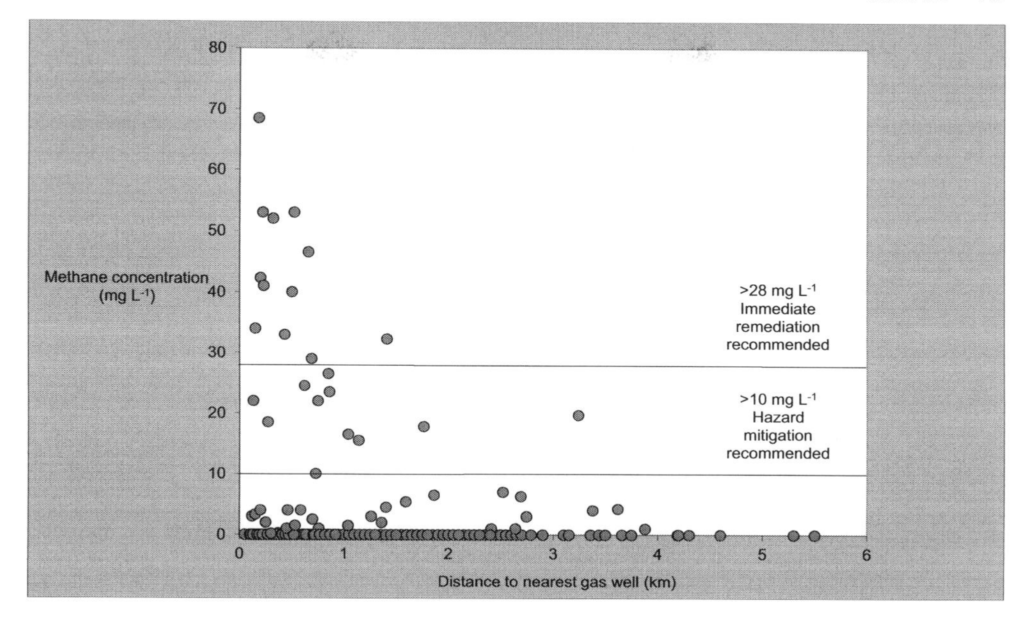

Figure 2.16 Dissolved methane concentrations in drinking waters plotted against distance from the nearest gas well in the Marcellus shale gas area, Pennsylvania, USA (n = 141). Methane (p = 0.007) and ethane (p = <0.005) were significantly correlated with distance from gas wells. The horizontal lines indicate US Department of the Interior recommendations for methane-contaminated drinking waters.

Source: Adapted from Jackson et al. (2013).

Impacts of water pollution: the oceans

Pollution of the marine environment is caused by some of the same processes described above for freshwaters, particularly where the pollutants described enter the oceans via riverine inputs; this is the case for ROM, toxic organic compounds and endocrine-disrupting chemicals. Alternatively, inputs may be direct. For example, ROM, in the form of human waste, is sometimes emitted directly into coastal waters via sewage outflows. Ships wash out their empty fuel and cargo tanks at sea using seawater and this can lead to pollution by residues of oil and other wastes; for example, in 2013 a material called polyisobutene, a lubricant additive, caused the deaths of hundreds of seabirds on the southern coast of England. The deliberate dumping of wastes at sea is banned under the international 'Convention on the Prevention of Marine Pollution by Dumping of Wastes and Other Matter' (1972), but the Convention allows permits for the disposal of some wastes under certain conditions, including dredged material and sewage sludge.

Accidental spillages of oil at sea can cause particularly serious pollution, the main effects being damage to the plumage of sea birds (inhibiting buoyancy, flight and insulation) and to their digestive and other organs, as a result of self-cleaning. Affected bird populations typically recover over time by natural re-population but short- to medium-term damage is often severe. Other affected organisms include coastal-dwelling mammals and shoreline invertebrates, populations of which can take many years to recover, sometimes leading to seaweed proliferation in the absence of grazers. A portion of the spilled oil is volatilised, diluted and/or biodegraded by bacteria, but some fractions may persist in sediments and beach sands for many years. Oxygen depletion of affected waters can also occur as the oil is broken down by aerobic bacteria. Economic damage, mainly related to fishing and/or tourism industries, can also be serious after a major oil spill. Some notable oil spills are listed in Table 2.14 and the most recent of these is outlined in more detail in Box 2.6.

Table 2.14 Notable oil spills.

Year	*Tanker/Rig*	*Location*
1967	Torrey Canyon	United Kingdom
1978	Amoco Cadiz	France
1979	Atlantic Empress	Trinidad & Tobago
1989	Exxon Valdez	USA
2010	Deepwater Horizon	USA

Box 2.6. The Deepwater Horizon oil spill.

On 20 April 2010, oil began to leak into the Gulf of Mexico following an explosion on the Deepwater Horizon oil drilling rig. An estimated 6.5×10^8 L of oil was spilled over the following 3 months until the leak was sealed. The spill covered a large area of the Gulf (Fig. 2.17) and >1500 km of coastline were contaminated by oil (~350 km heavily contaminated). Approximately 2.9×10^6 L of dispersant chemicals were injected at the source of the leaking oil by BP, the owners of the drilling rig, and a further 4.5×10^6 L were applied to the surface oil slick. Seabirds, fish, marine mammals and turtles were directly exposed to the oil. Estuarine oyster reefs were also exposed to the oil as it drifted towards the Gulf shoreline. On the Gulf coast, more than 8000 dead or potentially injured seabirds were collected by response teams, around 1400 of which were treated and released. Approximately 1000 oiled turtles were also collected (most died), as well as 171 marine mammals, of which 153 were dead. Reduction in benthic invertebrate biodiversity in the deep ocean was observed over an area of 148 km^2 and researchers estimate that recovery of the affected sea bed will take decades. Injury to other specialised deep sea creatures including bacteria, coral, jellyfish and fish, is still being evaluated by marine experts. Three years after the spill, hundreds of miles of coastline were still being cleaned or evaluated and large amounts of residual tar were still in evidence. The cost to the rig's owner, BP, has been at least $14 bn and court settlements are still being heard. Research teams continue to investigate the impacts of the spill on the natural environment.

Information: NOAA (2012) and Montagna et al. (2013).

Figure 2.17 Satellite image of the Gulf of Mexico and part of the south coast of the USA, 24 May 2010, showing the extent of the Deepwater Horizon oil spill one month after the explosion. The bright area at the centre is sunlight reflecting off the smooth surface of the main area of oil and numerous slicks can be seen (light grey) extending mainly to the north and east including the tip of the Mississippi Delta directly north of the main area of oil. The width of the image is approximately 500 km.

Credit: NASA (Wikimedia Commons).

Environmental impacts of insecticide use

Apart from organochlorine (OC) and organophosphate (OP) pesticides {→Cl →P}, the main insecticides of concern with regard to toxicity are carbamates (CB), pyrethroids (PT) and neonicotinoids (NN). These pesticides are insect pest neurotoxins and are generally preferred in agriculture to the OCs and OPs, because of their more rapid degradation in the environment (CB, PT), their lower toxicity to birds and mammals (CB, PT, NN) or their ability to be translocated by plants (NN); however, specific pesticides of these types can be toxic to non-target organisms.

The mode of toxicity in carbamates, such as aldicarb and carbaryl, is the inhibition of the acetylcholinesterase enzyme that is also observed in OP pesticides {→P} and that causes neurotoxicological damage in the target insect; the main difference with OPs is that carbamates have a more temporary effect and the most commonly affected non-target organisms, particularly invertebrate and fish species, can often recover relatively quickly.

Pyrethroids are synthetic (and more photostable) versions of pyrethrin, an insecticide that occurs naturally in plants of the *Chrysanthemum* genus. They are used in domestic products as well as agriculture and can enter the wider environment via urban and agricultural runoff, accumulating in sediments to some extent. Pyrethroids are also neurotoxins, paralysing the target insect, and they are toxic to most aquatic organisms including fish and invertebrates; however, they have relatively low toxicity to most birds and mammals, including humans.

In recent decades, neonicotinoid insecticides have been developed, based on the naturally occurring plant alkaloid nicotine, and are now very widely used; again they are neurotoxic to insects, causing persistent excitation of acetylcholine receptors, and they are targeted at crop pests. Serious concerns have been raised however, about their neurotoxicological and immunotoxicological effects on honey bees, as well as their potential toxicity to other wild insects, birds and aquatic invertebrates. Bees are vitally important in natural ecosystems and in the pollination of crops and it is feared that exposure to neonicotinoids may lead to reductions in bee populations. In 2013, the European Commission proposed a restriction on three neonicotinoids (clothianidin, imidacloprid and thiametoxam) because of the possibility of a high risk to bees as well as other organisms, including aquatic invertebrates. A well-publicised study (Lu et al., 2014) found that sub-lethal exposure of 12 honey bee (*Apis mellifera*) colonies to

Figure 2.18 The western (or 'European') honey bee, *Apis mellifera*. There are concerns that exposure to neonicotinoid pesticides may seriously harm these vitally important pollinators.

Photo credit: Jon Sullivan (pdpphoto).

clothianidin or imidacloprid eventually led to hive abandonment in six of them ('colony collapse disorder'), while a further six control hives had no abandonments. Neonicotinoids are not known to be toxic to vertebrates but concerns have been raised about the potential for food chain effects related to declines in non-target insects because of their relative stability in the environment. For example, imidacloprid pollution of freshwaters in the Netherlands was significantly associated ($p < 0.0001$) with declines in insectivorous bird populations following the introduction of imidacloprids in the 1990s (confounding factors corrected); average reductions of 3.5% per year were recorded where concentrations of imidacloprid were >19.4 ng L^{-1} (Hallmann et al., 2014).

Notes

1 The IPCC report considers natural variations including solar output and concludes that measured solar activity since 1986 has not contributed to the observed global warming; the IPCC does state however that natural variability in solar output over its 11-year cycle may have influenced climate in some regions.

2 The saturation scale for dissolved minerals in seawater is based on Ω (omega) values, where $\Omega<1$ = undersaturation and $\Omega>1$ = supersaturation. Deeper waters are naturally undersaturated, mainly as a consequence of higher pressure; in contrast surface waters are saturated (except where upwelling of deeper, undersaturated water occurs) and the boundary between the two, the saturation horizon, is becoming shallower in some areas.

References

Bergman, A., Heindel, J.J., Jobling, S., Kidd, K.A. and Zoeller, R.T. (Eds). 2012. *State of the Science of Endocrine Disrupting Chemicals*. WHO and UNEP, Geneva.

Boden, T.A., Marland, G. and Andres, R.J. 2014. *Global, Regional and National Fossil Fuel CO_2 Emissions*. Available at: http://cdiac.ornl.gov

Bopp, L., Resplandy, L., Orr, J.C., Doney, S.C., Dunne, J.P., Gehlen, M., Halloran, P., Heinze, C., Ilyina, T., Seferian, R., Tjiputra, J. and Vichi, M. 2013. Multiple stressors of ocean ecosystems in the 21st century: projections with CMIP5 models. *Biogeosciences* 10, 6225–6245.

CDIAC (Carbon Dioxide Information Analysis Center). 2014. *Index of Trends: CO_2*. Available at: http://cdiac.esd.ornl.gov/ftp/trends/co2/siple2.013

Hallmann, C.A., Foppen, R.P.B., van Turnhout, C.A.M., de Kroon, H. and Jongejans, E. 2014. Declines in insectivorous birds are associated with high neonicotinoid concentrations. *Nature* 511, 341–343 (doi:10.1038/nature13531).

IGBP/IOC/SCOR (International Geosphere-Biosphere Programme, Intergovernmental Oceanographic Commission and Scientific Committee on Ocean Research). 2013. *Ocean Acidification: Summary for Policymakers – Third Symposium on the Ocean in a High-CO_2 World*. International Geosphere-Biosphere Programme, Stockholm.

IPCC (Intergovernmental Panel on Climate Change). 2013. *Climate Change 2013: The Physical Science Basis. Working Group II Contribution to the Fifth Assessment Report of the IPCC*. IPCC, Geneva.

IPCC (Intergovernmental Panel on Climate Change). 2014. *Climate Change 2014: Impacts, Adaptation and Vulnerability. Volume 1: Global and Sectoral Aspects. Working Group II Contribution to the Fifth Assessment Report of the IPCC*. IPCC, Geneva.

Jackson, R.B., Vengosh, A., Darrah, T.H., Warner, N.R., Down, A., Poreda, R.J., Osborn, S.G., Zhao, K. and Karr, J.D. 2013. Increased stray gas abundance in a subset of drinking water wells near Marcellus shale gas extraction. *Proceedings of the National Academy of Sciences* 110(28), 11250–11255.

Johnson, A.C., Dumont, E., Williams, R.J., Oldenkamp, R., Cisowska, I. and Sumpter, J.P. 2013. Do concentrations of ethinylestradiol, estradiol and diclofenac in European rivers exceed proposed EU environmental quality standards? *Environmental Science and Technology* 47, 12297–12304.

Lu, C., Warchol, K.M. and Callahan, R.A. 2014. Sub-lethal exposure to neonicotinoids impaired honey bees winterization before proceeding to colony collapse disorder. *Bulletin of Insectology* 67(1), 125–130.

Montagna, P.A., Baguley, J.G., Cooksey, C., Hartwell, I., Hyde, L.J., Hyland, J.L., Kalke, R.D., Kracker, L.M., Reuscher, M. and Rhodes, A.C.E. 2013. Deep-sea benthic footprint of the Deepwater Horizon blowout. *PLoS ONE* 8(8): e70540 (doi:10.1371/journal.pone.0070540).

NASA/GISS (National Aeronautics and Space Administration/Goddard Institute for Space Studies). 2014. Giss Surface Temperature Analysis (GISTEMP). Raw data updated monthly and available at: http://data.giss.nasa.gov/gistemp/

NOAA (National Oceanic and Atmospheric Administration). 2012. *Natural Resource Damage Assessment for the Deepwater Horizon Oil Spill*. Available at: http://www.gulfspillrestoration.noaa.gov/wp-content/uploads/FINAL_NRDA_Status Update_April2012.pdf

NOAA/ESRL (National Oceanic and Atmospheric Administration/Earth System Research Laboratory). 2014. *Trends in Atmospheric Carbon Dioxide*. Available at: www.esrl.noaa.gov/gmd/ccgg/trends/

RCP (Representative Concentration Pathways) Database. 2014. Available at: http://tntcat.iiasa.ac.at/RcpDb

SEPA (Scottish Environment Protection Agency). 2014. *Prevention of Environmental Pollution from Agricultural Activity: Code of Good Practice*. Available at: http://www.sepa.org.uk/land/guidance.aspx

UNEP-CEP (United Nations Environment Program: Caribbean Environment Program). 1998. *Appropriate Technology for Sewage Pollution Control in the Wider Caribbean Region, Technical Report 40*. UNEP-CEP, Kingston.

USEPA (United States Environmental Protection Agency). 2012. *Study of the Potential Impacts of Hydraulic Fracturing on Drinking Water Resources: Progress Report*. USEPA, Washington DC.

WHO (World Health Organization). 2011a. Tackling the global clean air challenge. WHO News Release. Available at: http://www.who.int/mediacentre/news/releases/2011/air_pollution_20110926/en/

WHO (World Health Organization). 2011b. *Urban Outdoor Pollution Database*. WHO, Geneva.

3 Nitrogen

Environmental reservoirs and chemical forms

Atmosphere

Most of the nitrogen (N) within the Earth system is contained in the atmosphere. More than 99.9% of atmospheric N is present as nitrogen gas (N_2), which comprises 78% of the atmosphere by volume. Its atmospheric abundance is attributable to its stability: the triple covalent bond between the two nitrogen atoms in the N_2 molecule requires a large energy input to be broken (Fig. 3.1). For this reason, photolysis of N_2 molecules to monatomic N occurs only in the upper atmosphere, in the presence of sufficiently strong short-wave radiation.

Other nitrogen-containing gases are present in trace amounts, collectively accounting for less than 0.001% of the atmosphere by volume. Of these trace gases, nitrous oxide (N_2O), a greenhouse gas, is by far the most abundant, mainly because it is relatively stable and has an atmospheric lifetime of >100 years. The other main N gases are more short-lived in the atmosphere: they include ammonia (NH_3), nitric oxide (NO) and nitrogen dioxide (NO_2). The chemistry of the latter two oxides of N is closely linked and they are collectively known as 'NO_X'. Other atmospheric components include:

(i) nitric acid (HNO_3), which is present as a gas or within aerosols/cloud droplets;
(ii) ammonium nitrate (NH_4NO_3), which is produced via reaction of NH_3 and HNO_3 (equation 3.1); it occurs either as a solid (as shown in the equation), forming cloud condensation nuclei, or as an aqueous species;

$$NH_3\ (g) + HNO_3\ (g) \rightarrow NH_4NO_3\ (s) \qquad [3.1]$$

(iii) amines (derivatives of ammonia), where one or more of the H atoms in NH_3 is replaced by an organic group such as an aromatic or an alkyl (e.g. methylamine, CH_3NH_2 and trimethylamine, $(CH_3)_3$-N) or by elements such as chlorine (e.g. chloramine, $NClH_2$).

Lithosphere

Nitrogen is present at high concentrations in the interior of the Earth, but it is not abundant in the near-surface rocks of the Earth's crust, in contrast to the other major

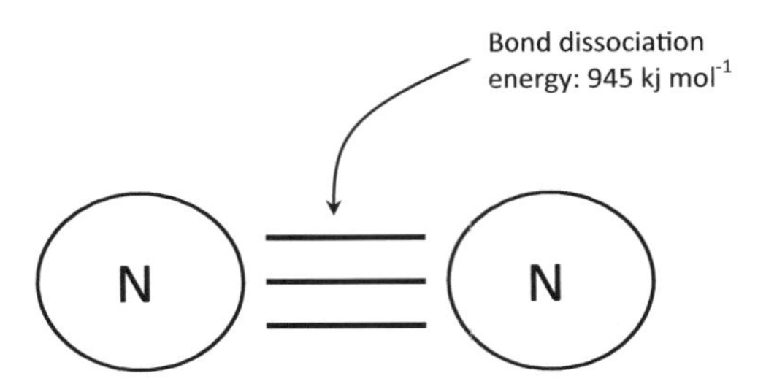

Figure 3.1 The N_2 molecule is very stable, with a high bond energy (bond enthalpy).

Figure 3.2 Rare deposits of sodium nitrate in the Atacama Desert, Chile were mined for export as fertiliser.

Credit: Carlos Varela (Flickr).

elements. It is present in fossil fuels (up to 2% of the contents of coal, for example) and organic-rich shales, reflecting the role of N in the biosphere. It is distributed finely, as NH_4, in components of igneous rocks, such as feldspars and micas. Evaporite nitrate formations are rare, but a notable deposit of sodium nitrate exists, as the mineral nitratine, in the Atacama Desert of Chile and part of Peru (Fig. 3.2); deposits of the highly soluble nitratine are found in smaller quantities in other arid environments, including Death Valley, USA and the deserts of Egypt. In soils, nitrogen is present in organic matter and, in a less-available form, as adsorbed NH_4 in clay minerals.

Hydrosphere

Nitrogen is mainly present in the oceans as unreactive and non-bioavailable dissolved N_2, comprising 95% of the total N. Dissolved N_2O is also present in seawater. Freshwaters and marine waters contain dissolved ions of nitrate (NO_3), nitrite (NO_2) and ammonium (NH_4) as well as organic N, in dissolved and particulate forms.

Biosphere

Nitrogen is an essential element, being present in the fundamental building blocks of living tissues. It mainly exists in biomass as amino N (i.e. in the amine functional group, NH_2) in protein structures and amino acids and in nucleotides in DNA and RNA. It is excreted in urea, $(NH_2)_2CO$. Because of the non-reactive nature of its dominant N_2 form, nitrogen is one of the key limiting factors in primary productivity and, as such, limited availability of N is a control on growth at the base of food chains.

The natural nitrogen cycle

The key reservoirs and fluxes of N are detailed in Fig. 3.3. Atmospheric N_2 is very abundant, but nearly all organisms are unable to metabolise it. Lightning or high-temperature biomass fires have sufficient energy to split the molecule into the individual N atoms required for life, but they provide relatively small amounts of available N; however, specialised N-fixing micro-organisms in soils and waters have the ability to metabolise N_2, and their activities provide available forms of N for the food chain (note that the internal N cycles in soils and waters represent the majority of the total flux). The budgets (inputs minus outputs) for the atmospheric and soil reservoirs can be balanced out within the ranges given, but there are net inputs to the hydrosphere and rocks, the latter via permanent burial of N in ocean sediments. This is not unexpected, given the excess (anthropogenic) fixation of atmospheric N_2 into soils (see below) and the subsequent transfer of excess soil nitrate to rivers and seas. Anthropogenic N fixation also implies a net atmospheric output, which is borne out by the negative atmospheric budget that ensues when either the minimum, mid-range or maximum values of inputs and outputs are used.

Nitrogen is found in several chemical forms and oxidation states, ranging from –3 to +5 (Table 3.1), and transformations between the various forms underlie the key processes within the N cycle. These are described in detail below, but in summary they

Table 3.1 Oxidation states of important nitrogen species and compounds mentioned in the text (with typical chemical state in brackets).

Oxidation state	*Species*
–3	Ammonia, NH_3(g) Ammonium, NH_4^+ (aq) Ammonium salts, e.g. $(NH_4)_2SO_4$ (s) Methylamine, CH_3NH_2 (g)
0	Nitrogen, N_2 (g)
+1	Nitrous oxide, N_2O (g)
+2	Nitric oxide, NO (g)
+3	Nitrous acid, HNO_2 (g/aq)
+4	Nitrogen dioxide, NO_2 (g) Nitrite, NO_2^- (aq)
+5	Nitric acid, HNO_3 (g/aq) Nitrate, NO_3^- (aq) Nitrate salts, e.g. $NaNO_3$ (s)

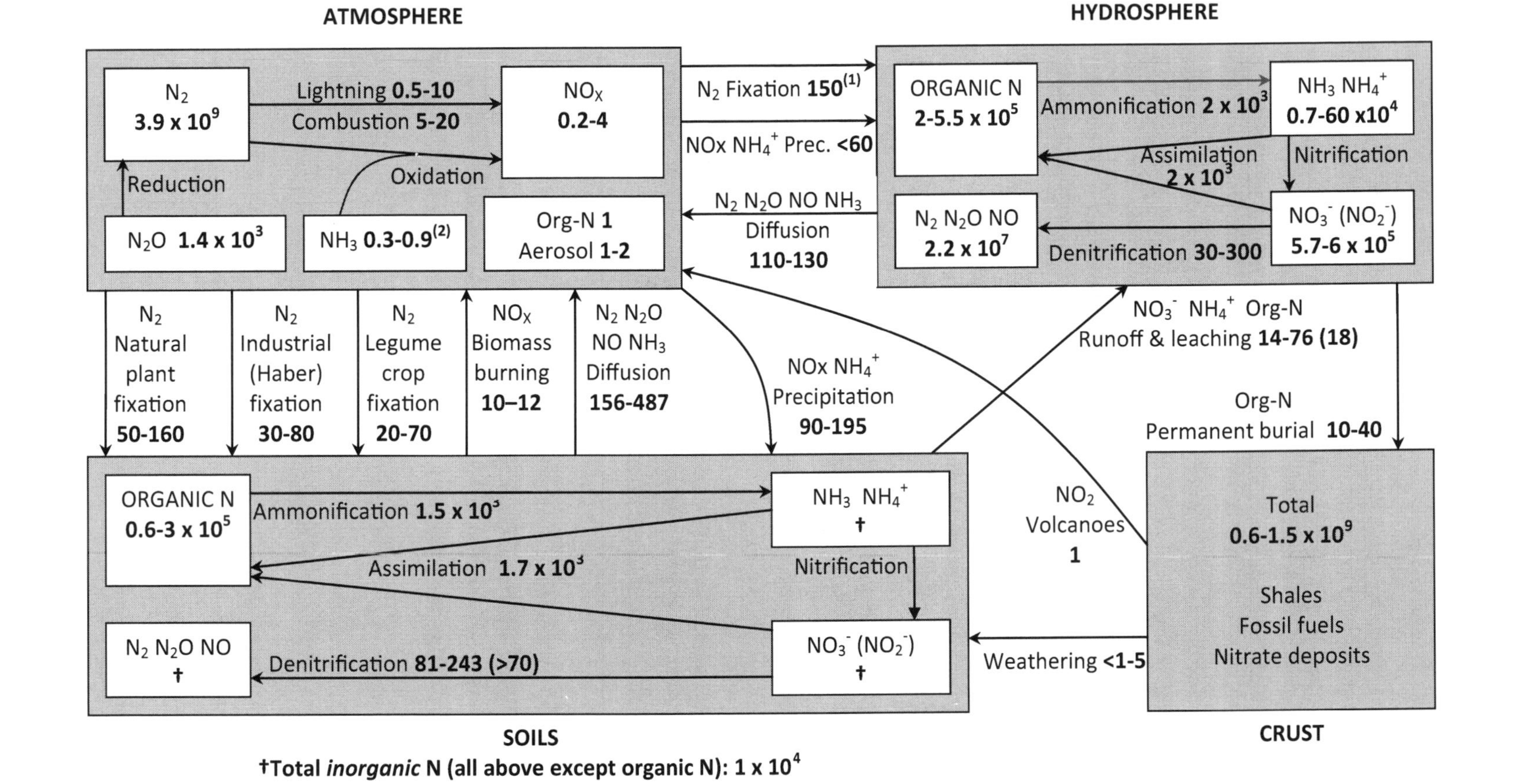

Figure 3.3 The nitrogen cycle. Units for estimated reservoirs (in boxes) and fluxes (adjacent to arrows) are Mt N and Mt N a^{-1}, respectively. Figures in brackets show pre-industrial fluxes, where estimated. The ranges of values demonstrate the considerable uncertainties in their estimation, although only one range covers much more than one order of magnitude.

Notes:

(1) Estimates of oceanic N fixation vary widely, but Schlesinger and Bernhardt (2013) point out that recent estimates of ~150 Mt a^{-1} (used here) are 10 times higher than they were two decades ago

(2) atmospheric ammonia (NH_3) is precipitated as ammonium (NH_4).

Key data sources: Jaffe (1992) and Schlesinger and Bernhardt (2013).

are: N fixation from the atmosphere into soils and waters; assimilation of fixed N into proteins by soil and aquatic organisms; breakdown of organic N to inorganic forms by ammonification; nitrification of ammonia to bioavailable ammonium and nitrate ions; and denitrification of nitrate to N_2 and other gases that are returned to the atmosphere.

Abiotic cycling

Nitrogen fixation (atmosphere → lithosphere/hydrosphere)

Some N_2 is 'fixed' to NOx via its oxidation in the atmosphere, the energy being supplied by lightning. A lightning bolt has sufficient energy to split N_2 and O_2 molecules into monatomic N and O, which recombine as nitric oxide, NO (equation 3.2). Natural fires can also drive this reaction but high temperatures of at least 1500°C are required. The NO is oxidised to NO_2 (equation 3.3), which dissolves in atmospheric water to form nitric acid, HNO_3 and nitrous acid, HNO_2 (equation 3.4). The component nitrate (NO_3), nitrite (NO_2) and hydrogen ions of these acids (see equations 3.5 and 3.6) return to the surface in rainfall providing an abiotic source of available N to plants.

$$N_2 + O_2 \rightarrow 2NO \qquad [3.2]$$

$$2NO + O_2 \rightarrow 2NO_2 \qquad [3.3]$$

$$2NO_2 + H_2O \rightarrow HNO_3 + HNO_2 \qquad [3.4]$$

$$HNO_3 \rightarrow H^+ + NO_3^- \qquad [3.5]$$

$$HNO_2 \rightarrow H^+ + NO_2^- \qquad [3.6]$$

Other N gases that are naturally present in the atmosphere, particularly ammonia, are also water soluble and may similarly be returned to the surface via rainfall as well as by dry deposition.

Biotic cycling

Nitrogen fixation (atmosphere → lithosphere/hydrosphere)

Although the atmosphere is dominated by N_2, this gas cannot be used directly by plants; they can only take up N through the roots as NO_3 and NH_4. The gap is bridged by specialised soil and aquatic micro-organisms called diazotrophs, which are able to undertake N fixation – i.e. the transformation of atmospheric N_2 into NH_3/NH_4, which higher organisms are able to assimilate. Although the process requires energy (to split the strong triple bond in N_2), such N-fixing species gain an advantage in environments that have low levels of available N, in low-humus soils for example. The first diazotrophs, cyanobacteria, evolved in the primitive oceans that contained little if any available N; this is because abiotic N fixation requires an oxygenated atmosphere (see equation 3.2), which was not present before photosynthetic organisms had evolved. Nitrogen fixation supplies an estimated 12% of the N input to biomass each year; the remainder is provided by the biodegradation of organic matter, which releases previously fixed N (see ammonification, below).

The energy required for N fixation is provided by oxidation of organic molecules, with adenosine triphosphate (ATP) and electron transfer being integral to the process. The N is fixed as ammonia:

$$N_2 + 8H^+ + 8e^- + 16ATP \rightarrow 2NH_3 + H_2 + 16ADP + 16P \quad [3.7]$$

The fixed N is then assimilated into amino acids such as glycine, CH_2NH_2COOH:

$$NH_3 + 2CO_2 + H_2O \rightarrow 1.5O_2 + CH_2NH_2COOH \quad [3.8]$$

In the pedosphere, N-fixers are either free-living in the soil (e.g. *Azotobacter* spp.) or exist in symbiotic association with plants (e.g. *Rhizobium* spp.). The latter live in nodules that form on the roots of leguminous plants (legumes) such as peas, beans and clover (Fig. 3.4). Symbiosis occurs also in some non-legumes, such as alder (*Alnus* spp.) and gut bacteria of termites fix N, making it available to their hosts. Symbiotic N-fixers are host-specific. For example, *Rhizobium japonicum* bacteria are symbiotic with soybean plants (*Glycine max*) and *Rhizobium trifola* bacteria with clover (*Trifolium* spp.). These bacteria obtain the energy required for N fixation by oxidising sugars that are provided by the plant; free-living bacteria obtain their energy from respiration of organic molecules in the rhizosphere.

In the aquatic environment, cyanobacteria are the main N-fixers, mainly as free-living or filamentous cells but also in symbiotic association with coral organisms and aquatic ferns. To obtain the energy for N fixation, they oxidise the sugars they manufacture via photosynthesis. Cyanobacteria also undertake N fixation on land as symbiotic partners with fungi in lichens, which grow on rocks, trees and other surfaces, and with some plants.

N-fixers are highly specialised and adapted. They link N fixation to their cycling of other elements such as C, O, S and P, providing a net surplus of energy and giving an advantage in environments with low levels of available N. Second, they possess nitrogenase, an enzyme that is essential for fixation; it contains a molybdenum-iron protein, which holds an N_2 molecule in place, and a separate Fe protein, which supplies the electrons required for the reductive breakdown of N_2 into its constituent atoms. Growth of diazotrophs can be limited in locations where these elements are scarce; particularly in soils without Mo and in seawater containing relatively low amounts of Fe. Finally, because N fixation is a reductive process, N-fixers must also be anaerobes or have the ability to create anoxic conditions, either by the use of specialised cells called heterocysts (in cyanobacteria) or metabolic processes (in *Azotobacter* spp. for example). In legumes, a molecule called leghaemoglobin, which binds oxygen in the same manner as haemoglobin in blood, is present in the root nodules; leghaemoglobin regulates oxygen release in the nodule in precise amounts so that the symbiotic bacteria can respire without disrupting the anaerobic N-fixation process. While specialised N-fixers accomplish the highly important task of providing N to the food chain, several other biological processes are involved in the cycling of N in soils and waters, as detailed below (and summarised in Fig. 3.5).

Figure 3.4 Rhizobium nodules on the roots of cowpea (*Vigna unguiculata*), an important leguminous crop plant in tropical areas.

Credit: Harry Rose (Flickr).

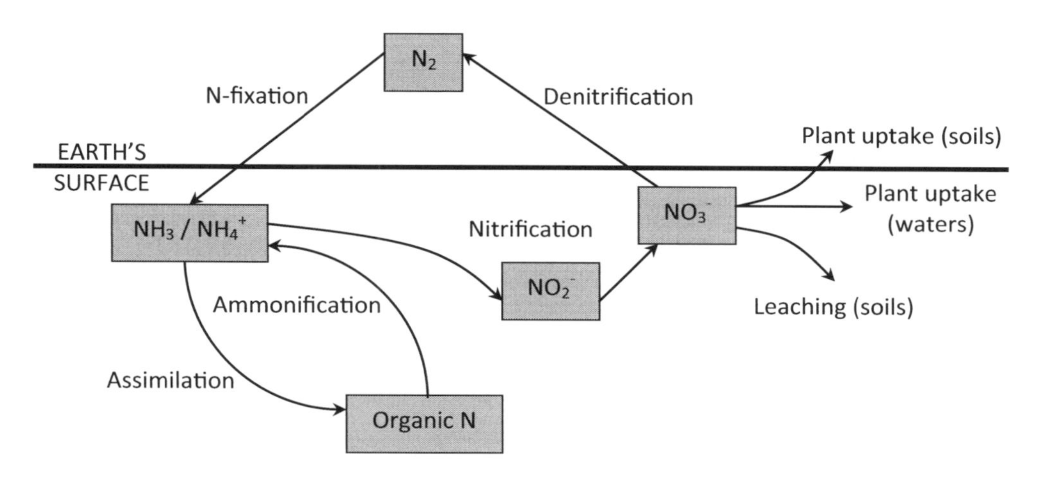

Figure 3.5 A simplified summary of the internal nitrogen cycle of soils and waters showing key features of interest.

Ammonification (transfer within biosphere)

Soils and waters constantly receive an influx of organic matter, from wastes excreted by living organisms and from their bodies when they die. As dead organic matter is biodegraded by saprophytic fungi and bacteria, organic N compounds are released and transformed to ammonia via the process of ammonification. For example, in the case of glycine, the amino acid formed in equation 3.8, the reaction is as follows:

$$CH_2NH_2COOH \text{ (glycine)} + 1.5O_2 \rightarrow 2CO_2 + H_2O + NH_3 \quad [3.9]$$

The nitrogenous component of waste excretions is urea and this too undergoes ammonification:

$$(NH_2)_2CO \text{ (urea)} + H_2O \rightarrow CO_2 + 2NH_3 \quad [3.10]$$

Some of the excess ammonia diffuses out of soils and waters into the atmosphere, where a proportion dissolves in atmospheric moisture, hydrolyses to aqueous ammonium ions, NH_4, and returns to Earth in wet deposition. Non-diffusing ammonia is also hydrolysed to NH_4, which is taken up by plants, or adsorbed (in soils). If the NH_3/NH_4 released (equation 3.9 and 3.10) is all used by microbes, it is said to be immobilised. Immobilisation typically occurs where the organic matter or wastes being biodegraded have a carbon to nitrogen (C:N) ratio of >25:1. Biodegradation of wastes with lower C:N ratios (i.e. relatively nitrogenous organic matter such as dead legumes) is likely to result in excess N being produced over and above microbial requirements. The resulting input of inorganic forms of N into soils and waters is termed 'N mineralisation'.

Nitrification (transfer within biosphere)

In soils and waters, reduced forms of N produced by ammonification are oxidised by nitrifying micro-organisms to nitrite and nitrate, as long as the environment is sufficiently oxic and has a pH of >5.5. Nitrification is a two-stage process and is mainly mediated in soils by two different types of bacteria, *Nitrosomonas* spp. and *Nitrobacter* spp., which both obtain energy from the oxidation. These bacteria are also present in fresh waters, and in the oceans where *Nitrosococcus* spp. and *Nitrococcus* spp. are also important nitrifiers.

Nitrosomonas spp. perform the oxidation of ammonium:

$$2NH_4^+ + 3O_2 \rightarrow 2NO_2^- + 4H^+ + 2H_2O + \text{energy } (\Delta G° = -290 \text{ kJ mol}^{-1}) \quad [3.11]$$

Nitrobacter spp. oxidise the nitrite produced in equation 3.11:

$$2NO_2^- + O_2 \rightarrow 2NO_3^- + \text{energy } (\Delta G° = -82 \text{ kJ mol}^{-1}) \quad [3.12]$$

If the initial oxidation of NH_4 is incomplete, gaseous nitrous oxide (N_2O) is also produced in relatively small amounts. The second reaction proceeds more rapidly than the first, so nitrite does not normally accumulate in soils and waters. Nitrate in soil solution is taken up readily by plant roots. Excess nitrate does not accumulate in temperate soils; being both soluble and anionic (and thus not adsorbed in soils with a high cation exchange capacity), it typically leaches into ground or surface waters if not taken up by plants.

Denitrification (lithosphere/hydrosphere → atmosphere)

In reducing environments like poorly drained soils (Fig. 3.6), wetlands and anoxic water bodies (including deep marine waters), where the supply of oxygen is limited, NO_3 is utilised for respiration by specialised bacteria such as *Thiobacillus denitrificans*, *Micrococcus denitrificans* and *Psuedomonas denitrificans*, which all possess the nitrate reductase enzyme. Denitrifying bacteria are facultative anaerobes, meaning they are able to also respire aerobically in oxygenated conditions.

The process of *denitrification*, the breakdown of organic matter using NO_3 as the oxidation source, is exemplified here by the oxidation of a simple carbohydrate molecule, CH_2O:

$$4NO_3^- + 5CH_2O + 4H^+ \rightarrow 2N_2 + 5CO_2 + 7H_2O \quad [3.13]$$

Figure 3.6 Denitrifying bacteria are able to use nitrates for respiration when aeration levels are low, in waterlogged soils for example.

Credit: David Morris (Flickr).

The N_2 produced in this reaction degasses from soils and waters into the atmosphere, effectively closing the cycle which began with the fixation of N_2 from the atmosphere. Nitrogen gas accounts for >90% of the N gases produced by denitrification; smaller percentages (but not insignificant amounts) of N_2O and NO are also produced when reduction is incomplete. In addition to denitrification, N_2 (with some NOx) is also released to the atmosphere from natural biomass burning.

It should be noted that, annually, approximately 10–40 Mt of fixed N leaves the cycles described in this section to be incorporated into sedimentary rocks, mainly within residual organic matter that has settled on the sea bed.

Anthropogenic disturbance of the nitrogen cycle

The natural N cycle is altered by human activity primarily by the industrial fixation of the element from the atmosphere and the subsequent application of the fixed form to the Earth's surface as fertiliser. Table 3.2 summarises some of the most important uses of anthropogenically fixed nitrogen; manufacture of fertiliser is the dominant use, but N is also utilised in several other products and processes. Fuel combustion provides the other major human disturbance to the N cycle (see below).

Industrial nitrogen fixation

Despite the abundance of atmospheric N_2, natural fixation transfers only a small fraction of N_2 to soils each year. This is problematic for modern agriculture because

Table 3.2 Uses of nitrogen.

Use	*N compound(s)*
Fertilisers	Ammonium nitrate, NH_4NO_3 Urea, $CO(NH_2)_2$ Ammonium sulfate, $(NH_4)_2SO_4$ Ammonium phosphate, $(NH_4)_3PO_4$ Anhydrous ammonia
Explosives[a]	Potassium nitrate, KNO_3, in gunpowder and fireworks Ammonium nitrate, NH_4NO_3 Nitroglycerine (glyceryl trinitrate) in dynamite Trinitrotoluene (TNT)
Medicine	Organic nitrates in treatment of heart disease Nitrous oxide, N_2O, as an anaesthetic in dentistry Amines in antihistamines and other drugs
Food preservatives	Sodium nitrate, $NaNO_3$ Sodium nitrite, $NaNO_2$ Nitrous oxide, N_2O
Refrigeration	Ammonia, NH_3
Aerosol propellant	Nitrous oxide, N_2O, in food products such as whipped cream
Dyes	Amines
Household cleaners	Ammonium hydroxide
Nylon	Amines

a The explosive power of these N compounds is derived mainly from the sudden expansion of dense solid material to N_2 and other gases.

N is required by all life forms and it is a key limiting nutrient, meaning that additional inputs are necessary. In traditional farming systems, supplementary N is provided by animal manures and other organic wastes, such as crop residues and human wastes. Crop rotation, using leguminous plants, is also utilised widely. However, the nutrient requirements of modern humanity and the rapid loss of soil NO_3 (via leaching) and NH_3 (via volatilisation) mean that significant additional N inputs to soils are required. This is coming into sharper focus as the human population continues to grow and as per capita consumption of food increases, particularly that of meat, which requires the largest areas of agricultural land. Furthermore, urbanisation will continue to swallow existing agricultural land; maintaining global productivity and crop yields on the diminishing land area will therefore require further inputs of fertiliser, as will declines in soil N levels, which can be caused by modern agricultural practices and the clearance of natural ecosystems for cropland.

The additional soil N required by the current human population is provided by industrial N fixation. By the 19th century, western countries were importing natural $NaNO_3$ deposits from South America to provide additional fertiliser inputs to their croplands; however, in the global political turmoil of the early 20th century, Atlantic shipments of this mineral fertiliser could not be relied on. During this period, industrial N fixation was developed by German scientists, Fritz Haber and Carl Bosch (Box 3.1). At first, the synthesised ammonia was used solely in the manufacture of explosives but, following the First World War, production of mineral fertilisers began. Gradual increases in the efficiency and affordability of the Haber-Bosch process resulted in widespread use of mineral N fertilisers by the second half of the 20th century.

Box 3.1 The Haber-Bosch process.

The synthesis of ammonia from nitrogen gas and hydrogen gas in the Haber-Bosch process (Fig. 3.7) is shown in the simple equation:

$$N_2 + 3H_2 \rightarrow 2NH_3 \quad [3.14]$$

Prior to this, however, other processes are required to provide the two reactants shown in the equation. Nitrogen gas, N_2, is sourced from the atmosphere; it is separated from the other components of the air by liquefaction and distillation. Hydrogen gas, H_2, is provided by the reaction (called 'steam reforming') of natural gas (methane, CH_4) and water:

$$CH_4 + H_2O \rightarrow CO + 3H_2 \quad [3.15]$$

In the Haber-Bosch process, manufacturing plants utilise high temperatures and pressures, in the ranges of 300–550°C and 150–250 atmospheres (15–25 MPa), respectively, to synthesise NH_3 from N_2 and H_2; the process is energy intensive, adding to the financial and environmental costs of N fertilisers. The reaction proceeds in the presence of an Fe catalyst that also contains small amounts of K, Ca and Al. Ammonia is removed from the equilibrium mix of reactants and products shown in equation 3.14 by condensation, and is subsequently processed. Most is converted to HNO_3, in the 'Ostwald Process', which involves the oxidation of NH_3 to NOx followed by dissolution of NOx in water. Most HNO_3 is then converted to ammonium nitrate fertiliser by reaction with further ammonia:

$$HNO_3(aq) + NH_3(l) \rightarrow NH_4NO_3(aq) \quad [3.16]$$

Similarly, synthesised NH_3 can also be reacted with phosphoric acid to form ammonium phosphate fertiliser, or with sulfuric acid to produce ammonium sulfate fertiliser. Another widely used N fertiliser is urea, $CO(NH_2)_2$. Granules of urea are produced by reacting NH_3 with CO_2 at high pressure. Urea granules decompose gradually, giving a relatively slow release N fertiliser. Ammonia may also be used as a direct fertiliser by being injected in liquefied form directly into soils; this readily leads to the formation of NH_4 in the soil and the ions are then taken up by plant roots.

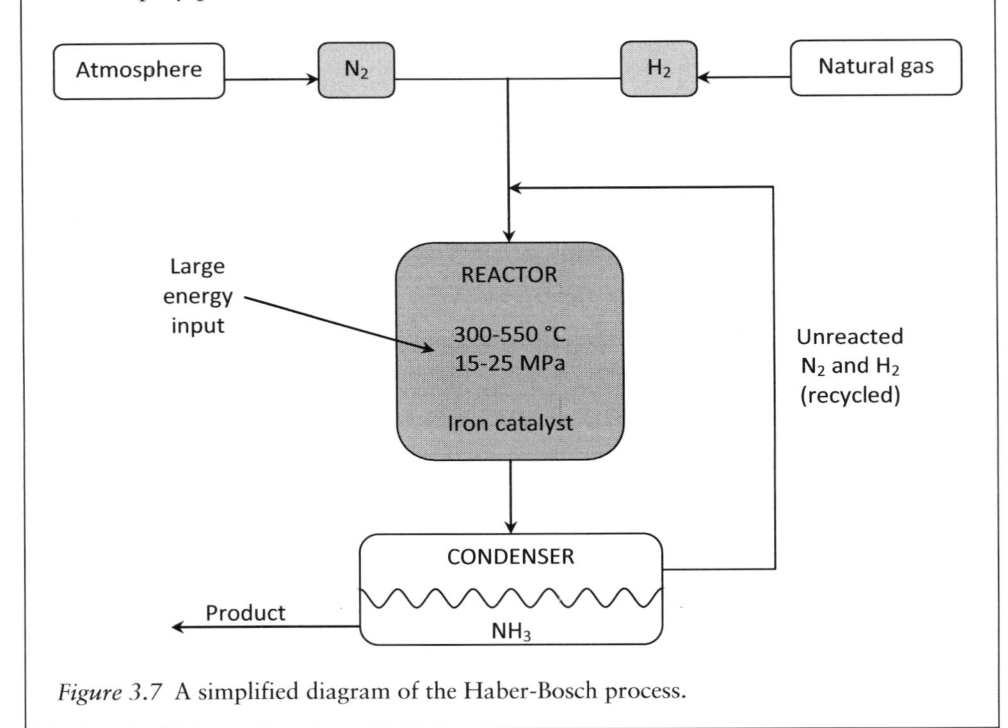

Figure 3.7 A simplified diagram of the Haber-Bosch process.

Farmers are understandably keen to apply sufficient amounts of fertiliser (inorganic and organic) to their fields to ensure that the potential yield of their crop is fully realised; however, over-application, above an 'optimum rate' (Fig. 3.8), does not increase N uptake further and incurs unnecessary financial costs for the farmer as well as environmental costs when excess N is leached into water bodies by rainfall (see below). In practice, the optimum application rate is not easily determined because of uncertainty over future rainfall rates and farmers are likely to err on the side of caution, over-applying to compensate for potential future loss of soil N.

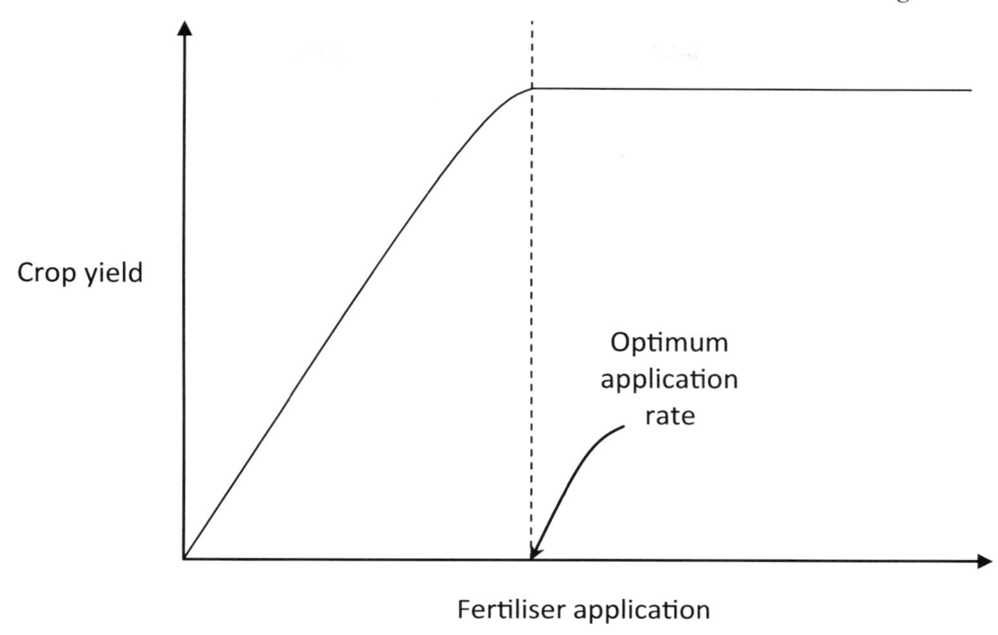

Figure 3.8 The optimum application rate.

Table 3.3 Nitrogen supply and uptake and residual nitrogen in fourteen European countries in 2009. Approximately 77–93% of N supply in these countries in 2005–08 was from mineral fertiliser and manure, with a higher contribution from the former in most countries (the remaining percentages are mainly attributed to atmospheric deposition and natural fixation). Nitrogen uptake is via crop harvest and grazing. The significant amounts of residual N are highly vulnerable to leaching, runoff and volatilisation, potentially creating pollution problems, particularly in waters.

Country	*N supply ($kg\ ha^{-1}$)*	*N uptake ($kg\ ha^{-1}$)*	*Residual N ($kg\ ha^{-1}$)*	*Residual N (kt)*
Austria	115	93	22	70
Czech Republic	144	83	61	214
Denmark	197	121	76	202
Estonia	68	48	20	16
Finland	114	72	42	95
Germany	208	133	75	1273
Ireland	205	154	51	214
Norway	182	87	95	96
Poland	132	78	54	841
Portugal	81	67	14	52
Slovakia	102	69	33	64
Spain	80	50	30	701
Sweden	110	80	30	93
United Kingdom	200	104	96	1185

Source: Data from Eurostat (2014).

Taken together, industrial N fixation and direct fixation by leguminous crops appear to be matching or even exceeding the amount of atmospheric N_2 that is fixed by natural vegetation each year (Fig. 3.3). The main biotic process that removes N from soils (denitrification) cannot always keep pace with inputs of mineral fertiliser and manure, leading to excess soil N, mainly in the form of nitrates. Nitrates are water-soluble and, being anionic, are not adsorbed by cation exchange, so are able to pollute water bodies via runoff and leaching (Table 3.3).

In addition to fertiliser use, another major source of NO_3 pollution of water bodies is land-use change (Box 3.2), particularly the conversion of pasture to arable land, which reduces soil organic matter and releases mineralised N to waters. Also important in this context is improved drainage, which transports excess nitrates to rivers; furthermore, denitrification is reduced in the drier soils so more N is left in the ground (and therefore vulnerable to leaching) instead of being returned to the atmosphere as stable N_2 (Howden et al., 2010). Some of the suggested and applied responses to N loss are listed in Table 3.4, but their efficacy may be limited if long-term land-use changes are not addressed.

Box 3.2 Nitrates in drinking waters in England and Wales.

Approximately two-thirds of public drinking water supplies in England and Wales are derived from surface waters, the other third being sourced from underground aquifers. At times, NO_3 concentrations in both surface and ground waters exceed the WHO drinking water guideline for nitrate of 50 mg L^{-1}, necessitating expensive treatment, usually dilution with 'cleaner' water from elsewhere. Taking the River Thames as a long-term example (Fig. 3.9), NO_3 concentrations increased in the 1940s by approximately 50% over the decade and again, but more steeply (by another 70%), between 1968 and 1974. Howden et al. (2010) note that there were no sudden changes in climate or population in the Thames catchment area at these times, indicating that agricultural activities, including land-use changes, were the main causes. The 1940s increase is attributed to surface runoff of mineralised N from the conversion of pasture to arable land in World War Two (see black line in Fig. 3.9); there is no evidence of increased fertiliser use at this time. This activity is likely to have also been a causal factor in the longer-term (decadal) increase in river N levels because of slow transport of leached N via groundwaters. The later (1960s–70s) increase is concurrent with a period of further agricultural intensification, including additional (but smaller scale) pasture-to-arable conversions and, more particularly, a substantial increase in the use of mineral fertilisers (light grey line, Fig. 3.9) together with uptake of grant-aided land drainage schemes, which facilitate the rapid transport of soluble nitrates from field to river. Despite a general decline in fertiliser use since the 1980s, nitrate concentrations in the River Thames have decreased to a lesser degree, further indicating the likely importance of long-term land-use changes.

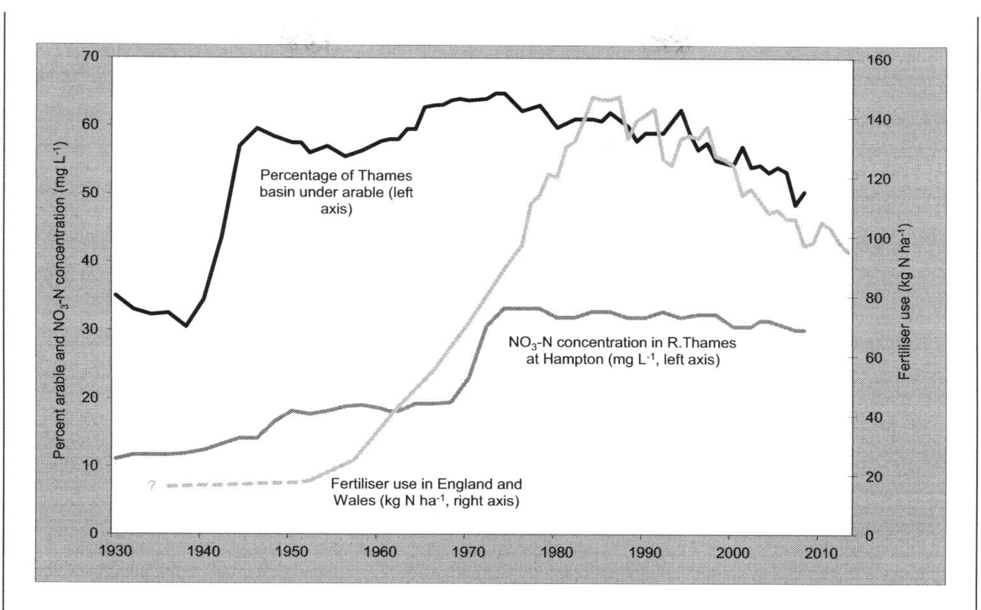

Figure 3.9 Nitrate concentrations in the River Thames at Hampton (upstream of London), 1930–2008. Also shown are the extent of arable land in the Thames catchment area and fertiliser use in England and Wales. Note: Comprehensive records for fertiliser use start in 1974; the period from 1952 to 1973 is interpolated from data for the years 1952, 1957, 1962, 1966 and 1970.

Source: Adapted from Howden et al. (2010), AIC (2013) and DEFRA (2014a).

Table 3.4 Responses to losses of nitrogen in farming.

Response	*Rationale*
Organic farming	Less N input (but more land required for same food output)
Lagoons	To intercept N-polluted runoff, allowing for denitrification
Timely application	So more of the nutrient is taken up by a growing crop
Soil/plant testing	To better predict the most appropriate application rate
Winter cover crops	To take up residual N and minimise runoff from bare soils
Genetic engineering	N fixation by non-legume crops (not currently viable)

Combustion processes

Upon combustion, most of the N contained in fuel (comprising approximately 1.5% of the mass of coal, for example) is released to the atmosphere, mostly as NOx ('fuel NOx') but with appreciable amounts of NH_3 as well. In addition to fuel NOx, combustion processes at high temperatures (particularly >1500°C) produce 'thermal NOx', where N_2 and O_2 in the combustion air are separated into monatomic N and O, subsequently recombining as NO (equation 3.2). The important point is that the N source in this case is the atmosphere, not the fuel. Coal burning generally produces more fuel NOx, whereas NOx emissions from the combustion of natural gas, which has a lower N content, are mainly of the thermal type. While fossil fuels are the major sources of NOx from combustion processes, there are also significant additions to the atmosphere from the burning of biomass; for example, when wood is used as fuel and where forested land is cleared for agricultural use.

While the atmospheric lifetime of NOx is quite short, it is an important air pollutant, particularly in urban areas, where vehicles are major sources (Table 3.5); the NOx is mostly emitted by vehicles that do not carry a three-way catalytic converter (mainly diesels). In contrast, NOx that is produced by industrial combustion processes, including power stations, is usually dispersed more widely into the atmosphere from tall stacks. Both fuel NOx and thermal NOx, together with NH_3, ultimately return to the Earth's surface in wet (and some dry) deposition. Once deposited, thermal NOx represents an additional flux of inert N_2 from the atmosphere to available forms (particularly NO_3) in soils and waters where pollution impacts may ensue (see below); similarly, fuel NOx represents a transfer, via the atmosphere, of inert N from long-term storage in the lithosphere to soils and waters.

Attempts to minimise NOx emissions from combustion sources have mainly involved, in large combustion plants, 'low NOx burners' and similar techniques to reduce combustion temperatures so that thermal NOx emissions (equation 3.2) are minimised. Reduction of NOx from vehicles has been attempted by techniques including greater fuel efficiency and the use of catalytic converters, which were introduced on a wide scale in the 1980s. A catalytic converter contains a ceramic honeycomb structure coated with a precious metal, such as rhodium or platinum; the metal (shown as M in equations 3.17–3.19) catalyses the reduction of NOx by hydrogen (e.g. equation 3.17) or by carbon monoxide (e.g. equation 3.18). The hydrogen gas on the left side of equation 3.17 is generated by reaction on the catalyst surface of hydrocarbons and water, both of which are present in the exhaust fumes. Carbon monoxide, a reactant in equation 3.18, is also present in the exhaust gases. The reduced forms in the examples above are N_2 and N_2O, but NH_3 may also be formed in the catalytic converter (equation 3.19).

$$2NO + 2H_2 + M \rightarrow N_2 + 2H_2O + M \qquad [3.17]$$

$$2NO + CO + M \rightarrow N_2O + CO_2 \qquad [3.18]$$

$$2NO + 5H_2 + M \rightarrow 2NH_3 + 2H_2O + M \qquad [3.19]$$

Table 3.5 Typical atmospheric nitrogen dioxide concentrations.

Location type	*Concentration ($\mu g\ m^{-3}$)*
Rural background	<1–10 (annual mean)
Urban areas	20–90 (annual mean)
Urban areas	75–1015 (hourly mean)

Source: Data from WHO (2006).

Human and animal wastes

Human and animal wastes are significant sources of nitrogen in the environment. Water bodies may receive organic and inorganic forms of N (e.g. urea and NH_4, respectively) from manure and sewage, while gases such as N_2O and NH_3 are released from such wastes to the atmosphere. Such pollution is typically managed by storage and digestion of animal wastes and sewage treatment of human wastes (Table 3.6). For example, in the UK, areas with elevated levels of NO_3 in surface waters are designated as 'Nitrate Vulnerable Zones' and, in practice, this means that farmers in these zones must ensure adequate storage of nitrogenous wastes, in addition to controlling their use of fertilisers.

Table 3.6 Example techniques for the management of human and animal wastes.

Example	*Method*	*Outcomes*
	Animal waste	
Capture and storage	Waste composted or anaerobically digested	Waste prevented from entering waterways; volume reduced and nutrients stabilised; odours from NH_3 and other gases controlled
	Human waste (sewage and effluent)	
Reaction chamber	Use of nitrifying bacteria and reducing agents	NO_3 generated by nitrification is reduced to N_2 by reducing agents (e.g. methanol); N_2 degasses into the atmosphere
Liming	Effluent pH increased to 11	NH_4 converted to gaseous NH_3, which diffuses out of the effluent on subsequent aeration

Environmental impacts arising from disturbance of the nitrogen cycle

Impacts on waters: eutrophication

Excess soil N, mostly in the form of soluble nitrates from inorganic and organic fertiliser applications, enters rivers and estuaries in many agricultural areas. Ammonia, nitrates and other nitrogenous compounds also reach waterways in animal waste and sewage effluent and as soil particles transported in surface runoff. Direct deposition to waters (from the atmosphere) of pollutant NOx and NH_3 and nutrient release from aquaculture are important additions. Such influxes, particularly those from agriculture, increase the levels of available N in water bodies. Excess inputs to waters of N and other nutrients, particularly phosphorus (P), can induce rapid population growth of primary producers that are normally kept in check by lower concentrations of the limiting nutrient. The resulting 'algal bloom' is symptomatic of 'eutrophication', from the Greek *eutrophos*, meaning 'well-nourished'.

Eutrophication is, in fact, a natural process, whereby nutrients, particularly P, accumulate in lakes and slow-flowing, low-lying rivers over time, while moorland streams, for example, remain 'oligotrophic' (low in nutrients). Human sources of

Figure 3.10 Bloom of the macroalga, *Ulva lactuca* (sea lettuce) on Oxford Beach, Chesapeake Bay, USA.

Credit: Emily Nauman (Flickr).

excess nutrients cause 'cultural eutrophication' or 'hypertrophication' (meaning 'excess nourishment'). However, these latter terms are seldom utilised and eutrophication is generally used, as here, to describe the effects of pollutant inputs. 'Algal bloom' is in many cases also a misnomer as other primary producers than algae proliferate in eutrophic waters. In marine areas, blooms are often of dinoflagellates, such as *Karenia* spp. or *Noctiluca* spp., and in freshwaters the flourishing organisms are typically cyanobacteria, which are not algae, despite being commonly known as blue-green algae. In other cases, macroalgae (colloquially 'seaweed'), rather than planktonic organisms, proliferate.

Eutrophication of estuaries and coastal waters is generally attributed to inputs of N rather than other nutrients such as P, which is thought to be a more important cause of freshwater eutrophication {→P} (although there is some debate about their relative importance). For example, the macroalgal blooms that often occur on beaches, for example in the USA (Fig. 3.10) and Brittany, France, are attributed to nitrate pollution from intensive agriculture. This is partly because soluble nitrate is more likely to reach coastal waters, and remain in dissolved form, than P, more of which is held in sediment particles. In the case of cyanobacteria blooms, another important factor is that in the alkaline waters they are less able to fix N (than in most freshwaters), so N is more likely to be the limiting nutrient for seawater cyanobacteria, which will then bloom if excess N is supplied.

Eutrophication is problematic for a number of reasons (Table 3.7). In particular, rapid biodegradation of the dead tissues of an 'algal bloom' by aerobic micro-organisms causes hypoxia, the depletion of dissolved oxygen (DO) from the water body; nitrogenous effluents can also reduce DO levels of freshwaters more directly by the reaction of NH_4 with oxygen. This causes stress to other aerobic organisms in the water, including fish, because the amount of DO held by fresh water at saturation point (e.g. 11 mg L^{-1} at 10°C and less in warmer waters) is not much higher than the oxygen requirements of such organisms. Depletion of DO to <5 mg L^{-1}, for example,

Table 3.7 The main impacts of eutrophication.

Hypoxia and anoxia	Biodegradation of phytoplankton blooms depletes oxygen in the water column.
Release of toxins	Some phytoplankton species release toxic compounds such as neurotoxins into the water body. Human impacts from direct exposure and via diarrhoeic and paralytic shellfish poisoning.
Biodiversity	Biodiversity is decreased as certain species become dominant, including fast-growing aquatic plants, which can take advantage of the excess nutrients, and organisms such as coarse fish, which have a greater tolerance to lower DO levels.
Turbidity and blanketing	Increased numbers of plankton or macroalgae reduce the amounts of light required by other photosynthesising organisms in the water body. Proliferation of phytoplankton and/or higher plants also has a blanketing effect.
Tissue damage	Excess plankton concentrations in waters can physically damage sensitive tissues in aquatic organisms, including fish gills.
Nuisance	Blooms can have a nuisance value in terms of visual amenity and odour.
Economic damage	Phytoplankton blooms can block water treatment filters.

Figure 3.11 In summer 2003, >1 million menhaden fish were killed by anoxia in Greenwich Bay, Rhode Island, USA, resulting in the implementation of nitrogen pollution controls in sewage treatment plants, which were recognised as the main sources of nitrogen inputs to the bay.

Credit: Chris Deacutis (Flickr).

is known to affect the function and survival of many aquatic organisms. In extreme cases, waters become very hypoxic or even anoxic (zero DO), leading to 'fish kills' where aquatic organisms die from asphyxia (e.g. Fig. 3.11). A separate concern is that many cyanobacteria species emit toxic organic chemicals into the water, including neurotoxins, hepatotoxins and dermatoxins. In a phytoplankton bloom the concentrations of toxins in the water body can become high enough to poison aquatic organisms and cause illness in humans eating the affected seafood; there are also records of fatalities in dogs and cattle that have drunk eutrophic waters affected by such toxins.

In eutrophic coastal regions, areas of hypoxia or anoxia due to organic matter biodegradation are often described as marine 'dead zones' because of the serious effects on aerobic organisms. More than 400 dead zones have been reported worldwide, covering a total area of 245,000 square km and the number has increased rapidly in recent decades, lagging behind increases in fertiliser use by about 10 years (Diaz and Rosenberg, 2008) (Fig. 3.12). Most of the hypoxic areas associated with fertiliser pollution (and smaller inputs from deposition of NOx emissions) have so far been recorded in the coastal areas and seas of Europe, China and the USA, and in some of the North American Great Lakes. The persistently hypoxic dead zones of the Baltic Sea in Europe are estimated to have a 'missing biomass' each year equivalent to 2.4×10^5 t C; while there is partial compensation because of increased secondary production outside of the dead zones, there is a substantial deficit in the Baltic Sea overall (Diaz and Rosenberg, 2008). Management of nutrient inputs to selected rivers has led to recovery in some hypoxic areas, but in severe cases full recovery has still not occurred after 10 years.

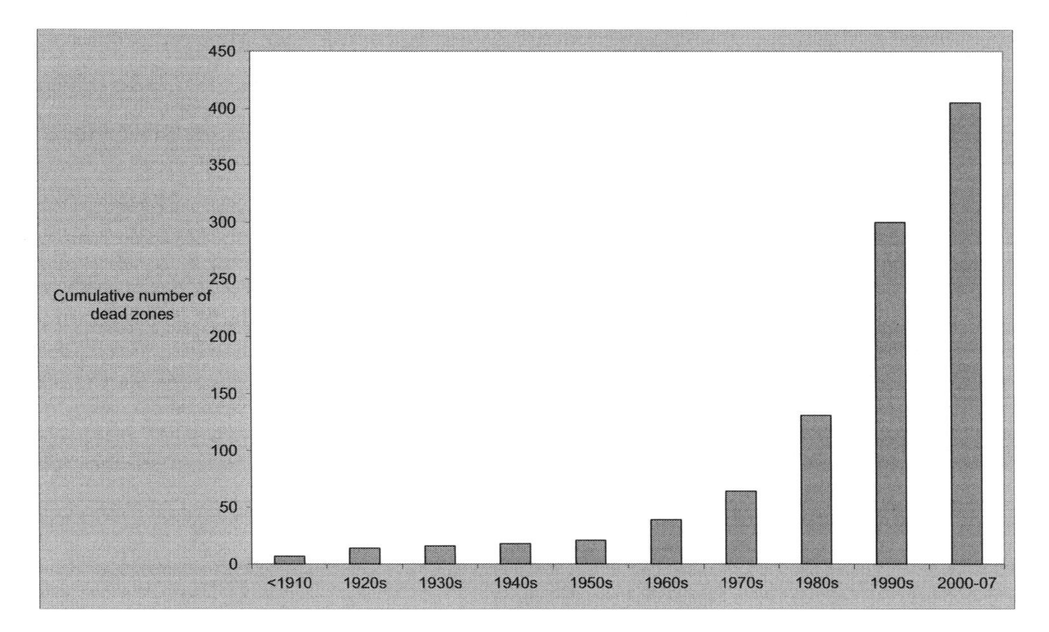

Figure 3.12 Cumulative increase (global) in marine dead zones over time.

Source: Adapted from Diaz and Rosenberg (2008).

Marine hypoxia is also caused naturally by seasonal episodes of coastal upwelling, which bring nutrient-rich organic matter into the photic zone of surface waters causing blooms known as 'red tides' or 'green tides'. For example, in 2012 a widely-reported red tide occurred in New South Wales, Australia of *Noctiluca scintillans*, a dinoflagellate (a single-celled heterotroph). Such organisms are common globally but bloom when nutrients become plentiful. Their colour appears to depend on their main food source. The reason for the red colour seen in Australia and elsewhere (e.g. China) is that the main food source is red algae (which contain only chlorophyll a). The dinoflagellates are green in other regions, where mainly green algae (containing both chlorophyll a and b) are consumed. Some dinoflagellates produce saxitoxin, a potentially lethal neurotoxin that is implicated in paralytic shellfish poisoning. In addition, high concentrations of NH_3 may appear in the affected waters; the NH_3, which is produced by the organisms as they consume the algae, is toxic to fish and other marine organisms.

Impacts on waters: nitrates in drinking water

Concerns about nitrate contamination of drinking water supplies mainly relate to methaemoglobinaemia (MHA), a condition that occurs when ingested nitrates are reduced in the gastrointestinal tract to nitrites, NO_2. Nitrites enter the bloodstream and transform oxygen-carrying ferrous Fe^{2+} in the haemoglobin molecule to ferric Fe^{3+}. The resultant Fe^{3+} form of the molecule (called methaemoglobin, MH) does not readily release O_2 to tissues and can cause cyanosis (skin discoloration) and other disorders if it comprises >3% of total haemoglobin (Table 3.8). The World Health Organization declared a guideline for nitrates in drinking water of 50 mg L^{-1} to protect against MHA in infants. Nitrate ingestion is not unusual because it is present in some vegetables (indeed, maximum concentrations are sometimes established for such foods) and adult intakes are estimated to be in the range of 50–150 mg per day; however, where intake is increased because of contaminated drinking water, there is the potential for MHA risk to increase. There is some uncertainty in the medical literature about the exact role of drinking water nitrates in MHA; most incidences of water-related MHA have involved private wells, which are often contaminated with bacteria (gastrointestinal infection is a known causative factor in elevated nitrite synthesis).

Infants up to 4 months of age are the most vulnerable to MHA, partly because they have a higher stomach pH than adults; this favours the nitrate-reducing bacteria, which are suppressed in the very acidic adult stomach. Young infants also carry residual foetal haemoglobin, which is more readily oxidised to MH, and they also

Table 3.8 Effects of methaemoglobinaemia.

Methaemoglobin content of blood (% of haemoglobin)	*Symptom*
3–15	Mild skin discoloration
15–20	Cyanosis
25–50	Headache, weakness, chest pain
50–70	Cardiac arrhythmia, seizure, delirium
>70	Death likely

have less of the enzyme responsible for converting any MH formed in the body back to haemoglobin. For these reasons, and the characteristic cyanosis it causes, MHA is also known as 'blue baby syndrome'. Babies can be exposed to excess nitrate via contaminated drinking water (when used to make up baby formula) and they have a relatively large fluid intake, relative to their body mass. MHA can cause similar problems in livestock, which, like infants, carry more of the nitrate-reducing bacteria in their stomachs.

Nitrites in the human body, mainly from the ingestion of excess nitrates, can also react with amino acids to form nitrosamines, which are known carcinogens. The role of nitrate-contaminated drinking water in this context is contentious, however. Some researchers have also shown associations between ingestion of nitrate-contaminated water and incidence of goitre.

Impacts on waters: acid rain

Deposition of acidifying substances, chiefly compounds of sulfur {→S} and nitrogen can cause ecological damage. Nitrogen dioxide, produced by combustion processes, oxidises in the atmosphere to nitric acid:

$$4NO_2 + O_2 + 2H_2O \rightarrow 4HNO_3 \rightarrow 4\{H^+ + NO_3^-\} \text{ (aq)} \qquad [3.20]$$

Nitric acid is also a product of the reaction of NO_2 with free radicals such as hydroxyl:

$$NO_2 + {}^{\cdot}OH + M \rightarrow HNO_3 + M \qquad [3.21]$$

Nitric acid is subsequently deposited to the Earth's surface in rainfall where it contributes to acidification of soils and water bodies. Acidification can also be caused by the nitrification of excess NH_4 in soils (equation 3.11).

Measurements of wet deposition of acidifying substances in the Tatra Mountains of Europe show that while S deposition has decreased significantly in recent decades, N deposition has remained unchanged and has become the main form of acid deposition (Rzychon and Worsztynowicz, 2008) (Fig. 3.13); decreased N levels were observed in Tatra lakes, however, suggesting that deposited N is being retained in the terrestrial ecosystem. A separate study of soil acidification in the Tatra Mountains found that historic and ongoing atmospheric N deposition has depleted soils of the important nutrients calcium and magnesium and led to the leaching of toxic aluminium; reduction of plant biomass has been recorded, associated with lower Ca and Mg levels in plant shoots (Bowman et al., 2008).

Impacts on waters: ammonia contamination

Ammonia is directly toxic to aquatic organisms. In fish, dissolved NH_3 is transported across the gills to a greater extent than NH_4, and is therefore more bioavailable. Inside the body NH_3 is converted to NH_4, which can have acute toxic effects on the central nervous system, causing death in extreme cases. Some species of freshwater fish are affected at NH_3 concentrations of 0.1 mg L^{-1}; levels exceeding 1 mg L^{-1} can be fatal. Fish kills have resulted from ammonia-rich slurry waste being accidentally or negligently released into waters. Another common source of large-scale, sudden fish

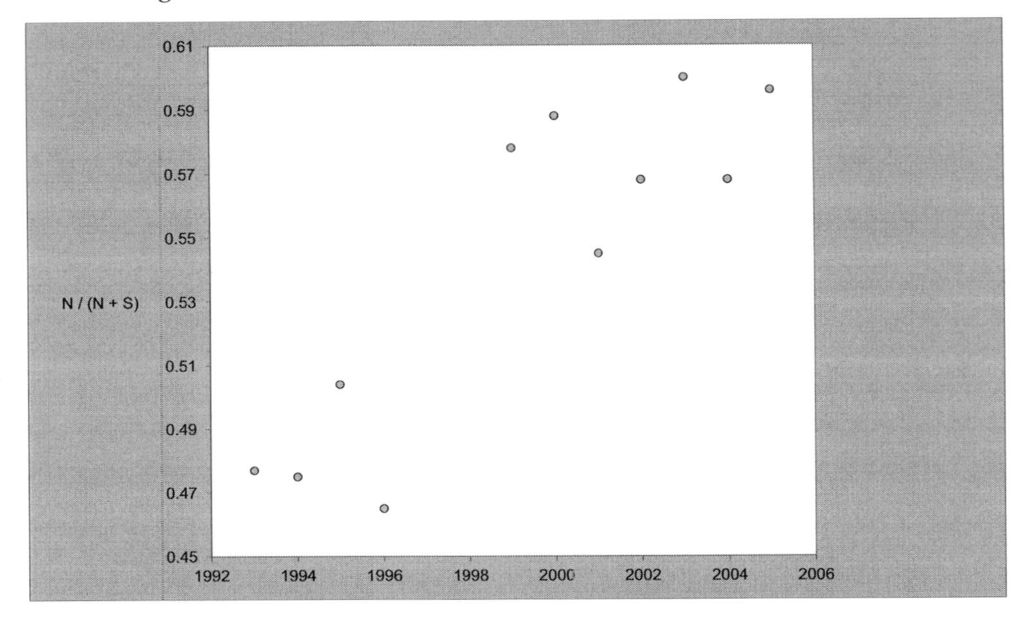

Figure 3.13 Nitrogen deposition as a fraction of total nitrogen and sulfur deposition in the Tatra Mountains, Poland, a region historically affected by acid deposition. Nitrogen compounds accounted for the minority of acid deposition in the 1990s (ratio <0.5) but had become the main acidifying substances by the 2000s, following controls on sulfur emissions.

Adapted from Rzychon and Worsztynowicz (2008).

kills is the leakage of ammonia refrigerant from industrial units. Ammonia is generally more toxic to marine fish, but the latter are less likely to be exposed to the pollutant discharges described above.

Atmospheric impacts: photochemical smog

Nitrogen oxides (NOx) are the main precursors of photochemical smog. Nitric oxide is produced in combustion processes (equation 3.2) and is rapidly oxidised in air to NO_2 (equation 3.3). Nitrogen dioxide, via a series of photochemical and chemical reactions, is the main source of tropospheric ozone, a gas that causes respiratory problems and is the chief component of photochemical smog. Pollutant NO_2 can also form particulate nitrates, such as NH_4NO_3, which are an important fraction of ultra-fine particulate pollution ($PM_{2.5}$); NOx are thus known as primary pollutants (i.e. emitted directly from pollutant sources). Secondary pollutants, such as tropospheric O_3, are those created from reactions of the primary pollutants. The other main category of primary pollutants involved in the formation of photochemical smog is the reactive organic gases (ROGs); these compounds are sometimes referred to as unburnt hydrocarbons (HCs).

The generation of photochemical smog pollution begins when NO_2 is photolysed in UV light (and some of the shortest wavelengths of visible, violet light):

$$NO_2 + h\nu\ (\lambda < 420\ nm) \rightarrow NO^{\cdot} + O^{\cdot} \quad [3.22]$$

Once formed, the $O^{\cdot}$ radical reacts readily with other atmospheric components. First, its reaction with O_2 produces O_3 (in the presence of a third body, M):

$$O_2 + O^{\cdot} + M \rightarrow O_3 + M \quad [3.23]$$

Equations 3.22 and 3.23 can be combined as equation 3.24 (note that this reaction is reversible as the products NO and O_3 react with each other):

$$NO_2 + O_2 \leftrightarrow NO + O_3 \quad [3.24]$$

Second, the $O^{\cdot}$ radical can react with water vapour to generate $^{\cdot}OH$ radicals:

$$H_2O + O^{\cdot} \rightarrow 2^{\cdot}OH \quad [3.25]$$

And third, $O^{\cdot}$ radicals can react with ROGs producing another group of compounds called organic free radicals (generic symbol $R^{\cdot}$, where R = an alkyl group, C_nH_{2n+1}):

$$ROG + O^{\cdot} \rightarrow R^{\cdot} + \text{other products} \quad [3.26]$$

For instance, using methane, CH_4, as a simple example of an ROG:

$$CH_4 + O^{\cdot} \rightarrow {}^{\cdot}CH_3 + {}^{\cdot}OH \quad [3.27]$$

Equation 3.27 shows methane reacting with an $O^{\cdot}$ radical to produce a hydroxyl ($^{\cdot}OH$) radical and an organic free radical $R^{\cdot}$, in this case methyl, $^{\cdot}CH_3$.

The organic free radicals ($R^{\cdot}$) formed in this way are subsequently converted in the atmosphere to an important category of secondary pollutants called peroxy radicals, RO_2. For example, the methyl produced in equation 3.27 subsequently reacts with molecular oxygen (equation 3.28) to produce a methoxyl radical, CH3OO$^{\cdot}$, which is just one example of a peroxy radical.

$${}^{\cdot}CH_3 + O_2 \rightarrow CH_3OO^{\cdot} \quad [3.28]$$

The peroxy radicals are important because they go on to react with primary pollutant NO to produce NO_2 (equation 3.29), which generates further O_3 (equations 3.22 and 3.23) adding to the smog.

$$NO + RO_2 \rightarrow NO_2 + RO\ \text{(oxy radical)} \quad [3.29]$$

Crucially, this reaction also removes NO, which could otherwise destroy O_3, as seen in equation 3.24.

The resulting photochemical smog contains a complex mixture of pollutants: (i) the primary pollutants, NOx and ROGs; (ii) the secondary pollutants seen above – O_3 (comprising up to 90% of the main smog gases) and peroxy radicals; (iii) further secondary pollutants, such as aldehydes and peroxyacetyl nitrate (PAN), chemical formula $CH_3COOONO_2$. PAN is a powerful lachrymator (tear-inducer), being extremely irritating to the eyes. Being fairly stable in the atmosphere, it can be a long-range pollutant and may release NO_2 back into the atmosphere far away from where it was first formed.

The complex series of reactions depicted above typically takes place over several hours subsequent to the initial photolytic reactions and are favoured when ROG to NOx ratios are in the range 4:1 to 10:1. Because they are driven by photolysis, photochemical smogs are particularly prevalent in the sunnier cities of the tropics and sub-tropics. Photochemical smogs are a *seasonal* occurrence in temperate climates and are sometimes referred to, in this context, as summer smogs; however, peaks are often observed in springtime as well, as increasingly strong sunlight begins to act on primary pollutants that are newly emitted and/or have accumulated on a hemispherical basis over the winter months. The distinctive dirty-orange haze that characterises a photochemical smog results from the absorbance of blue and green wavelengths of visual light by the main smog pollutants, leaving behind reds and yellows.

In stable, windless conditions, primary pollutants emitted in an urban area, during the morning rush hour, for example, will tend to remain there all day, creating urban smog and potentially exposing a large number of people to unhealthy air. On breezier days, however, the polluted air will have moved out into the surrounding suburbs or countryside by the time secondary smog pollutants have been formed by sunlight acting on the moving air mass. Furthermore, O_3 levels are typically lower in urban areas because of ongoing emissions of NO, which reacts with O_3 (equation 3.24); Fig. 3.14 illustrates this effect. For these reasons tropospheric O_3 is unusual as a pollutant that, despite being caused by urban and industrial emissions, tends to be at higher concentrations in rural locations (see examples in Table 3.9 and Fig. 3.14) and is also a transboundary pollutant, crossing national boundaries. Additional tropospheric O_3 is derived from natural sources and from intrusions of stratospheric O_3 during periods of extreme atmospheric turbulence.

Despite technologically driven reductions in primary pollutant emissions, hemispheric O_3 has not decreased in recent decades because growing population and consumption levels have increased the total number of industrial and transport sources. Meanwhile, the typically lower O_3 concentrations in many urban centres have risen as NO emissions from vehicles (which normally consume urban O_3) have gradually decreased because of technological improvements. This has resulted in a convergence of rural and urban O_3 levels in some areas as urban concentrations gradually rise towards the higher, hemispheric levels (Fig. 3.15).

Table 3.9 Example ozone concentrations in urban and rural areas.

Location	*Urban*	*Rural*	*Note*	*Source*
UK	40–60	65–78	Daily maximum 8-hour running mean; mean of all urban or rural sites, 2011	AEA Group and DEFRA (2012)
Turkey	16–46	40–80	Mean hourly concentration, all days during 2007 and 2008	Im et al. (2013)

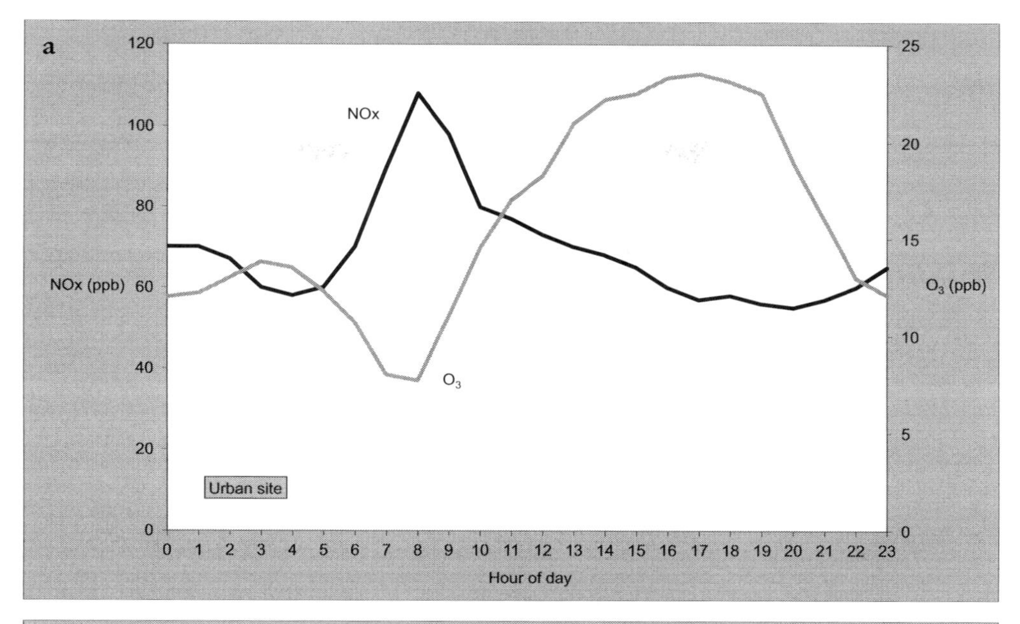

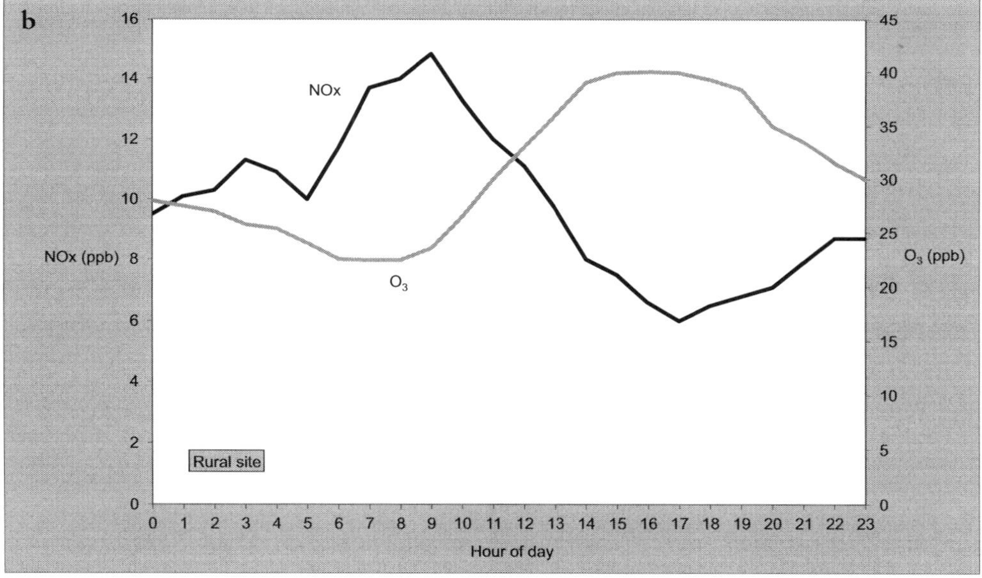

Figure 3.14 Nitrogen oxides and ozone concentrations in: (a) urban Istanbul, Turkey (10 m from a major highway with 2×10^5 vehicles d^{-1}) and (b) a rural location ~5km to the south and affected by dominant northerly winds from the city. Both graphs show diurnal variation in average hourly concentrations (collected over a 2-year period, 2008–2009). Comparison of the two graphs shows that, while NOx concentrations are at their highest in the city, O_3 levels are more elevated at the rural site. This is partly because of the photochemical formation of O_3 as the air mass (containing primary pollutants, mainly NOx) moves out of the city. It is also because O_3 is consumed by fresh vehicle emissions of NOx (specifically NO) in the city; this effect can be seen in both graphs. NOx is emitted in the morning rush hour and levels peak at around 8 am in the city (about an hour later at the rural site); O_3 concentrations gradually rise during the morning as the photochemical reactions take place and peak in the afternoon; again the rural site has a time-lag of about 1 hour, presumably as the air mass slowly moves out of the city to the south. Vehicular NOx emissions in the evening rush hour are not translated into a second NOx peak because of reaction with the O_3 that has accumulated during the day.

Source: Adapted from Im et al. (2013).

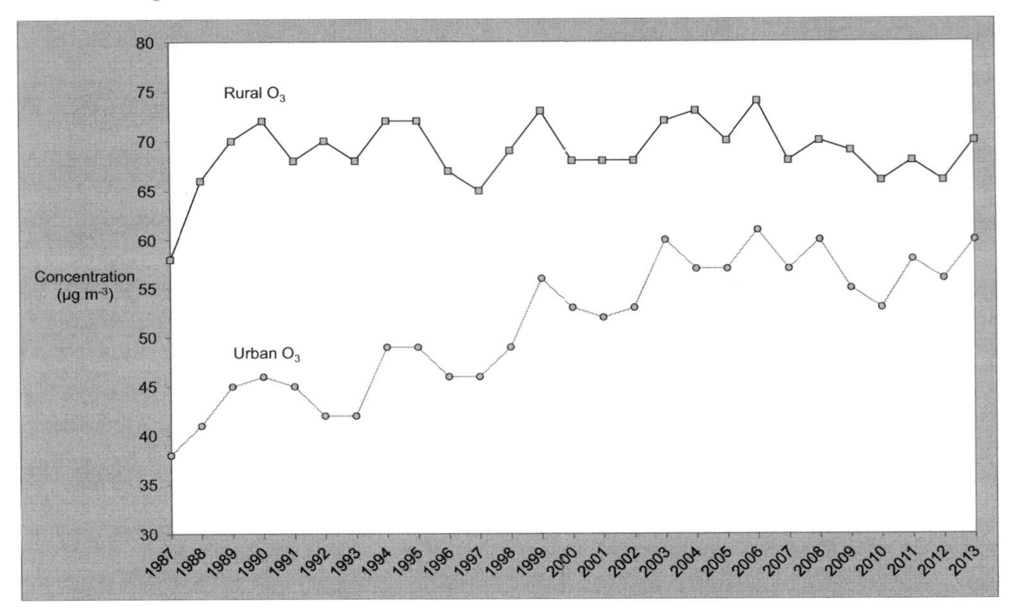

Figure 3.15 Tropospheric ozone concentrations in UK urban and rural areas, 1987–2013.
Source: Data from DEFRA (2014b).

Most of the key primary and secondary pollutants involved in photochemical smog formation have human health impacts. Ambient levels of nitric oxide are not thought to damage human health but exposure to pollutant levels of NO_2 can affect individuals with asthma and pulmonary disease; very high levels can also affect healthy subjects (Table 3.10). It constricts the airways and may increase susceptibility to infections of the respiratory system. High concentrations can cause pulmonary oedema (accumulation of fluid in the air sacs of the lungs) and more serious cases of airway constriction can lead to bronchiolitis obliterans, a potentially fatal condition. Populations living and working in close proximity to busy roads are at particular risk of exposure. In the general population, asthmatics, infants and some elderly people are most vulnerable to NO_2 pollution. The WHO guideline levels for NO_2 are 40 µg m^{-3} (21 ppb) as an annual mean and 200 µg m^{-3} (105 ppb) as a 1-hour mean. These guidelines are typically used as the basis for national air quality standards, but NO_2 concentrations in many cities around the world exceed these levels, at least some of the time (Table 3.5).

Table 3.10 Human health impacts of nitrogen dioxide. Summary table based on information from World Health Organization (2006) and Helleday et al. (1995).

Concentration (µg m^{-3})	*Effects*
>200	Bronchial constriction increases in asthmatic subjects
560	Decreased lung function in asthmatic subjects
6580	Temporary decreased lung function (mucociliary clearance of particles) in healthy subjects (20 min exposure)

Ozone, an oxidising agent, reacts with compounds lining the airways of the respiratory system, reducing its ability to withstand foreign bodies such as allergens and bacterial infections. The body's natural defence mechanism produces mucus, leading to inflammation and breathing difficulties. Chronic exposure can permanently scar lung tissue. Ozone exposure can also exacerbate respiratory diseases such as asthma, bronchitis and emphysema. Table 3.11 illustrates the health effects that can be expected at various concentrations; the WHO guideline value for O_3 is 100 μg m^{-3} (50 ppb), averaged over 8 hours. To provide an example of how often such guideline levels are breached, in summer 2013, all 28 member states of the EU exceeded the EU air quality standard for O_3 (120 μg m^{-3}, averaged over 8 hours) on at least one occasion and in 19 member states the standard was breached on 25 days or more (Fig. 3.16). Epidemiological studies indicate that each 10 μg m^{-3} increase in O_3 above background levels may lead to a daily mortality increase of 0.3% in an exposed population (WHO, 2006). Because O_3 concentrations tend to be higher away from urban centres, pollutant monitoring is also undertaken in suburban and rural areas in some countries, but this does not appear to be a widespread practice.

Ultra-fine particulate matter ($PM_{2.5}$) is an additional health risk in areas suffering from photochemical smogs {→C}. Reaction of NOx with NH_3, emitted mainly from animal wastes and sewage, leads to the production of minute ammonium nitrate particles (equation 3.1). Such particles can comprise significant percentages of total $PM_{2.5}$ load; in Los Angeles, for example, the classic photochemical smog location, the proportion is approximately one-third.

Table 3.11 Human health effects of ozone based on 8 hours of exposure.

Concentration (μg m^{-3})	*Effects*
100	Effects on respiratory systems of vulnerable individuals; 1–2% increase in 'brought-forward' deaths among those vulnerable to respiratory damage.
160	Transient lung inflammation among healthy adults with intermittent vigorous exercise under test conditions; 3–5% increase in 'brought-forward' deaths among those vulnerable to respiratory damage.
240	Airway inflammation and decreased lung function in healthy adults; 5–9% increase in 'brought-forward' deaths among those vulnerable to respiratory damage.

Source: Based on information from the World Health Organization (2006).

Atmospheric impacts: effects of air pollution on plants and materials

Tropospheric ozone, generated by photochemical NOx reactions, can enter plants through the stomata (leaf openings) and dissolve in water in the plant tissues. Here ozone's oxidative capacity can form compounds that disrupt vital plant functions. Brief exposure to slightly elevated levels of the gas reduces the rate of photosynthesis in some plants. Visible signs of ozone damage are chlorosis (reduced chlorophyll), interveinal necrosis (cell death), pigmentation (e.g. leaf spots), depressed bud/flower development, reduction in stomatal aperture and reduced growth. Tissue damage and reduction in photosynthetic rate also occur in plants exposed to NOx. The most

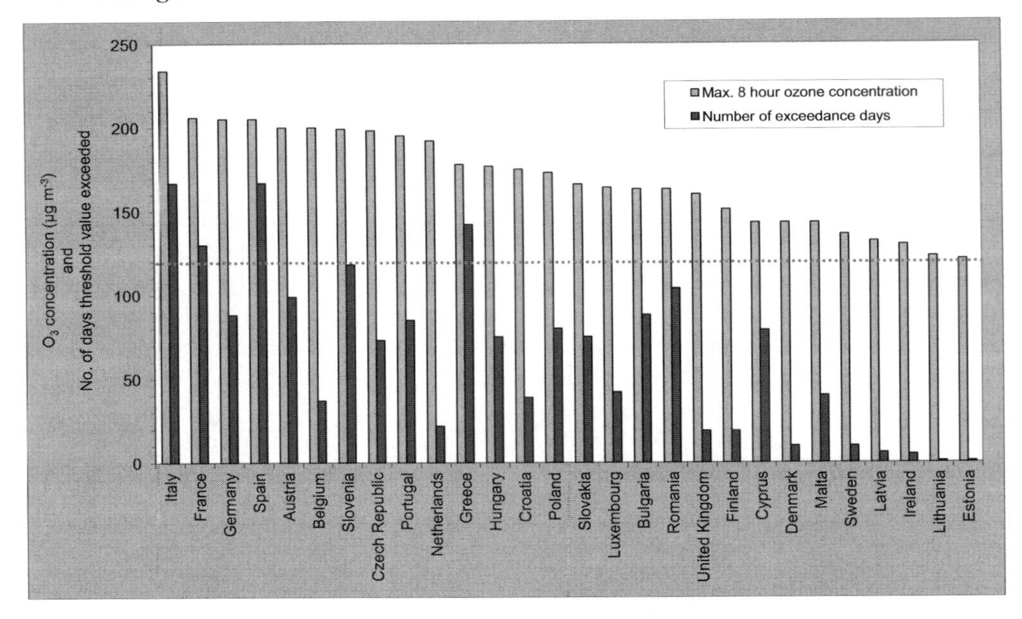

Figure 3.16 Maximum 8-hour averaged ozone concentrations (µg m^{-3}) in the 28 EU member states in summer (April–September) 2013; the horizontal line denotes the EU environmental quality standard for O_3 (120 µg m^{-3}). Also shown for each country is the number of calendar days this EQS was breached during the period.

Source: Data from EEA (2014).

toxic of the secondary pollutants, PAN, is more toxic to plants than O_3, but is present in smaller concentrations so its overall impact is lower. It reacts with the sulfhydryl groups of proteins and phospholipids, key components of cell membranes. A less direct impact of N deposition (both as NOx and NH_3) is biodiversity loss in semi-natural habitats as a result of fast-growing plants becoming dominant over less competitive species, including N-fixers, that flourish when nutrient levels are lower. Particularly affected are habitats such as nutrient-poor meadows, peatlands and bogs.

Fading of dyes in textiles and paints may be caused by NOx and tropospheric O_3, and the latter is also known to damage some rubber and plastic objects. Nitrogen compounds also directly damage building materials via acid rain formation.

Atmospheric impacts: nitrous oxide pollution and the global atmosphere

Natural atmospheric concentrations of N_2O are significantly enhanced by anthropogenic emissions from agricultural soils and wastes, vehicles and industrial sources. Agriculture is particularly important, as artificial N fixation significantly raises the amounts of N in both soils and waters, increasing the rate of denitrification and associated N_2O releases. Such emissions are of concern because N_2O, which has an atmospheric lifetime of 120 years, is a greenhouse gas and is a (stratospheric) ozone-depleting chemical (ODC).

As a greenhouse gas, N_2O is much more potent molecule-for-molecule than CO_2, having a 100-year global warming potential (GWP) of 265. Importantly, it absorbs terrestrial radiation of wavelengths 3–5 μm and 7.5–9 μm, the latter being within the atmospheric window, i.e. infrared radiation that is not absorbed by either CO_2 or H_2O. It accounts for approximately 5% of the overall greenhouse effect.

Because of its long atmospheric lifetime, N_2O reaches the stratosphere. Here, while some molecules are destroyed by photolysis, it persists long enough to reduce levels of stratospheric ozone. Monatomic oxygen, which is relatively abundant in the stratosphere due to photolysis of O_2 molecules, reacts with N_2O yielding NO (equation 3.30), which can destroy O_3 (equation 3.31). The NO_2 created in the latter reaction can react with further monatomic oxygen to regenerate the NO molecule (equation 3.32), which can thus destroy thousands of O_3 molecules in a continuous cycle.

$$N_2O + O \rightarrow 2NO \quad [3.30]$$

$$NO + O_3 \rightarrow NO_2 + O_2 \quad [3.31]$$

$$NO_2 + O \rightarrow NO + O_2 \quad [3.32]$$

Prior to the significant O_3 depletion observed in the 1980s, scientists had expressed concerns about the ozone-depleting potential of N_2O emissions. Following the phasing-out of CFCs and other halogenated ODCs, attention has once again switched to N_2O. For a detailed explanation of stratospheric ozone depletion, including the key role of polar stratospheric clouds, see {→Cl}.

References

AEA Group and DEFRA (Department for Environment, Food and Rural Affairs, UK). 2012. *Air Pollution in the UK 2011*. DEFRA, London.

AIC (Agricultural Industries Confederation). 2013. *Fertiliser Statistics 2013*. AIC, Peterborough.

Bowman, W.D., Cleveland, C.C., Halada, L., Hres, J. and Baron, J.S. 2008. Negative impact of nitrogen deposition on soil buffering capacity. *Nature Geoscience* 1, 767–770.

DEFRA (Department of Environment, Food and Rural Affairs). 2014a. *The British Survey of Fertiliser Practice*. DEFRA, York.

DEFRA (Department of Environment, Food and Rural Affairs). 2014b. *DEFRA National Statistics Release: Air Quality Statistics in the UK, 1987–2013*. DEFRA, London.

Diaz, R.J. and Rosenberg, R. 2008. Spreading dead zones and consequences for marine ecosystems. *Science* 321, 926–929.

EEA (European Environment Agency). 2014. *Air Pollution by Ozone across Europe during Summer 2013*. EEA Technical Report No. 3/2014. EEA, Copenhagen.

Eurostat. 2014. *Nitrogen Balance in Agriculture*. Available at: http://epp.eurostat.ec.europa.eu/statistics_explained/index.php/Nitrogen_balance_in_agriculture

Helleday, R., Huberman, D., Blomberg, A., Stjernberg, N. and Sandstrom, T. 1995. Nitrogen dioxide exposure impairs the frequency of the mucociliary activity in health subjects. *European Respiratory Journal* 8, 1664–1668.

Howden, N.J.K., Burt, T.P., Worrall, F., Whelan, M.J. and Bieroza, M. 2010. Nitrate concentrations and fluxes in the River Thames over 140 years (1868–2008): are increases irreversible? *Hydrological Processes* 24, 2657–2662.

Im, U., Incecik, S., Guler, M., Tek, A., Topcu, S., Unal, Y.S., Yenigun, O., Kindap, T., Odman, M.T. and Tayanc, M. 2013. Analysis of surface ozone and nitrogen oxides at urban, semi-urban and rural sites in Istanbul, Turkey. *Science of the Total Environment* 443, 920–931.

Jaffe, D.A. 1992. The nitrogen cycle. In: Butcher, S.S., Charlson, R.J., Orians, G.H. and Wolfe, G.V. (Eds.). *Global Biogeochemical Cycles*. Academic Press, London.

Rzychon, D. and Worsztynowicz, A. 2008. What affects the nitrogen retention on Tatra Mountains Lakes' catchments in Poland? *Hydrology and Earth System Sciences* 12, 415–424.

Schlesinger, W.H. and Bernhardt, E.S. 2013. *Biogeochemistry: An Analysis of Global Change*. Academic Press, Waltham.

WHO (World Health Organization). 2006. *Air Quality Guidelines: Global Update 2005*. World Health Organization Regional Office for Europe, Copenhagen.

4 Phosphorus

Environmental reservoirs and chemical forms

Phosphorus (P) is an essential element, but available forms are very limited in the natural environment because it exists primarily in insoluble solids. It is concentrated in the rocks of the Earth's crust, with much smaller concentrations in other parts of the environment (Table 4.1).

Phosphorus generally occurs in the natural world in the +5 oxidation state, as phosphate. Elemental P is unstable and not found in nature. Phosphorus halides and oxides, with oxidation states of +2, +3 and +5, are synthesised for use in chemistry and industry, but are not present naturally. Free phosphate (PO_4^{3-}) is the fully dissociated anion of phosphoric acid (H_3PO_4) and forms bonds with many different cations. The anhydride of H_3PO_4 is phosphorus pentoxide (P_2O_5), a white powder used in the production of many P compounds, including fertilisers. Phosphorus also occurs in organophosphates, in which organic molecules are bonded to one, two or three of the oxygen atoms in the PO_4^{3-} ion.

Lithosphere

Phosphorus is mainly present in the Earth's crust as insoluble calcium phosphate minerals, called apatites, which account for approximately 95% of crustal P. Apatites have the formula $Ca_5(PO_4)_3X$, where X can be OH (hydroxyapatite), F (fluorapatite) and, less commonly, Cl (chlorapatite). This is something of a simplification because there is generally a high degree of isomorphic ('equal form') substitution in phosphate

Table 4.1 Typical concentrations of phosphorus in the environment.

Environmental reservoir	*Typical concentration*
Crust	Basic igneous rock: up to 2500 mg kg^{-1} Other rock types: 300–1000 mg kg^{-1}
Soils (total P)	200–600 mg kg^{-1}
Seawater (surface)	1.5 µg L^{-1}
Seawater (deep)	60–90 µg L^{-1}
Freshwater	<100 µg L^{-1}
Atmosphere (particulate)	<1 µg m^{-3}
Atmosphere (gas, mainly phosphine)	<100 ng m^{-3}

minerals, particularly of Ca by Fe and Al. In a mass of apatite, many of the PO_4^{3-} anions may also be replaced by other anions such as sulfate and arsenate. Apatites are found most commonly in igneous rocks, but also occur in metamorphic and sedimentary rocks. Extensive sedimentary deposits of phosphate, termed phosphorite or rock phosphate, are found particularly in China, Russia, the USA and parts of northern and southern Africa and the Middle East.

Rarer phosphates include those of Al, Fe, Cu (including turquoise) and rare earth elements (monazite). Most other rocks do not have a high P concentration, but it is widely dispersed in the crystal lattices of many other minerals and clays. A more concentrated form, albeit uncommon compared with phosphorites, is found in thick guano deposits. There are large deposits of bird guano along parts of the South American west coast, where seabirds are attracted in huge numbers by the rich upwelling waters of the Humboldt Current and where the guano is not washed away because of arid conditions (Fig. 4.1). In the 19th century, deposits of >10 m depth were observed in this area. In other areas of the world, thick deposits of bat guano can build up in caves.

Phosphorus is held strongly in soils, mainly because of the reaction of phosphates with common soil cations to form compounds of varying, but generally low, solubility. In alkaline soils a wide variety of Ca phosphate minerals are present, from apatites to monocalcium phosphate. In more acidic soils, P is associated with Fe and Al hydroxyl phosphates such as strengite, $FePO_4.2H_2O$. In near-neutral soils (pH 6–7), P is more

Figure 4.1 Birds fill the landscape and the skies, Ballestas Islands, Peru. The accumulations of their guano (excrement) is so rich in phosphates and other nutrients that it is collected and sold as fertiliser to this day, mainly to organic farmers.

Credit: Lisa Weichel (Flickr).

available, existing as hydrogen phosphate (HPO_4^{2-}) and dihydrogen phosphate ($H_2PO_4^-$) ions released from mineral weathering; the concentrations of these two ions are approximately equal at pH 7 and increase, relative to each other, in alkaline and acidic soils, respectively. However, even at neutral pH, overall concentrations of available P remain low; liberated ions are likely to be adsorbed on Fe and Al oxide particles, for example. As P is an essential element, it is not surprising that it is also present in soil organic matter; most soils appear to have 25–75% of their P content in the organic fraction. Most of the *available* P in soils originates from the decomposition of organic matter.

Atmosphere

The atmosphere is a relatively insignificant part of the P cycle. Most P in the atmosphere is contained in particulate matter originating from soil dust, pollen and sea spray. Phosphine, PH_3, is the only gaseous form of any note in the atmosphere but is rare and oxidises readily to phosphoric acid (H_3PO_4), which may then act as a minor source of cloud condensation nuclei. Phosphine may be formed naturally by the microbial reduction of phosphate in anoxic environments.

Hydrosphere

Because of the low solubility of mineral phosphate, freshwaters contain very small amounts of dissolved P. The vast majority of P present in the hydrosphere exists within suspended or settling particles derived from the erosion of rock formations. No more than 10% of the particulate P is thought to be released into solution. Where dissolved forms are present, the most common in freshwaters (with pH levels of 6–7) is $H_2PO_4^-$, while in seawater (pH 8) HPO_4^{2-} is more dominant, together with $MgHPO_4$, $NaHPO_4^-$ and $CaPO_4^-$. Dissolved organic forms, released from the decomposition of organic matter, are also present in both fresh and marine waters, as are polyphosphates (phosphate groups in chains or cyclic compounds). Available P in surface seawater is readily taken up by photosynthesisers and, as a result, most oceanic P is present at depth, carried there by the settling of solid particles. In certain geological circumstances these sediments may subsequently become P-bearing sedimentary rocks.

Biosphere

Phosphorus is essential for life. In particular, it is an integral component of DNA and RNA (in the phosphorus-sugar backbone) and of the adenosine mono-, di- and tri-phosphates (AMP, ADP and ATP; Fig. 4.2) that are involved in energy transfer. It is also a key element in phospholipids (phosphoglycerides) in cell membranes and in bones and teeth (in hydroxyapatite). It is also present in nerve and brain tissue and in many organic molecules in the form of phosphate ester bonds. The biosphere is an important reservoir of available P; it is efficiently recycled by organisms because there is no atmospheric source and very little is released from its main reservoir (rocks). Its essentiality for life coupled with its low levels of availability mean that P is perhaps the greatest limiting factor on primary productivity and thus the proliferation of life.

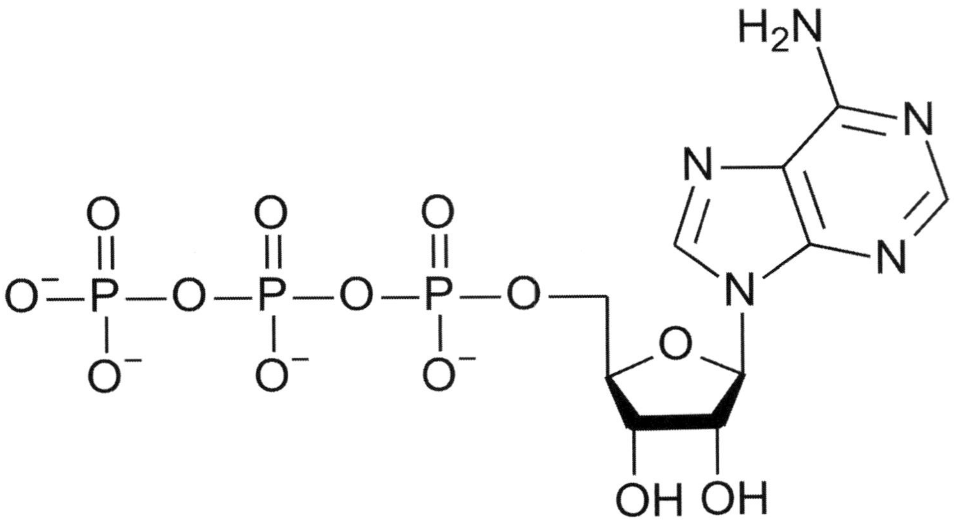

Figure 4.2 Adenosine triphosphate (ATP), a metabolic molecule present in cells, which transfers energy obtained from respiration. Energy is released by the removal of one of the three (tri)phosphates, on which ATP is converted to adenosine diphosphate, ADP. ATP is regenerated again in the cell by the enzyme, ATP synthase.

Credit: Mysid (Wikimedia Commons).

The natural phosphorus cycle

The P cycle (Fig. 4.3) mainly centres on the transfer of weathered rock particles to the sea floor and their lithification in sedimentary rocks, followed by uplift at a later stage in geological time. Cycling of available forms of P occurs over shorter time periods and is dominated by the biosphere.

Atmospheric fluxes

Phosphorus is unusual among the major elements in that the atmosphere plays a very small role in its natural cycle. Phosphate minerals are non-volatile and levels of P gases are very low. The flux of the main gaseous form, phosphine, is thought to be less than 0.04 Mt a^{-1}. However, larger amounts of particulate P are transported in the atmosphere, mainly sourced from soil dust and sea spray.

Biospheric fluxes

Phosphorus in the biosphere is recycled very efficiently. On land, only a miniscule proportion of total P in soils is available for uptake; plant roots take up scarce $H_2PO_4^-$ and HPO_4^{2-} ions from the soil. The diffusion of these anions to root surfaces is very slow and, to combat this, symbiotic fungi channel the phosphate ions to plant roots. In response to the low levels of available P in the soil, many plants exude acid phosphatase, an ectoenzyme, from their roots to liberate additional P from soil organic matter. Bacterial cells are known to store P in case of later decreases in environmental

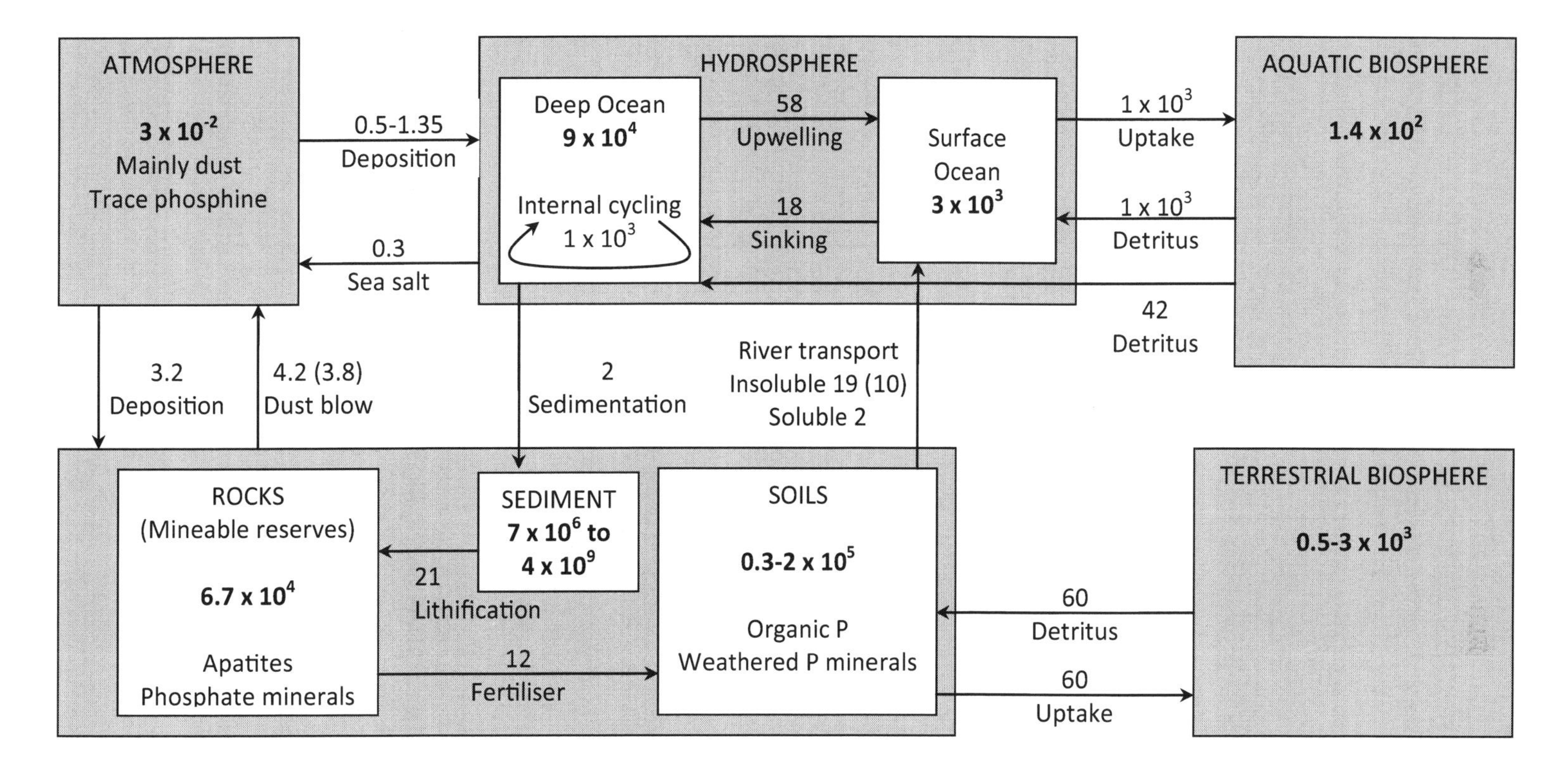

Figure 4.3 The phosphorus cycle. Units for estimated reservoirs (in boxes) and fluxes (adjacent to arrows) are Mt P and Mt P a^{-1}, respectively. Figures in brackets show pre-industrial fluxes, where estimated. The ranges of values for some fluxes and reservoirs demonstrate the uncertainties in their estimation.

Key sources: Jahnke (1992); O'Neill (1998); Schlesinger and Bernhardt (2013); United States Geological Survey (2013); and Yang et al. (2013).

levels. In the oceans, marine organisms may cycle up to 1×10^3 Mt a^{-1} of P, many times that of terrestrial biomass. Annually, the oceans may receive only 2 Mt of available P from rivers, implying that nearly all uptake by marine biomass is from P that has been released in the oceans by microbial biodegradation of organic matter.

Hydrospheric fluxes

Approximately 90% of the P carried by rivers is in the form of soil particles (runoff); most of this P is non-available, being locked up in mineral fragments that settle out in rivers and lakes and sink to the seabed, particularly on continental margins. The much smaller, available fraction is present in the water as dissolved phosphates. In the oceans, and in freshwater bodies such as lakes, photosynthesising organisms near the surface take up available P from the water. When such organisms die, most of the P is re-released by decomposition; however, any P contained in organic matter that escapes biodegradation remains in particulate form and is transported to the lake or sea bed in the sinking detritus, together with the settling mineral particles. Most of the released (or already free) phosphate ions are taken up again by new biomass, but many are immobilised by interaction with sinking mineral particles (e.g. sorption on calcium carbonates and iron oxyhydroxides) or by the formation of insoluble phosphate minerals on reaction with free Fe^{2+}, Al^{3+} and Ca^{2+} cations. Most of this inorganic P remains unavailable; however, in the reducing conditions encountered at depth in lakes and oceans, phosphates held in Fe oxide precipitates may be released when solid ferric iron (Fe^{3+}) is reduced to aqueous ferrous iron (Fe^{2+}).

The stratification of oceans, and lakes in summertime (see Chapter 1), limits the cycling of deep P back to the surface, where it could be used by photosynthesisers. In lakes, stratification typically breaks down in winter, but P reaching the surface is then of limited use because photosynthesis is limited by light levels. In the oceans, however, upward convection of P occurs at other times, both in high latitudes, where surface waters are colder resulting in less stratification, and in regions where upwelling of deep seawater occurs, such as along parts of the west coast of South America.

Lithospheric fluxes

Weathering and erosion liberate P from the lithosphere, largely within solid particles that are carried to varying degrees by streams and rivers. Input of P to the lithosphere is via sedimentary rock formation; seabed sediments comprising the newly formed rocks may include P from organic and, mainly, inorganic particles.

Production and uses of phosphorus

The main disturbance to the natural cycle of phosphorus arises from its extraction from the Earth's crust for fertiliser manufacture, the subsequent application of P fertiliser to farmlands around the world and its accumulation in soils and sediments. Hundreds of millions of tonnes of P have been transferred to the Earth's surface in this way over the past century.

Extraction and processing of phosphate rock

Extraction of phosphate (Table 4.2) increases year on year, driven currently by increasing demand for fertilisers in China, India and Latin America. Rock phosphate is generally extracted at or near the Earth's surface by destructive strip-mining techniques and most is converted industrially to phosphoric acid (H_3PO_4) by dissolution in sulfuric acid. Phosphoric acid is the main feedstock for phosphate fertiliser manufacture; it is also used to produce anti-corrosion and rust-proofing compounds, while purer grades are used in food additives and cleaning products. Other uses of phosphates include organophosphate pesticides.

Table 4.2 Global phosphate reserves and extraction in 2013.

Country	*Reserves (Mt × 10^3)*	*Extraction (Mt a^{-1})*
Morocco and West Sahara	50	28
China	3.7	97
Algeria	2.2	1.5
Syria	1.8	0.5
South Africa	1.5	2.3
Russia	1.3	12.5
Jordan	1.3	7
United States	1.1	32.3
Global total[a]	67	224

a Including countries not shown.

Source: USGS (2014).

Agriculture: fertilisers

Phosphorus is mainly used as a fertiliser and has been employed as such throughout recorded history; for example, bonemeal (hydroxyapatite) was used in ancient China and guano was also used extensively. Manure is another traditional source of P. In the more recent past, ground rock phosphate was commonly used as a mineral fertiliser, but because of its relative insolubility and consequent slow release of P (and because of high transport costs), it is used less today. Most P fertilisers are in the form of single or triple 'superphosphate', which is a relatively soluble Ca phosphate. Single superphosphate is formed, together with gypsum (hydrated calcium sulfate), by reaction of mineral phosphate with sulfuric acid (equation 4.1). Triple superphosphate is produced when phosphoric acid (H_3PO_4), is used as the reactant instead of sulfuric acid (equation 4.2). Another common phosphate fertiliser is ammonium phosphate, which also supplies N to crops; it is created by the addition of ammonia to phosphoric acid. Much of the P that is not used in fertilisers is employed as animal feed in intensive livestock operations and in some countries this is the second most common use of the element.

$$Ca_3(PO_4)_2\,(s) + 2H_2SO_4\,(aq) + 2H_2O\,(l) \rightarrow Ca(H_2PO_4)_2\,(s) + 2CaSO_4.2H_2O\,(s) \quad [4.1]$$

$$Ca_3(PO_4)_2\,(s) + 4H_3PO_4\,(aq) \rightarrow 3Ca(H_2PO_4)_2\,(s) \quad [4.2]$$

Agriculture: organophosphate pesticides

Organophosphate (OP) compounds comprise one of the main categories of pesticides in current use (Table 4.3). Together with carbamates and pyrethroids {→C} they have largely replaced organochlorine (OC) pesticides {→Cl}. The main advantage over OCs is their non-persistence in the environment and some are broken down in a matter of days. Others are longer-lived; for example, the OP dimethoate persists in soils and waters for several months. Most OPs are narrow-spectrum (selectively toxic) insecticides, mainly because the active toxic groups can be broken down by enzymes possessed by mammals but not insects; however, some OPs are broad-spectrum and OPs in general are much more acutely toxic than OCs to individuals exposed to them, including non-targeted animals and humans.

The toxicity of OPs (and carbamates) to insect pests is caused mainly by their irreversible binding to acetylcholinesterase, an important enzyme in the nervous system of nearly all animals; the enzyme is inactivated by the addition of a phosphate group to a chemical group present at the enzyme's active site, a process called phosphorylation. Acetylcholinesterase reverses the acetylation of choline to acetylcholine; this reversal controls the transmission of nerve impulses between synapses (nerve cells). Alteration of the enzyme by OP prevents the breakdown of acetylcholine, causing malfunctioning (continuous stimulation) of nerves, which can lead to death. The OP herbicide, glyphosate, inhibits an enzyme called EPSP synthase that plants need to synthesise amino acids such as phenylalanine that are vital for protein production and continued growth.

Table 4.3 Some commonly used organophosphate pesticides.

Pesticide	*Typical uses*
Chlorpyrifos[a]	Insecticide
Coumaphos	Insecticide: agriculture, residential
Diazinon[a]	Insecticide: sheep dip, residential
Dichlorvos	Insecticide: livestock, greenhouses
Dimethoate	Insecticide: agriculture, residential
Fenamiphos	Nematicide
Glyphosate	Herbicide: weedkiller
Isofenfos	Insecticide
Malathion	Insecticide: agriculture, horticulture, malaria control, residential
Parathion[b]	Insecticide: agriculture, horticulture
Parathion methyl	Insecticide: agriculture
Phosmet	Insecticide: fruit crops
Profenofos	Insecticide: agriculture

a Banned for residential use in some countries.
b Banned or restricted in many countries.

Phosphate detergents and other uses

Sodium tripolyphosphate, $Na_5P_3O_{10}$, is added to detergents to remove Ca^{2+} and Mg^{2+} ions from water because they react with detergent molecules and prevent them from removing grease and dirt particles; it also acts as a pH buffer. Phosphorus is used widely in anti-corrosion, anti-rusting products to protect metal objects against the elements. The object is exposed to an acidic solution containing phosphate, which then forms an insoluble precipitate on the metal surface by reacting with metal cations released by the acid. Additionally, phosphoric acid is used in the chemical polishing of alloys. A common historic use of P was in the heads of matches. In the 19th century, white phosphorus was used, but was eventually discontinued because it caused a serious bone disorder, called phossy jaw, in workers. Modern safety matches use red phosphorus, which is contained in the striking surface; the match head is mainly potassium chlorate. Phosphorus has seen military use, as white phosphorus in explosives and as organophosphates in nerve agents, such as Sarin and Tabun. Nerve agents have also been used in terrorist attacks. The use of Sarin on the Tokyo underground in 1995 hospitalised thousands of people and caused 12 fatalities, illustrating its toxicity to humans. Some other common uses of P compounds are shown in Table 4.4.

Table 4.4 Uses of phosphorus compounds not mentioned in the text.

Use	*Phosphorus compound(s)*
Oil additives	P sulfides
Light-emitting diodes (LEDs)	Gallium phosphides
Flame retardants	Organophosphates, phosphite esters and phosphine gas
Medicine (bone disease)	Bis-phosphonate
Food additives	E.g. phosphoric acid as an acidity regulator in carbonated drinks; calcium phosphate in baking powder
Toothpaste	Calcium phosphates and hydroxyapatite (abrasive agents and enamel enhancers); sodium monoflurophosphate (F additive); sodium polyphosphate (tartar reduction)
Ceramics	Disodium phosphate
Water treatment	Various ortho- and poly-phosphates for reduction of impurities (e.g. Fe, Mn, Pb, Cu), scale and corrosion
Fuel cells	Phosphoric acid fuel cells (PAFCs): Among the first fuel cells to be developed commercially; used today in power generators and public transport

Phosphorus pollution

Agricultural sources

Mineral and organic P fertilisers that are spread onto agricultural land can be washed directly into rivers if conditions are rainy, but otherwise PO_4^{3-} ions generally adsorb readily onto soil particles or form precipitates of insoluble solids. Therefore, in contrast to soluble nitrates, little of the added P leaches into ground or surface waters; however, this more intransigent behaviour means that more applications are required for crop uptake and the overall phosphate burden of the soil gradually rises. When heavy rains lead to soil erosion and runoff, P-laden soil particles are transported to rivers and other water bodies (Fig. 4.4). Such soil erosion is exacerbated by agricultural and arboricultural land clearances that expose areas of bare soil.

Figure 4.4 Soil erosion in a wheat field, USA. Because phosphate has limited solubility in soils, particulate runoff is an important transport pathway for P, particularly in fertilised and/or cleared land.

Credit: Jack Dykinga, US Dept. of Agriculture (Wikimedia Commons).

When applied to agricultural crops, OP pesticides can reach the wider environment via atmospheric transport during spraying and percolation through soil to ground and surface waters. There have also been cases of criminal or negligent disposal of OP pesticides (including used sheep dip) to soils and/or freshwaters.

Sewage treatment as a source

Domestic sewage and wastewaters contain high levels of phosphorus and are therefore potential sources of P pollution in receiving waters. The tripolyphosphates in detergents are broken down in wastewater and treatment is therefore necessary to minimise the release of PO_4^{3-} ions into freshwaters. Primary and secondary sewage processes transfer approximately 20% of phosphates from wastewater into sewage sludge, which may subsequently be spread on agricultural fields. Where a greater level of P removal from wastewater is necessary, tertiary treatment is undertaken; this involves the formation of insoluble phosphates by reaction with added Fe, Al or Ca, which are introduced, respectively, as ferric chloride (equation 4.3), aluminium sulfate (alum, equation 4.4) or slaked lime (equation 4.5). Such reactions remove approximately 95% of phosphate from wastewaters, leaving a proportion to enter the receiving watercourse; for higher removal rates of >99%, adsorption of PO_4^{3-} ions on activated alumina, Al_2O_3, may be used.

$$FeCl_3\,(s) + PO_4^{3-}\,(aq) \rightarrow FePO_4\,(s) + 3Cl^-\,(aq) \qquad [4.3]$$

$$Al_2(SO_4)_3.14H_2O\,(s) + 2PO_4^{3-}\,(aq) \rightarrow 2AlPO_4\,(s) + 3SO_4^{2-}\,(aq) + 14H_2O\,(l) \qquad [4.4]$$

$$5Ca(OH)_2\,(s) + 3HPO_4^{2-}\,(aq) \rightarrow Ca_5OH(PO_4)_3\,(s) + 6OH^-\,(aq) + 3H_2O\,(l) \qquad [4.5]$$

Because of the pollution risk from P in wastewaters, its use in washing powder has been banned in the USA for many years, although it is still used there in dishwasher powder. The European Union has proposed that the amount of phosphate allowed in detergents will be limited in future to a fraction of 1%. The main alternatives to phosphates in detergents are zeolite and nitriloacetic acid (NTA). Organophosphate flame retardants have also been detected in wastewaters.

Impacts of phosphorus pollution

Eutrophication

Eutrophication of freshwaters, particularly lakes and slow-flowing rivers, is typically attributed to P as the main limiting nutrient. Nitrogen, the main cause of estuarine and coastal eutrophication {→N}, is naturally provided to lakes (but much less to saline waters) by aquatic N-fixers and is therefore less likely to be the limiting nutrient, despite the fact that phytoplankton require up to 20 times more N than P. Furthermore, nitrate is much more soluble and bioavailable in freshwaters than phosphate, which is bound in sediment particles and unavailable, in contrast to saline waters where phosphates can desorb into the water column. If P is the limiting nutrient in many freshwater bodies, therefore, it is clear that additional inputs of P to freshwaters will be the prime cause in them becoming eutrophic.

In the natural P cycle, soil-bound P typically remains in lakes and slower rivers as bottom sediments. Naturally occurring soil erosion will therefore provide a reservoir of P in such water bodies; however, considering the low solubility of P, this will not normally amount to a significant input of soluble P to the water column at any one time. On the other hand, human activities add significantly to the input of P to freshwater bodies. The extraction of crustal P to produce fertiliser, its surface application to farmland and subsequent runoff result in an increased riverine flux of particulate P, as does enhanced soil erosion from bare fields. This adds to the P that flows into rivers in sewage effluent. The excess P can render water bodies eutrophic (e.g. Box 4.1), which causes problems of hypoxia, toxin release by cyanobacteria and turbidity as aquatic organisms proliferate in response to the added nutrient {see →N for detailed explanation of eutrophication}. Hypoxia can cause the release of further phosphate into the water column from the sediment reservoir (see 'Hydrospheric fluxes' sub-section, above). Freshwaters tend to become eutrophic when available-P levels reach 0.03 mg L^{-1}.

Organophosphate pesticides

The main exposure routes of OP pesticides in humans are: (i) inhalation (agricultural workers, rural populations, residential use); (ii) ingestion (traces on foodstuffs, deliberate ingestion in suicides); and (iii) dermal absorption (unprotected workers, residential use). To assess the risk of human ingestion of pesticide residues on foods, the authorities in some countries take samples of home-grown and imported food and drink products that are on sale to the public and analyse them for pesticide concentrations. In the UK, for example, the Committee on Pesticide Residues in Foods reports annually; a typical report from a recent year shows that 3% of more than 3000 analysed samples had pesticide residues in excess of established maximum guideline levels. The residues found on foodstuffs included OP pesticides such as dimethoate, chlorpyrifos, glyphosate, malathion and profenofos, on both homegrown and imported foodstuffs. The committee recommend actions to be taken where the residues have the potential to affect human health; for example in samples where more than one OP or carbamate pesticide is detected (because the effects of such acetylcholine disrupters may be additive).

Organophosphate insecticides work by disrupting acetylcholinesterase in insect pests (see above); however, this can occur in exposed humans and non-target organisms. In exposed human subjects, acetylcholinesterase disruption can cause a range of outcomes including sweating, lachrymation, salivation, diarrhoea, vomiting, bronchial spasm, hypotension, respiratory failure and death. Chronic effects include depression, drowsiness and memory loss; studies have suggested links to low birthweight, carcinogenicity (parathion) and an increase in attention deficit hyperactive disorder (ADHD) in children. Published claims that thousands of people die each year from acute exposure to OPs appear to relate to deliberate ingestion of OP pesticides as a form of suicide, particularly in rural areas of less developed countries.

There are concerns about the ecological toxicity of OP pesticides, particularly because of spray drift and the broad spectrum nature of some of the compounds used. Aquatic invertebrates appear to be particularly affected by chlorpyrifos, a commonly used, broad spectrum OP insecticide with a 'no observed effect level' (NOEL) of <0.1 μg L^{-1}. Effects on organisms are also seen with other OPs, including dimethoate

Box 4.1 Freshwater eutrophication.

In Lake Erie, North America, there have been major blooms of *Microcystis* sp. cyanobacteria in recent years, attributed primarily to P runoff from farmland, although an additional factor may be the invasive zebra mussel (*Dreissenid polymorpha*), which feeds on green algae that compete with *Microcystis*. In the worst years, one-third of the lake bottom has been classed as a 'dead zone' (anoxic) as a result of aerobic micro-organisms digesting the dying bloom. In addition, the cyanobacteria release microcystin, a liver toxin, to levels well in excess of the guideline level of 1 $\mu g\ L^{-1}$. In 2011 the lake's largest ever *Microcystis* bloom was recorded, covering an estimated area of >5000 km^2 (Fig. 4.5); this was attributed in large part to a >200% increase in dissolved P inputs to the lake from the Maumee River, Ohio, USA, from 1995 to 2011 (Michalak et al., 2013). Long-term changes in agricultural practices in Ohio, including increases in surface fertiliser application (rather than soil injection) were hypothesised to be the main source of the extra P, coupled with heavy spring rainfall of the kind that is predicted to increase with future climate change.

Figure 4.5 A large *Microcystis* sp. bloom (light grey areas) in Lake Erie in summer, 2011, caused primarily by agricultural inputs of phosphorus. The areas at the top and bottom of the image are Ontario, Canada and Ohio, USA, respectively.

Credit: NASA Earth Observatory (Wikimedia Commons)

(Fig. 4.6). Organophosphate pesticides have been shown to affect behaviour in birds and have been implicated in the deaths of wild birds in France, Greece and elsewhere; there may also be teratogenic effects in birds, causing skeletal defects in offspring.

By-products of fertiliser manufacture

The gypsum produced in superphosphate fertiliser manufacture (equation 4.1) typically has elevated levels of radioactivity because of the presence of uranium-238 and radium-226 in the phosphate rock used. Therefore, the 'phosphogypsum' by-product cannot all be used in normal gypsum products and has to be stockpiled above ground. In the phosphate-rich US state of Florida, at least 700 m tonnes are stored in this way, creating large tips in the landscape and there are concerns that the wastes are a source of acidity and trace elements, including radionuclides and fluoride.

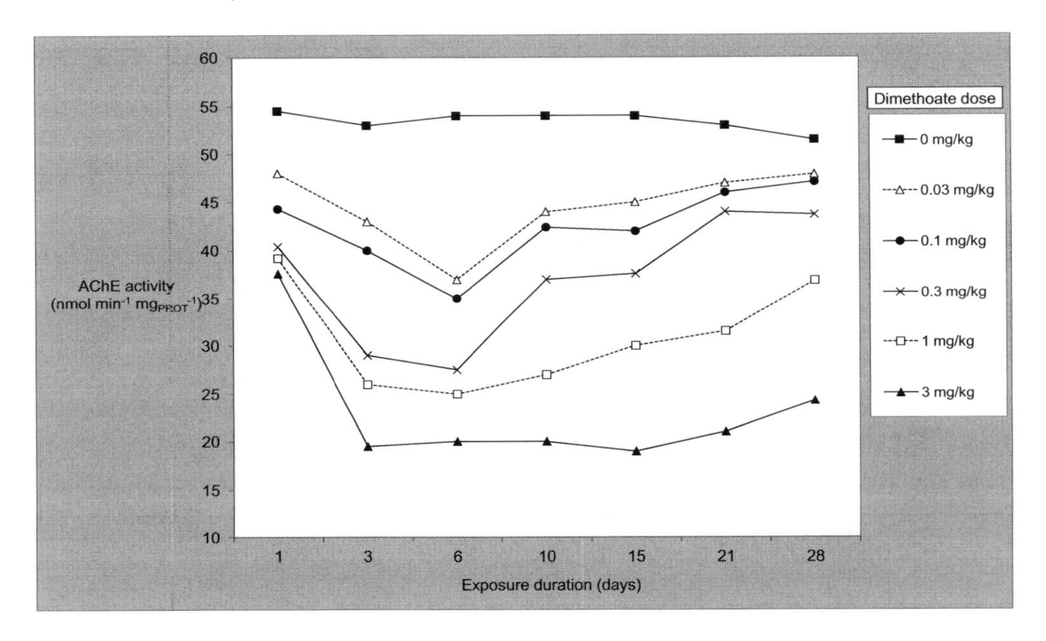

Figure 4.6 The effect of dimethoate in soil (dose and exposure time) on enzyme activity (acetylcholinesterase, AChE) in the earthworm *Eisenia andrei* (n = 21 earthworms for each dose and exposure time). The six lines on the graph, in order from the top downwards, relate to increasingly large doses of dimethoate (i.e. the top line is the control dose, see legend) and these progressively inhibit the earthworms' enzyme activity, as shown on the left axis. The authors used sublethal, 'environmentally relevant' concentrations (the recommended application rate on the pesticide label was 0.3 mg of active ingredient per kg of dry soil and they tested from 0.1 to 10 times this concentration). Varying degrees of recovery occurred after the low points of enzyme activity following 6 days of exposure; these were attributed to gradual dimethoate degradation in soil; despite this, significant ($p < 0.05$) inhibition of enzyme activity compared with the control treatment was still being observed, at doses of 0.3 mg kg^{-1} and higher, 28 days after the OP was added to the soil.

Source: Data from Velki and Hackenberger (2013).

References

Jahnke, R.A. 1992. The phosphorus cycle. In: Butcher, S.S., Charlson, R.J., Orians, G.H. and Wolfe, G.V. (Eds.). *Global Biogeochemical Cycles*. Academic Press, London.

Michalak, A. et al. 2013. Record-setting algal bloom in Lake Erie caused by agricultural and meteorological trends consistent with expected future conditions. *Proceedings of the National Academy of Sciences of the USA* 110(16), 6448–6452.

O'Neill, P. 1998. *Environmental Chemistry*. Blackie, London.

Schlesinger, W.H. and Bernhardt, E.S. 2013. *Biogeochemistry: An Analysis of Global Change*. Academic Press, Waltham.

USGS (United States Geological Survey). 2014. *Mineral Commodity Summaries: Phosphate Rock*. Available at: http://minerals.usgs.gov/minerals/pubs/commodity/phosphate_rock/mcs-2014-phosp.pdf

Velki, M. and Hackenberger, B.K. 2013. Inhibition and recovery of molecular biomarkers of earthworm *Eisenia andrei* after exposure to organophosphate dimethoate. *Soil Biology and Biochemistry* 57, 100–108.

Yang, X., Post, W.M., Thornton, P.E. and Jain, A. 2013. The distribution of soil phosphorus for global biogeochemical modelling. *Biogeosciences* 10, 2525–2537.

5 Sulfur[1]

Environmental reservoirs and chemical forms

Lithosphere

Sulfur (S) is one of the Earth's most abundant elements and, although most is concentrated in the planet's interior, the lithosphere is the major environmental reservoir (Fig. 5.1). Elemental or native S precipitates around volcanoes and hot springs and there are geological deposits in Japan, Mexico and parts of Europe. The element is also present in the crust in metal sulfide ores and there are deposits of gypsum and other sulfates (Table 5.1). The presence of S in fossil fuels is significant in the context of environmental pollution. The S content of coal can be up 5%; it is present in both the organic matter which formed the coal and within incorporated sulfide minerals, particularly pyrite. Oils also contain S; relatively sulfurous heavy fuel oils can contain 3–4% sulfur while refined products like petrol contain only a fraction of 1%. Natural gas also contains S, in the form of hydrogen sulfide (H_2S).

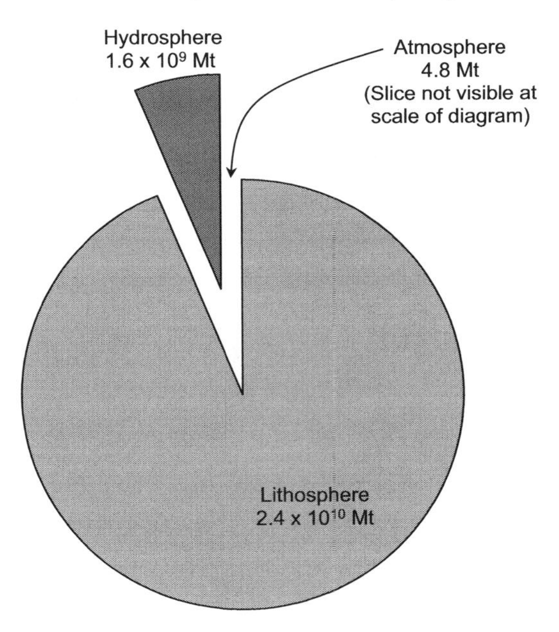

Figure 5.1 The main environmental reservoirs of sulfur.

Table 5.1 Forms of sulfur in the Earth's crust.

Sulfides		*Sulfates*	
Pyrite	FeS_2	Gypsum	$CaSO_4.2H_2O$
Galena	PbS	Epsomite	$MgSO_4.H_2O$
Sphalerite	ZnS	Barite	$BaSO_4$
Cinnabar	HgS	*Others*	
Chalcocite	Cu_2S	Native sulfur	S
Chalcopyrite	$CuFeS_2$	Fossil fuel	e.g. FeS_2

Atmosphere

The most abundant S gas in the atmosphere is carbonyl sulfide (COS). It is produced in smaller quantities than other S gases but is relatively resistant to oxidation and has a residence time of approximately 1 year (compared with days for other S gases), allowing it to accumulate to a concentration of ~1 ppb in the atmosphere. Millions of tons of sulfur dioxide (SO_2) enter the atmosphere each year but, because of its quick oxidation and its dissolution in atmospheric water, it has a relatively low background concentration (0.2 ppb). Other important sulfur gases include hydrogen sulfide (H_2S; 0.01 ppb); dimethyl sulfide (DMS; $(CH_3)_2S$; 0.01 ppb) and carbon disulfide (CS_2; 0.01 ppb). Sulfate, in compounds such as sulfuric acid (H_2SO_4) and ammonium sulfate [$(NH_4)_2SO_4$], is also an important atmospheric form of S.

Hydrosphere

Seawater is a large reservoir of S, nearly all in the aqueous sulfate (SO_4^{2-}) form, which is also present in fresh waters, much of it leached from soils. Other dissolved species include sulfite (SO_3^{2-}), bisulfite (HSO_3^-) and bisulfate (HSO_4^-) ions. The oceans also contain the aqueous form of H_2S. The gas DMS, $(CH_3)_2S$, is produced in the hydrosphere and subsequently diffuses into the atmosphere.

Biosphere

Sulfur is an essential element and is found in all living organisms as a constituent of two amino acids, cysteine and methionine. Plants take up the element in the SO_2 and SO_4^{2-} forms and are a key source of S in the food chain. Cysteine is present in keratin, the structural protein that forms hair and nails; in cysteine, disulfide (–S–S) linkages form between the –SH (sulfhydryl) groups in thiols (R–S–H, where R represents a carbon-containing group of atoms). These bridges link different parts of the peptide chains that make up the protein (Fig. 5.2), creating folds and other useful three-dimensional structures. The thiol groups in another cysteine-rich protein, metallothionein, are used by organisms to bind useful chalcophilic (S-loving) metals, particularly Zn, and subsequently release them for biological processes. Toxic metals like Cd and Hg may also be bound by metallothionein and safely stored in the liver. The other S-bearing amino acid, methionine, has a key role in the biological process of methylation. An important S-containing compound in the marine biosphere, and in

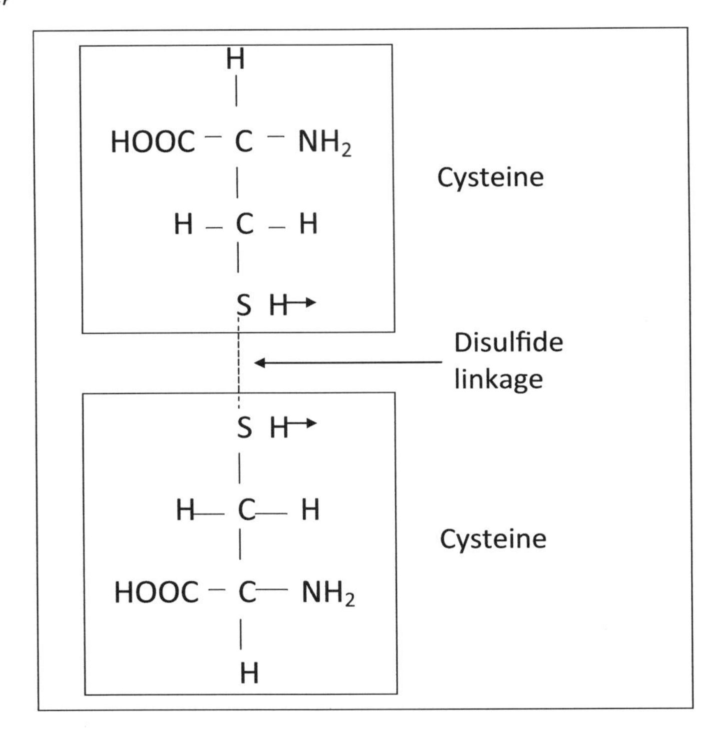

Figure 5.2 Disulfide linkages provide structure to keratin, a cysteine-based protein.

some plants, is dimethylsulfoniopropionate (DMSP), $(CH_3)_2S^+CH_2CH_2COO^-$. It is synthesised by marine algae, which use it to maintain osmotic balance with sea water.

The natural sulfur cycle

Figure 5.3 summarises the sulfur cycle, including the main S species, reservoirs and fluxes. Sulfur exists in several oxidation states and undergoes transformations between these states in the natural environment. In oxic conditions, reduced forms (upper rows of Table 5.2) are progressively oxidised to those forms listed further down. Conversely, in anoxic environments, oxidised forms such as sulfate may be converted to reduced forms like H_2S and metal sulfides. The resulting fluxes may be considered in terms of abiotic and biotic cycles (see below).

Abiotic cycling

Transfers from the lithosphere

Over time, weathering processes liberate S from rocks into percolating waters and much of this is transported to the world's oceans, mainly as sulfate. Similarly, S contained in the organic matter fraction of soils is slowly oxidised to sulfate, which is subsequently, if not taken up by plants, leached out of the topsoil into groundwater. Volcanoes transfer lithogenic S to the atmosphere in the forms of H_2S, COS and SO_2.

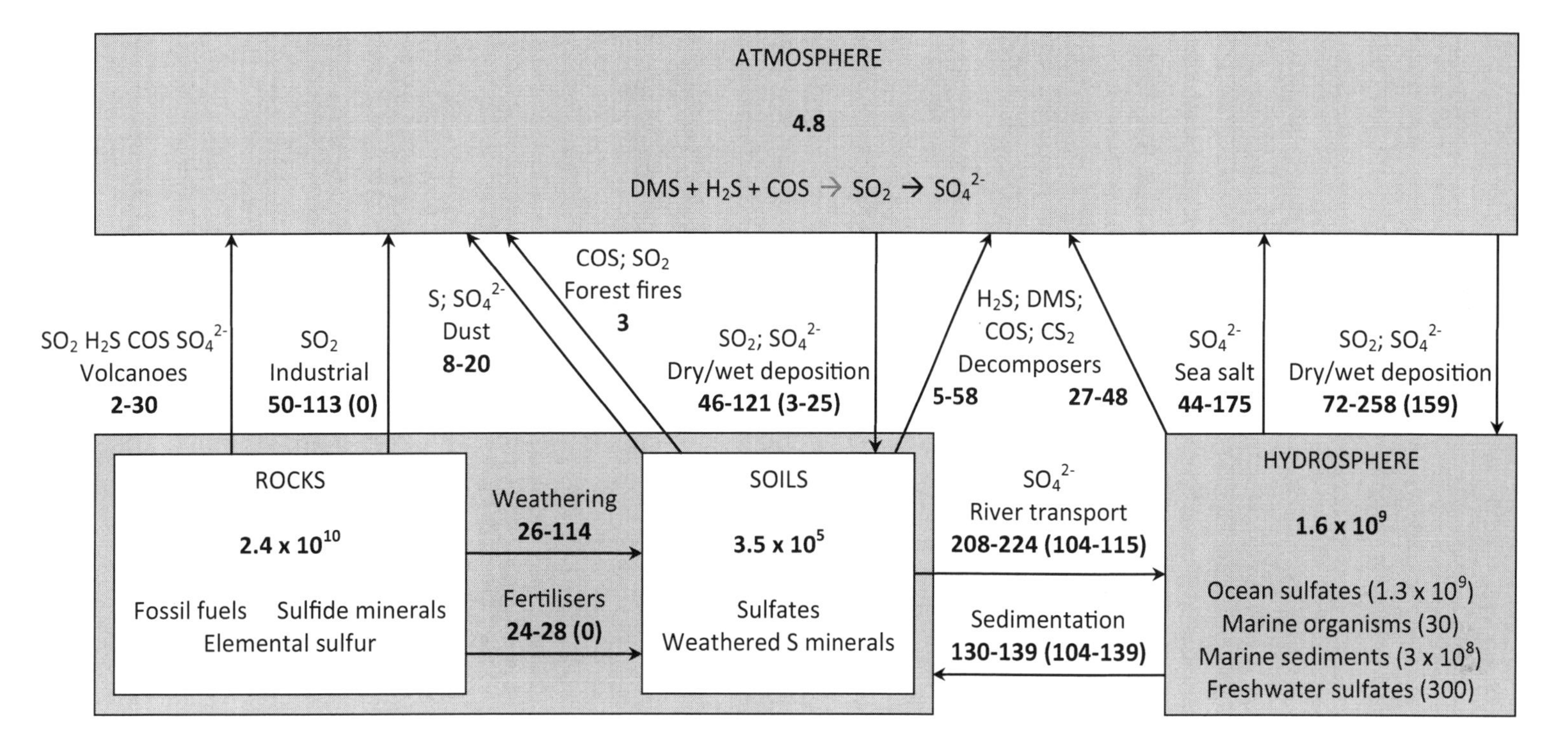

Figure 5.3 The sulfur cycle. Units for reservoirs (in boxes) and fluxes (adjacent to arrows) are Mt S and Mt S a^{-1}, respectively. The ranges of values for fluxes demonstrate the uncertainties in their estimation, but an annual flux through the atmosphere may be in the order of 250–300 Mt per year, with perhaps 20–30% of this attributable to human activity. In the latter context, comparison of current fluxes with bracketed (pre-industrial) figures shows that human activity has significantly increased the annual transfer of S from the lithosphere to the atmosphere as well as subsequent deposition and river transport (see later sections on pollution).

Key sources: Charlson et al. (1992); Schlesinger and Bernhardt (2013).

Table 5.2 Oxidation states of some important natural sulfur species and compounds (with typical state in parentheses).

Oxidation state	*Species*[a]
–2	H_2S (g); COS (g); CS_2 (g); DMS (g); sulfide minerals (s)
–1	Disulfide (s), e.g. iron disulfide, FeS_2
0	Native sulfur (s)
+4	SO_2 (g); HSO_3^- (aq)
+6	SO_3 (g); SO_4^{2-} (aq); H_2SO_4 (l); sulfates of Ca and Mg (s)

a SO_3 is sulfur trioxide; full names of all other compounds are given in the text.

For example, at least 17 Mt of SO_2 are estimated to have been emitted from the eruption of Mount Pinatubo in the Philippines in 1991. Hydrogen sulfide is also present in natural gas and can potentially seep from host rocks.

Transfers from the atmosphere

Gaseous forms of S have short residence times in the troposphere (e.g. DMS 1 day; H_2S 4 days; SO_2 14 days) because most are rapidly oxidised to other forms (e.g. equation 5.1) and/or undergo wet or dry deposition (e.g. SO_2, which is water soluble). Additionally SO_2 may be removed from the atmosphere by plant uptake via the stomata.

$$2H_2S + 3O_2 \rightarrow 2SO_2 + 2H_2O \quad [5.1]$$

In both the gaseous phase (equations 5.2a–c) and dissolved phase (equations 5.4a–b), SO_2 is ultimately converted to SO_4^{2-}. Gaseous-phase H_2SO_4 (formed in equations 5.2a–c) has a low saturation vapour pressure and therefore condenses readily onto water droplets, where it dissolves and dissociates to bisulfate, sulfate and hydrogen (H^+) ions (equation 5.3).

$$SO_2 + {}^{\cdot}OH \text{ (hydroxyl radical)} \rightarrow HSO_3^- \quad [5.2a]$$

$$HSO_3^- + O_2 \rightarrow HO_2 \text{ (perhydroxyl radical)} + SO_3 \quad [5.2b]$$

$$SO_3 + H_2O \rightarrow H_2SO_4 \quad [5.2c]$$

$$H_2SO_4\,(aq) \leftrightarrow H^+\,(aq) + HSO_4\,(aq) \leftrightarrow 2H^+\,(aq) + SO_4^{2-}\,(aq) \quad [5.3]$$

Alternatively, in the conversion of SO_2 to SO_4^{2-} in the aqueous phase, gaseous SO_2 first dissolves in a water droplet to become aqueous SO_2. This reacts with water to form sulfurous acid (H_2SO_3), the dissociation of which produces bisulfite and sulfite ions (equation 5.4a). Bisulfite reacts with hydrogen peroxide to form sulfate (equation 5.4b).

$$SO_2 + H_2O \leftrightarrow H_2SO_3 \leftrightarrow H^+ + HSO_3^- \leftrightarrow 2H^+ + SO_3^{2-} \quad [5.4a]$$

$$HSO_3^- + H_2O_2 \text{ (hydrogen peroxide)} + H^+ \rightarrow SO_4^{2-} + H_2O + 2H^+ \quad [5.4b]$$

The deposition of the dissolved species produced in equations 5.3, 5.4a and 5.4b is the main process of S removal from the atmosphere.

Biotic cycling

Transfers into and within the biosphere

In anoxic environments, such as lake sediments and deep oceans, sulfate reduction to sulfide, by specialist micro-organisms such as those of the *Desulfovibrio* genus (Fig. 5.4), regulates much of the anaerobic oxidation of organic matter. Some bacteria assimilate the S, while others reduce sulfate primarily to obtain energy via electron transfer (sulfate being used as a terminal electron acceptor) and the S is excreted as a waste product. Equation 5.5 shows sulfate reduction in the presence of acetate, CH_3COO^-, a known substrate; other substrates include alcohols and aromatic compounds (Sorensen et al., 1981).

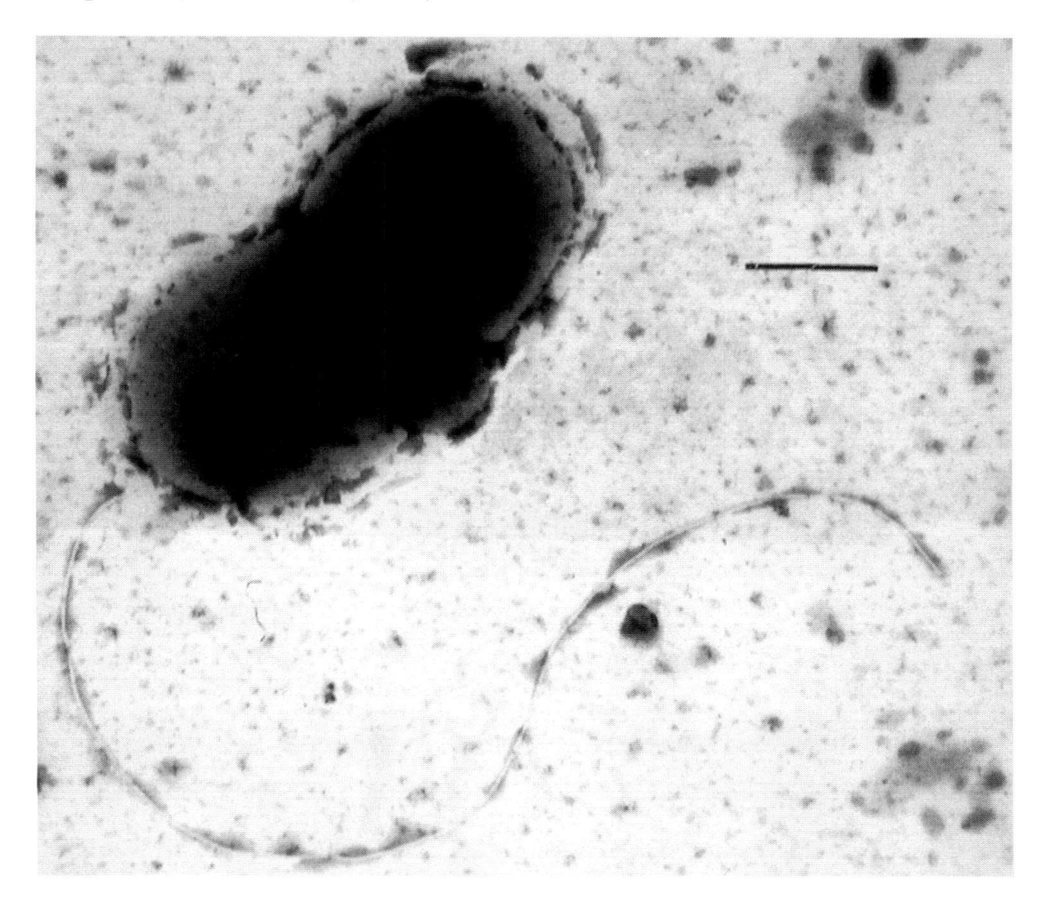

Figure 5.4 Desulfovibrio vulgaris, an anaerobic, sulfate-reducing bacteria. Bar length: 0.5 μm. Credit: Graham Bradley (Creative Commons).

$$SO_4^{2-} + CH_3COO^- + 3H^+ \rightarrow H_2S + 2CO_2 + 2H_2O \quad [5.5]$$

The H_2S produced in equation 5.5 may react with metal ions (e.g. Cu^{2+}, Pb^{2+}, Zn^{2+}) present in the anoxic sediments to form insoluble sulfides. For example:

$$Zn^{2+} + H_2S \rightarrow ZnS + 2H^+ \quad [5.6]$$

Chemosynthetic bacteria use H_2S as an energy source to manufacture sugars at hydrothermal vents in the sea floor, where sunlight, and therefore photosynthesis, is not available (equation 5.7). The bacteria may be free living or symbiotically hosted by animals such as the tubeworm *Riftia pachyptila*.

$$4H_2S + CO_2 + O_2 \rightarrow CH_2O \text{ (carbohydrate)} + 4S + 3H_2O \quad [5.7]$$

Specialist micro-organisms in anoxic zones of lakes where H_2S is abundant use dissolved H_2S, instead of water, for photosynthesis and produce elemental S rather than oxygen (equation 5.8). The species capable of this process are collectively known as purple sulfur bacteria. They produce grains of S that can be observed inside the bacterial cells.

$$H_2S + CO_2 + h\nu \rightarrow C_{org} + 2S + 2H_2O \quad [5.8]$$

Microbial processes are clearly an important part of S uptake into the biosphere but higher organisms provide an additional mechanism. To provide S for amino acids, plants take in sulfate from the soil (via the roots) and SO_2 (via stomata). There is evidence to suggest COS is also taken in by plants. The assimilated S is subsequently passed up the food chain to herbivores and carnivores.

Transfers out of the biosphere

Several S gases are produced from biodegradation of organic matter, including SO_2. In anoxic environments like swamps, H_2S, COS and CS_2 are produced. While some of the H_2S produced in such environments will remain there, by conversion to metal sulfides, much will diffuse out into the atmosphere along with the other gases. Carbonyl sulfide and SO_2 are also produced from natural biomass combustion.

Decomposition of marine algae and the dimethylsulfoniopropionate they contain (see above) annually transfers up to 30 Mt of S to the atmosphere as DMS. This is of interest to climate scientists because oxidation of DMS can ultimately produce cloud condensation nuclei (CCN) in the form of tiny droplets of sulfuric acid (Fig. 5.5). One tenet of James Lovelock's 'Gaia' hypothesis (of biosphere-regulating processes) is that this part of the natural S cycle may regulate climate: i.e. if natural increases in global temperature encourage algal growth and, ultimately, DMS production, the resulting rise in cloud cover (because of increased CCN) could reduce the overall warming. Lovelock (2010) states that the mechanism is observable in the southern hemisphere, but is obscured in the northern hemisphere by anthropogenic S emissions.

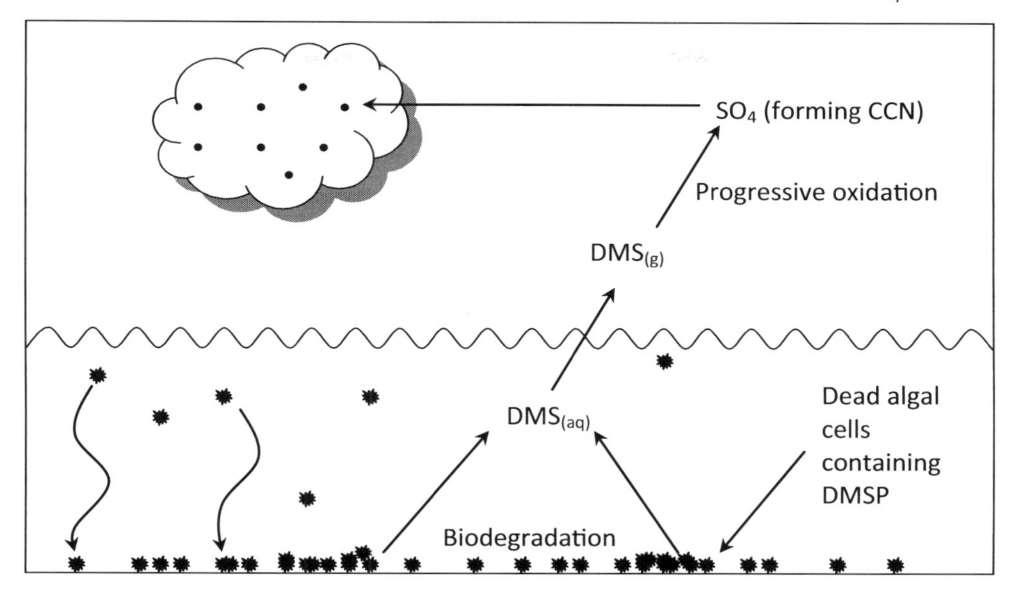

Figure 5.5 Biodegradation of marine algae, which contain dimethylsulfoniopropionate (DMSP) for osmotic regulation, releases DMS to the atmosphere; oxidation of DMS ultimately produces sulfates, which can act as cloud condensation nuclei (CCN).

Anthropogenic disturbance of the sulfur cycle

Uses of sulfur

Elemental S is used in vulcanised rubber and fungicides. Sulfur dioxide is used as a food preservative and in the production of sulfuric acid, the world's most heavily produced industrial chemical. Gypsum is used in plasterboard and barite is used as a weighting agent in oil drilling fluids. Sulfur, in its various forms, is used very widely and these are just a few examples.

Extraction and processing of lithospheric sulfur

The most significant anthropogenic disturbance to the natural S cycle is the extraction of long-term, geological reservoirs of S and subsequent releases to the hydrosphere and, in particular, the atmosphere. Native S is surface mined or extracted from underground deposits by the Frasch process, where superheated water and steam are pumped underground to melt the solid S, allowing it to be pumped to the surface. Today, most S is extracted, mainly as H_2S, from sulfurous fossil fuels, particularly natural gas and oil; elemental S is subsequently obtained via oxidation of the H_2S (equation 5.9 is a simplified summary).

$$2H_2S + O_2 \rightarrow 2S + 2H_2O \qquad [5.9]$$

The smelting of sulfide minerals and combustion of sulfurous fossil fuels transfers lithospheric S to the atmosphere, mainly as SO_2. H_2S is also emitted from industrial processes (mainly oil refining and paper production), but in much smaller quantities, and SO_2 is the form in which the vast majority of S is emitted. Equation 5.10 shows how the smelting of metal sulfide ores (denoted as MS) produces the desired metal (M), but also generates SO_2, which is emitted to the atmosphere as a waste product:

$$MS\ (s) + O_2\ (g) \rightarrow M\ (s) + SO_2\ (g) \qquad [5.10]$$

However, the combustion of fossil fuels, particularly in power stations, is the major source of atmospheric SO_2, accounting for 90% of emissions in the UK, for example. The S contained in the fossil fuel, partly as iron sulfide, FeS_2, reacts in air to produce SO_2:

$$S + O_2 \rightarrow SO_2 \qquad [5.11]$$

SO_2 emissions may be minimised to some extent by pollution control technologies such as flue gas desulfurisation (FGD) and fluidised bed combustion (FBC). These processes are based on the reaction of SO_2 with lime, limestone (equation 5.12) or dolomite. FGD, for example, can remove more than 90% of SO_2 emissions from coal-fired power stations by passing the flue gases through a slurry of finely ground limestone, $CaCO_3$, to produce calcium sulfite ($CaSO_3$) and carbon dioxide:

$$CaCO_3 + SO_2 \rightarrow CaSO_3 + CO_2 \qquad [5.12]$$

The anthropogenic input of S into the atmosphere is of the same order of magnitude as natural emissions. Global anthropogenic emissions peaked in the 1990s, since when developed countries have increasingly introduced air quality measures to reduce concentrations of SO_2 and other atmospheric pollutants. In the UK, for example, by 2010 SO_2 emissions had fallen to 11% of 1990 levels. Despite this, global SO_2 emissions remain considerable and have been significantly added to by economic growth in Asia in recent years. Coal consumption in China has risen significantly since 2000 and SO_2 emissions from Chinese power plants increased from ~11 Mt in 2000 to ~19 Mt in 2006 (Lu et al., 2010); however, China has been reducing its SO_2 emissions since 2006 due to the extensive adoption of FGD technology (Fig. 5.6).

Sulfur dioxide emissions from motor vehicles have decreased in most countries as a result of the widespread introduction of low sulfur petrol in recent times (the acceptable percentage of S in petrol in the European Union, for example, is 0.001%); levels of S in fuels are still high in some areas, particularly Africa. Shipping industries can still use fuel oil with up to 3.5% S, although the percentage allowed by the International Maritime Organization will be reduced to 0.5% by 2020 because of environmental concerns.

Pollution of freshwaters arises from the deposition of SO_4^{2-} and H^+ ions from the sources mentioned; however, the extraction and processing of lithospheric S can also directly pollute the hydrosphere. Oxidation of exposed metal sulfide mine workings and their associated waste heaps also results in the release of excess SO_4^{2-} and H^+ ions. The water carrying these ions is known as acid mine drainage (see below).

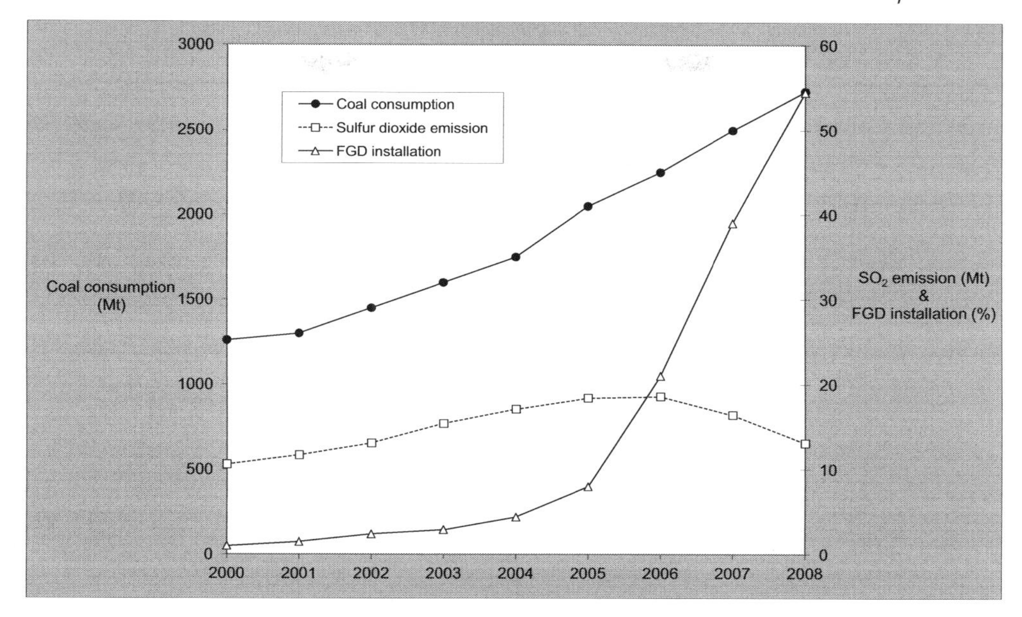

Figure 5.6 Trends of coal consumption, flue gas desulfurisation installation and estimated sulfur dioxide emissions in relation to power plants in China, 2000–2008. Despite rising coal consumption, the installation of FGD in >50% of power plants by 2008 resulted in declining SO_2 emissions.

Source: Data from Lu et al. (2010).

Environmental impacts arising from disturbance of the sulfur cycle

Sulfur deposition and acid rain

As we have seen, naturally produced SO_2 may be oxidised in the atmosphere to sulfuric acid, which is readily wet deposited. When excess SO_2 is produced, in power stations for example, its oxidation can lead to acid rain, an increase in the natural acidity of rainfall (nitrogen emissions also contribute to acid deposition {→N}). The normal pH of rainfall is a tolerable 5.6, due mainly to the dissolution of naturally abundant CO_2 molecules in raindrops creating a weak carbonic acid, but episodes of acid rain decrease the pH further. Although the residence time of SO_2 in the troposphere is relatively short, in the order of about 2 weeks, this is long enough for acid rain to travel across national boundaries and oceans and cause problems far from where the pollution was first produced. In catchments with low buffering capacities (i.e. where the rocks have limited acid neutralising capacity), the result is overly acidic natural waters and soils. Associated with the reduction in pH is an increase in the concentrations of aluminium (much of which is normally locked up in the soil) to potentially toxic levels in soil solution and fresh waters; the main biotic impacts of Al release are on plants and aquatic organisms (Box 5.1).

Long-term studies in Wales, UK provide a clear example of the impacts of acidic deposition on aquatic organisms. By the 1980s, the pH levels of half of the streams in Wales had decreased and this was having significant effects on aquatic invertebrates and riverine birds. Despite the controls on S emissions that have been enforced since

Box 5.1 Ecological impacts of acidification.

One of the main impacts on organisms in acidified catchments is exposure to elevated levels of Al in the bioavailable Al^{3+} form. Acidifying substances can release Al, which is naturally present in soil and sediment particles, into soil solution and surface waters. Aluminium is quite toxic to plants and elevated levels of bioavailable Al in soil solution can affect enzyme activity and cause root damage, leading to decreased water and nutrient uptake. Reproductive success and growth is affected in some amphibian species, including the Natterjack toad (*Bufo calamita*). In Al-enriched, acidic freshwaters, reductions are recorded in invertebrates such as mayflies, beetles and crustaceans, with consequent effects on their predators. The decline of the Dipper (*Cinclus cinclus*), an invertebrate-eating bird, has been noted in such waters, although low calcium concentrations may also have an effect in some cases; similarly, impacts on Pied Flycatcher (*Ficedula hypoleuca*) may be caused by decreased Ca intake affecting shell porosity in addition to Al-related prey effects. In fish, slight elevations in Al levels may affect osmoregulation (body salt level) because of damage to specialised cells in the gills that transfer salt ions; sufficient Ca^{2+}, which controls gill permeability, ensures adequate ion intake at low pH, but Al ions can displace Ca^{2+} in gill tissues, inhibiting ion uptake. At Al concentrations of >100 μg L^{-1}, respiration can be affected by the precipitation of gelatinous $Al(OH)_3$ on the gills. There are particular concerns about the impacts on Arctic char (*Salvelinus alpinus*), brown trout (*Salmo trutta*) and salmon (*Salmo salar*). Investigators of brook trout (*Salvelinus fontinalis*) mortality during acidic episodes in upland streams in the USA (Fig. 5.7) concluded that an Al concentration of >225 μg L^{-1} for at least 2 days was the primary determinant of mortality. Declines in smaller fish populations affect piscivorous birds like the Kingfisher (*Alcedo atthis*).

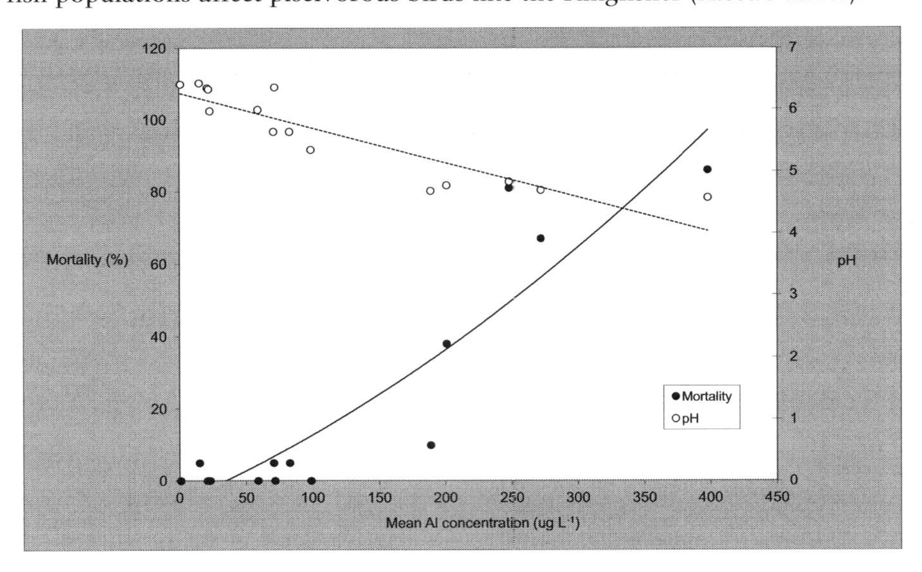

Figure 5.7 Relationship between inorganic aluminium concentration in river water and (i) brook trout mortality (left axis); (ii) pH of river water (right axis).

Source: Data from Baldigo and Murdoch (1997).

then (in Europe and elsewhere), the streams have not fully recovered. A 25-year study (1981–2005) of upland Welsh streams (Ormerod and Durance, 2009) showed that the mean winter pH of forest streams had increased by only 0.4 units over the period and that they were still too acidic to allow for recovery of invertebrates. Moorland streams had larger increases of approximately 1 pH unit, but were still at risk from 'episodic acidification' when heavy rainfall or snowmelt deposited anthropogenic S from the atmosphere; this is thought to have restricted the effectiveness of 'lime-dosing' of the streams in restoring invertebrate populations.

Plant damage is another symptom of SO_2 and acid deposition, perhaps best illustrated by forest dieback in regions where sulfurous coal has been used (e.g. the 'black triangle' in eastern Europe, where problems peaked in the 1980s) or around smelting complexes where metal sulfides are roasted (e.g. Sudbury, Canada and Norilsk, Russia {→Ni}). Elevated concentrations of SO_2, taken in via the stomata, can damage plants directly. In acute cases foliar necrosis (cell death) is observed, but chronic effects from long-term exposure to relatively low, but elevated, levels, are more typical. These include reduced growth, senescence (e.g. leaf-shedding) and chlorosis (yellowing), where chlorophyll production, and therefore photosynthesis, is affected and leaves become discoloured. Chlorosis is caused because SO_2 damages the light-sensitive components (thylakoids) of a plant's chloroplasts. The gas is also thought to affect the permeability of cell membranes, allowing plants' cells to leak essential elements. Another important effect of SO_2 is that of lowering the plant's response to other environmental stresses such as frost. Similarly, plants exposed to elevated levels of SO_2 have been found to accumulate higher levels of toxic metals like Cd (Li et al., 2011). Some crops, including barley, spinach and wheat, are more susceptible to SO_2 than others, such as asparagus and corn. Lichens and bryophytes are particularly susceptible because, unlike higher plants, they do not have a protective cuticle, or stomata that could be closed to prevent SO_2 intake. The gas may dissolve in moisture on the surface of these organisms, in some cases yielding toxic bisulfite ions, which can then directly enter individual cells. Documented effects include impairment of photosynthesis, respiration and nitrogen fixation.

In addition to ecological impacts, certain materials are corroded by acid rain, particularly ornamental building materials such as limestone, $CaCO_3$ (equation 5.13); however, such erosion also happens naturally and pollution may not always be the sole or main cause (see Fig. 5.8, for example).

$$2H^+ (aq) + CaCO_3 (s) \rightarrow Ca^{2+} (aq) + CO_2 (g) + H_2O (g) \qquad [5.13]$$

In Europe and the United States, acid deposition still occurs, although to a lesser extent than in past decades, and concerns remain because of sulfate retention in soils; for example, despite the closure of mining and smelting operations at Falun, Sweden in 1992, the acidity of nearby lakes has not increased according to diatom studies, probably because of long-term storage of deposited S in their catchment areas (Smol, 2008). Despite such remaining concerns, attention has now shifted somewhat to industrialised nations, particularly China, where rates of S-deposition in some areas now exceed those observed in the 'black triangle' of 1980s Europe.

Figure 5.8 Eroded limestone, Truro Cathedral, UK. This is a relatively modern cathedral, built from 1880 to 1910 when the main industrial period of this region had ended. Natural weathering in a maritime climate is likely to be the main cause of damage in this particular case, although the possibility of acid rain from further afield should perhaps not be discounted.

Acid mine drainage

The excavation of metal ores and coal deposits can expose large areas of underground workings and surface-tipped waste, leading to sulfide oxidation by air and water. This generates acid mine drainage (AMD), which can pollute local waterways with acidic waters. For example, oxidation of pyrite (FeS_2), which is ubiquitous in many mineral and coal deposits, may be summarised as follows:

$$4FeS_2 + 15O_2 + 14H_2O \rightarrow 4Fe(OH)_3 + 8H_2SO_4 \qquad [5.14]$$

The acidity is derived from the dissociation of the resulting sulfuric acid (H_2SO_4) into hydrogen cations (acidic) and sulfate anions (basic). The acidity typically leads to elevated concentrations of toxic metals dissolved in the mine drainage water, particularly in the vicinity of metal ore mines. The first product in equation 5.14 is iron hydroxide, or 'ochre', which is commonly evident as a vivid orange staining or blanketing of the affected watercourse (Fig. 5.9). Acid mine drainage is a common source of pollution at mine sites around the world. In the USA alone, 2×10^4 km of streams and rivers are estimated to have been degraded by AMD. Occasionally, large-scale incidents have occurred; a British example is an incident at Wheal Jane mine, Cornwall in 1992, when a bright orange plume of 50 million litres of acidic (pH 3) and metal-rich minewater was released into the local river and reached the nearby coast.

The acidity of AMD can cause direct ecological impacts, being particularly damaging to fish gills, and severe ochre deposits can smother benthic fauna and egg-laying habitats. The elevated concentrations in AMD of dissolved toxic metals such as Al, Cd and Pb are harmful to sensitive receptors such as aquatic invertebrates and young salmonid fish species. Higher species such as riverine birds and mammals are indirectly affected by the pollution because of declines in their prey.

Treatment of AMD is resource intensive, long term and expensive, and the costs often have to be borne by the state in the absence of the original mine owners. Treatment typically involves the gravitational passage of minewater through crushed limestone to neutralise the acidity and promote, together with water aeration, the removal of dissolved metals into solid precipitates. Artificial wetlands are also used in some situations; reed species such as *Phragmites australis* help to filter sediments, aerate the water and encourage metal precipitation.

Classic smogs and human health

When fossil fuels are burned or sulfide minerals roasted in, or close to, inhabited areas, human health impacts can result from inhalation of elevated levels of SO_2. The

Figure 5.9 The 'Red Waterfall' in Pennsylvania, USA. The image shows acid mine drainage from a coal mine, resulting from oxidation of iron disulfide (pyrite) in underground mine workings (see equation 5.14).

Credit: David Fulmer (Flickr).

gas stimulates nerves in the linings of the nose, throat and lung passages, causing irritation and constriction of the airways. The main effects are summarised in Table 5.3. A synergistic effect can occur when SO_2 is accompanied by fine particles, as is often the case when coal is burned; the particles provide a condensation surface for the formation of sulfuric acid (SO_2 combining there with moisture) and damage to the sensitive lung tissues ensues.

Using the UK as an example, maximum hourly concentrations of SO_2 today average approximately 200 $\mu g\ m^{-3}$, well within the UK limit value of 350 $\mu g\ m^{-3}$ and lower than any of the concentrations shown in Table 5.3. However, individual maximum hourly values may be higher; for example, 460 $\mu g\ m^{-3}$ and 293 $\mu g\ m^{-3}$ were recorded at industrial and urban sites, respectively, in 2011 (AEA Group and DEFRA, 2012). In past times, SO_2 concentrations in urban areas were much higher than they are today and the famous London smogs culminated in a fatal pollution episode in December 1952, when the combined and synergistic effects of SO_2 and fine particulate matter, both emitted in large quantities from domestic fires, were thought to have caused 4000 fatalities. Subsequent research has indicated that the number is more likely to have been 12,000, because many other deaths recorded at the time were attributed, it now seems incorrectly, to influenza (Bell and Davis, 2001). These fatalities are generally termed 'brought-forward' deaths, with elderly people most susceptible, along with individuals with pre-existing respiratory complaints, possibly caused by chronic exposure to poor air quality. In the UK, the Clean Air Acts that followed the 1952 incident decreased exposure to SO_2 and other pollutants from domestic coal fires. However, the ensuing move towards large power plants with tall stacks, emitting large volumes of flue gas, subsequently contributed to transboundary acid rain problems. Today, SO_2 concentrations are particularly elevated in industrialised parts of the world (see above).

Hydrogen sulfide is also a toxic gas, but is generally present in the atmosphere at much lower concentrations than SO_2. It irritates the throat and eyes and can be fatal if levels reach more than a few hundred parts per million. It is produced by anaerobic decomposition of sewage; therefore sewage workers wear personal monitors that sound an alarm when levels reach approximately 10 ppm. A well-documented case of H_2S pollution in recent years centres on the allegation that in 2006 an international shipping company exported toxic waste to the Republique de Cote d'Ivoire in Africa. The waste was said to be a S-rich oil residue produced from the treatment of fuel oil with caustic soda to produce cleaner oil for resale. The allegation is that H_2S emissions

Table 5.3 The human health effects of sulfur dioxide.

Concentration (ppm)	*Concentration ($\mu g\ m^{-3}$)*	*Effect*
0.25	6.7×10^2	Respiratory function of asthmatic individuals may be affected
<1	$<2.7 \times 10^3$	Detectable by taste/smell
1–2	$2.7–5.2 \times 10^3$	Can cause changes in lung function (increased airway constriction) in healthy individuals
10	2.7×10^4	Noticeable irritation of the respiratory tract
100	2.7×10^5	Considered to be a lethal dose

Source: Adapted from USDHHS (1998).

from the waste made thousands of local people ill and a UN report concluded that there seemed to be strong evidence linking the waste to a number of deaths. The company was later fined €1 million by a Dutch court (the waste had been exported from Amsterdam) and the government of the Republique have since compensated the families of 16 people who died at the time.

Global dimming and brightening

The SO_4 aerosol produced by the oxidation of SO_2 is known to reflect incident solar radiation, thus keeping the atmosphere and the planet somewhat cooler than it would otherwise be, an effect called 'global dimming'. This means that SO_2 emissions in Europe and North America in past decades may have been restricting the full extent of global warming. Climate scientists are concerned that, with the predicted CO_2 increase this century, current and future reductions in SO_2 pollution, required for the protection of human health and the environment, may conversely lead to 'global brightening', resulting in more accelerated warming than had been expected in the coming years and decades (Fig. 5.10). With the most recent reductions in SO_2 emissions, especially in China since 2006, some climate scientists predict that the rate of global warming this century may increase from the previously projected 0.2°C per decade to 0.3–0.4°C globally and 0.8°C in the northern hemisphere, where S emissions have already decreased (Raes and Seinfeld, 2009).

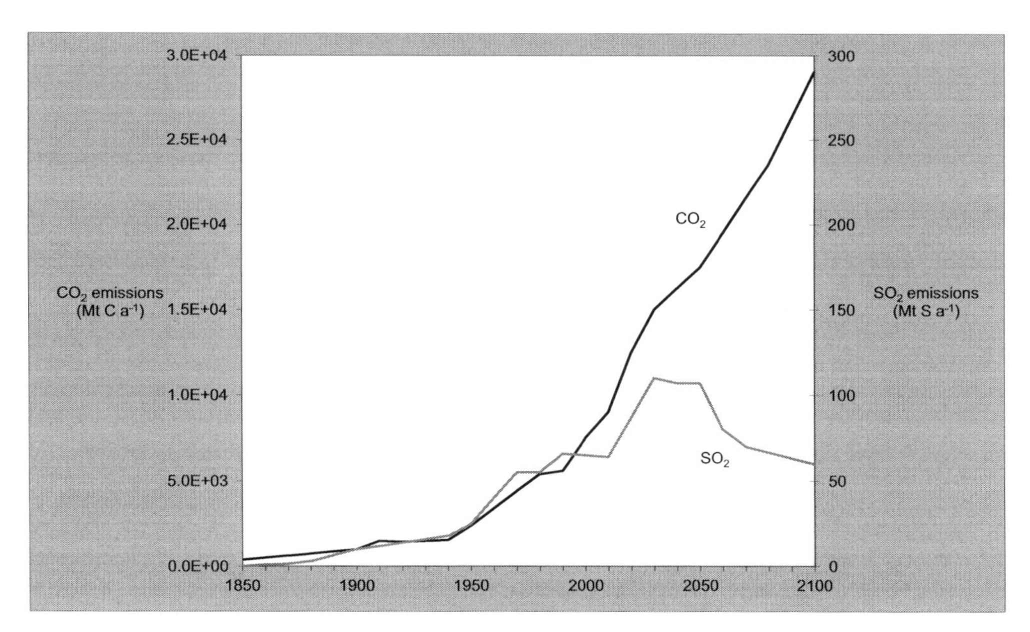

Figure 5.10 Global emissions of SO_2 and CO_2 in the past (1850–2000) and forecast for this century. The emissions reductions expected for SO_2 will decrease its atmospheric concentration relatively quickly because of its short atmospheric lifetime; meanwhile, CO_2 concentrations will remain high because of rising emissions (as shown) and because it has a much longer atmospheric lifetime. Consequently, the expected reduction in reflective sulfates without a concurrent fall in CO_2 levels may result in 'global brightening'.

Source: Adapted from Andreae et al. (2005).

Because of the global dimming effect of SO_4, injecting SO_2 into the stratosphere has actually been proposed as a potential technical fix for global warming. However, even putting aside concerns over the uncertain effects that such an undertaking might have on global climate and the environment, there are other reasons for caution. The injections would have to continue indefinitely (particularly if this 'fix' resulted in a continuance of greenhouse gas emissions), the costs would be large and the increased atmospheric SO_2 could clearly result in environmental impacts of the kind detailed in this chapter.

Note

1 The International Union of Pure and Applied Chemistry has formally adopted 'sulfur' as the accepted spelling of the element, in preference to 'sulphur'.

References

AEA Group and DEFRA (Department for Environment, Food and Rural Affairs, UK). 2012. *Air Pollution in the UK 2011*. DEFRA, London.

Andreae, M.O., Jones, C.D. and Cox, P.M. 2005. Strong present-day aerosol cooling implies a hot future. *Nature* 435, 1187–1190.

Baldigo, B.P. and Murdoch, P.S. 1997. Effect of stream acidification and inorganic aluminium on mortality of brook trout (*Salvelinus fontinalis*) in the Catskill Mountains, New York. *Canadian Journal of Fisheries and Aquatic Science* 54, 603–615.

Bell, M.L. and Davis, D.L. 2001. Reassessment of the lethal London fog of 1952: novel indicators of acute and chronic consequences of acute exposure to air pollution. *Environmental Health Perspectives* 109(S3), 389–394.

Charlson, R.J., Anderson, T.L. and McDuff, R.E. 1992. The sulfur cycle. In: Butcher, S.S., Charlson, R.J., Orians, G.H. and Wolfe, G.V. (Eds.). *Global Biogeochemical Cycles*. Academic Press, London.

Li, P., Wang, X., Allinson, G., Li, X., Stagnitti, F., Murray, F. and Xiong, X. 2011. Effects of sulfur dioxide pollution on the translocation and accumulation of heavy metals in soybean grain. *Environmental Science and Pollution Research* 18(7), 1090–1097.

Lovelock, J. 2010. *The Vanishing Face of Gaia*. Penguin, London.

Lu, Z., Streets, D.G., Zhang, Q., Wang, S., Carmichael, G.R., Cheng, Y.F., Wei, C., Chin, M., Diehl, T. and Tan, Q. 2010. Sulfur dioxide emissions in China and sulfur trends in East Asia since 2000. *Atmospheric Chemistry and Physics* 10, 6311–6331.

Ormerod, S.J. and Durance, I. 2009. Restoration and recovery from acidification in upland Welsh streams over 25 years. *Journal of Applied Ecology* 46, 164–174.

Raes, F. and Seinfeld, J.H. 2009. New directions: climate change and air pollution abatement: a bumpy road. *Atmospheric Environment* 43(32), 5132–5133.

Schlesinger, W.H. and Bernhardt, E.S. 2013. *Biogeochemistry: An Analysis of Global Change*. Academic Press, Waltham.

Smol, J.P. 2008. *Pollution of Lakes and Rivers: A Paleoenvironmental Perspective*. Blackwell, Malden.

Sorensen, J., Christensen, D. and Jorgensen, B.B. 1981. Volatile fatty acids and hydrogen as substrates for sulfate reducing bacteria in anaerobic marine sediment. *Applied and Environmental Microbiology* 42(1), 5–11.

USDHHS (US Department of Health and Human Services: Public Health Service, Agency for Toxic Substances and Disease Registry, ATSDR). 1998. *Toxicological Profile for Sulfur Dioxide*. ATSDR, Atlanta.

6 Arsenic

Arsenic in the natural environment

Lithosphere

Arsenic (As) is a metalloid, having properties of both metals and non-metals. It is a minor element in most rocks; the highest concentrations in the major rock types are observed in shales. Over 200 As minerals are known and, being a chalcophilic element, As is primarily concentrated in primary sulfide minerals, particularly arsenopyrite, but also minerals such as orpiment, realgar and enargite (Table 6.1). It is sometimes a component of pyrite (iron sulfide) and is often present as a 'guest element' in sulfidic ores of metals such as copper and lead. Arsenic is chemically similar to phosphorous and occurs in phosphate minerals at approximately 20 mg kg^{-1}, although nearly ten times this concentration has been reported in some phosphate deposits. Coal contains As, mainly in sulfide inclusions and, while concentrations are generally less than in phosphate minerals, some coals may contain in excess of 1000–2000 mg kg^{-1}. Precipitates of Fe oxyhydroxides, or 'ochre', in sediments and soils sometimes contain As. For example, in parts of southeast Asia, particularly Bangladesh and India, elevated levels of As (up to 100 mg kg^{-1}) are found in Quaternary alluvial deposits of the Ganges basin and the As is thought to be mainly associated with pyrite and/or Fe oxyhydroxides in the sediments.

In most soils, mean concentrations of As are typically <10 mg kg^{-1}, with maximum values of <100 mg kg^{-1} (Table 6.2); higher levels may be observed in heavily mineralised areas. Where As occurs in soils, it is typically present as arsenates in oxic conditions and as arsenites in waterlogged soils. These anionic species exist in the soil solution and are thus available for plant uptake.

Hydrosphere

Seawater and freshwaters contain traces of As, mainly as dissolved ionic species; particulates typically represent only 1% or so of the total As. In oxic waters, especially with higher pH levels, arsenate is prevalent; arsenite concentrations increase in less oxygenated waters, as do arsenic sulfides particularly at lower pH. Dissolved organoarsenic species also occur, mainly in the forms of monomethylarsonic acid (MMA) and dimethylarsinic acid (DMA). Groundwaters sourced from As-rich geological deposits can have greatly elevated levels of As. This is observed in parts of Argentina, Chile, China, Mexico and USA, but particularly in Bangladesh and India, where there are extensive deposits of As-rich alluvium. In Bangladesh, where thousands of boreholes have been drilled into the alluvial deposits to provide drinking water (Fig. 6.1), As concentrations of up to 4 mg L^{-1} occur, approximately three orders of magnitude higher than typical background concentrations. Natural arsenic enrichment of freshwaters to higher levels still is found in alkaline, closed-basin lakes fed by hot springs, such as Mono Lake (Fig. 6.2) and Searles Lake in California, USA.

Table 6.1 Summary of key minerals and chemical species of arsenic, including compounds in use today.

Minerals and compounds	*Chemical formula*
Minerals	
Arsenopyrite	FeAsS
Orpiment	As_2S_3
Realgar	AsS
Arsenolite	As_2O_3
Scorodite	$FeAsO_4$
Enargite	Cu_3AsS_4
Inorganic forms – pentavalent, As(V)	
Arsenate	AsO_4^{3-}
Arsenic acid	H_3AsO_4, $H_2AsO_4^-$, $HASO_2$
Chromated copper arsenate	$Cu_3(AsO_4)_2$
Inorganic forms – trivalent, As(III)	
Arsenite	AsO_3^{3-}
Arsenious acid	H_3AsO_3
Arsenic trioxide	As_2O_3
Arsine	AsH_3
Gallium arsenide	GaAs
Organic forms	
Monomethylarsonic acid (MMA)	$CH_3AsO(OH)_2$
Dimethylarsinic acid (DMA; 'cacodylic acid')	$(CH_3)_2AsO(OH)$
Salts of MMA: e.g. monosodium methylarsonate	CH_4AsNaO_3
Salts of DMA: e.g. 'Roxarsone' animal feed	$C_6AsNH_6O_6$
Trimethylarsine (TMA)	$(CH_3)_3As$
Tetramethylarsonium	$(CH_3)_4As^+$
Arsenobetaine	$(CH_3)_3As^+CH_2COO^-$
Arsenocholine	$(CH_3)_3As^+CH_2CH_2OH$
Arsenosugars (likely precursors to arsenobetaine)	As-containing carbohydrates

Table 6.2 Typical background concentrations of arsenic.

Environmental medium	*Typical background concentrations*[a]
Air	<3 ng m^{-3}
Soil	<100 mg kg^{-1}
Vegetation[b]	<400 µg kg^{-1}
Freshwater[c]	<10 µg L^{-1}
Sea water	<1.5 µg L^{-1}
Sediment	<15 mg kg^{-1}

a These are *typical* concentrations, encompassing the vast majority of reported values, based on a large number of literature sources. Higher natural concentrations occur in some cases; for example, in mineralised areas and in some saline lakes (see text).
b Including crops. Not including metal-tolerant species.
c Excluding areas with naturally high As in groundwater (see text).

Figure 6.1 Women accessing drinking water from a tubewell in Dhaka, Bangladesh. In many parts of Bangladesh and north-east India, waters from such wells are contaminated by arsenic from natural sources.

Credit: Development Planning Unit, University College London (Flickr).

Figure 6.2 Tufa formations on Mono Lake, California, USA. The lake is naturally enriched with arsenic.

Credit: Frank Kovalcheck (Flickr).

Atmosphere

The atmosphere is not considered a major reservoir for As. Microbial activity in soils and sediments produces volatile species and there are also inputs from volcanic activity and wind-blown dusts.

Biosphere

Although elevated levels of As are toxic to most forms of life, trace amounts are thought to be essential in some organisms. Its precise biological role is not clearly understood but studies have shown a wide range of effects of As deficiency in animals, including stunted growth and low birth weights. Many micro-organisms and plants can tolerate high levels of As and some micro-organisms use it as an energy source. Many plants and animals, such as molluscs and crustaceans, naturally contain inorganic (arsenate and arsenite) and organic forms of As. The latter are mainly arsenobetaine, particularly in marine species, but also arsenocholine and arsenosugars, such as glycerol ribose. The other main types of As found in biota are methylated compounds such as trimethlyarsine oxide and tetramethyl arsonium, as well as MMA and DMA and their salts.

Natural pathways and processes

There are four oxidation states of As (–3, 0, +3 and +5), but only the latter two generally occur in the environment. In environmental systems there are continuous transformations between the trivalent (+3) and pentavalent (+5) forms (mainly arsenite and arsenate, respectively), via oxidation, reduction and methylation. Key chemical forms of As are listed in Table 6.1 and the main transformations occurring in the environment, as detailed below, are summarised in Fig. 6.3.

The abiotic cycling of As can be mainly attributed to: (i) transfers to the atmosphere from volcanic emissions (thousands of tons per year); (ii) weathering and erosion of As-bearing rocks and subsequent inputs to the soil, atmosphere and hydrosphere; and (iii) atmospheric deposition to terrestrial and marine surfaces. Weathering of As-containing sulfide minerals produces secondary minerals such as scorodite and arsenolite. Weathering also releases dissolved species (mainly arsenates, or arsenites in anoxic environments), which can then enter soils, freshwaters and biota. Soil arsenic that is not taken up by plants or micro-organisms can accumulate in the soil or leach or runoff into freshwaters. Leaching of As is inhibited by sorption in the soil, associated with the presence of hydrated oxides of Fe and Al (especially for arsenates), clays and organic matter. Mobility and availability of As is dependent also on pH and Eh. For example, in very acidic ($pH<5$) and/or anoxic soils, Fe oxyhydroxides will more readily dissolve, releasing associated As, if present, from the solid phase. In water bodies, the concentration of As-containing ions may be decreased by complexation of arsenate (and some arsenite) by humic acids or by co-precipitation with iron and manganese oxyhydroxides.

Biotic pathways of As are well documented and are characterised by microbially mediated As metabolism in animals, water bodies and soils. Specialised micro-organisms, living in naturally As-enriched soils and water bodies, actively use As as an energy source (for cell growth). The key biotic processes (microbially mediated) are: oxidation of As(III) to As(V); reduction of As(V) to As(III); methylation of As(III) to MMA and DMA (in the human body, for example); and transformation of inorganic As to organic forms such as arsenocholine and arsenobetaine (in many marine species). Arsenobetaine appears to be the main end-product of As cycling in marine food webs. The organic forms of As are generally less toxic than the inorganic forms (particularly arsenite) and their production within organisms may be a detoxification measure. Arsenates in soils and sediments may also be microbially transformed (methylated) into volatile arsenic species such as trimethylarsine, which enter the atmosphere (and which ultimately may undergo oxidation back to As(V) before deposition to the Earth's surface). Plants can take up soil arsenate and arsenite; arsenate in particular is taken up as it competes with the same uptake carriers in the roots as those for phosphate, which is chemically similar.

In anoxic soils and sediments, the microbial reduction of arsenate to arsenite is coupled to the oxidation of organic matter and other reduced species such as hydrogen sulfide. In such conditions MMA and DMA may also be transformed to volatile methylarsines followed by their diffusion to the atmosphere. Oxidation of the methylarsine gases in the atmosphere forms DMA once more. The volatilisation of As via these processes, and those described for oxic soils, adds As to the atmosphere and decreases the residence time of As in soils.

Production and uses

Arsenic has been used in a wide variety of products. Many of its historical uses, in pigments, pesticides and medicines for example, have now diminished because of toxicity concerns, but the element is still used quite widely for other purposes. The

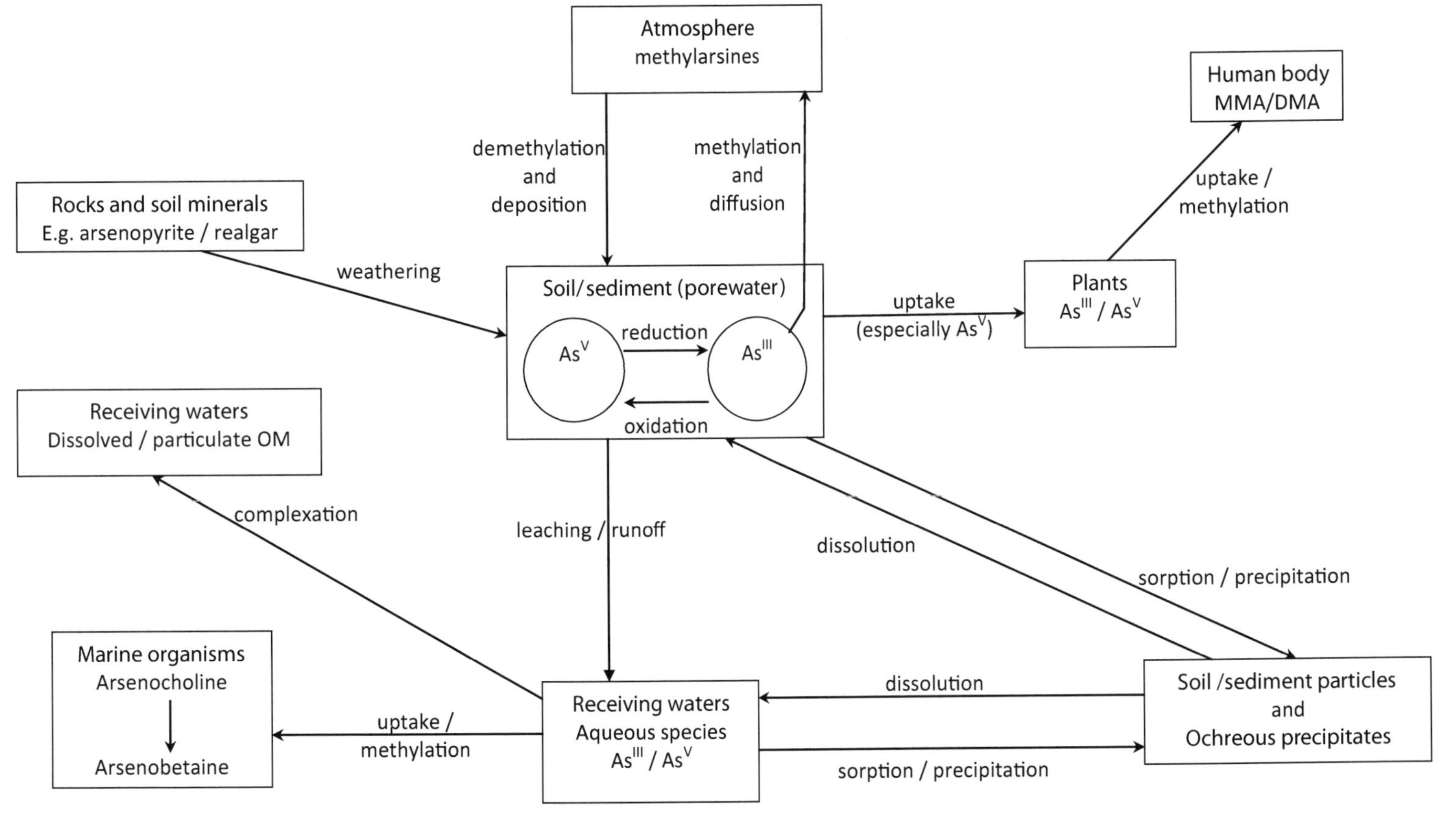

Figure 6.3 Typical environmental transformations involving arsenite (As^{III}) and arsenate (As^{V}). Abbreviations: MMA = monomethylarsonic acid; DMA = dimethylarsinic acid; OM = organic matter.

former use of copper arsenite pigments, such as Paris Green and Scheele's Green, was stopped because moulds, growing on wallpaper for example, were found to produce trimethylarsine. This gas was considered as toxic, and was even implicated in the death of Napoleon Bonaparte, but it is now thought that trimethylarsine (unlike arsine gas) is not particularly toxic by inhalation (Cullen and Bentley, 2005). Modern analytical techniques have shown that the hair of Napoleon and some other historical figures contain relatively high levels of As. The source may have been As medicines, such as Fowler's Solution (potassium arsenite), which were widely used in the past. For example, arsphenamine, an organoarsenic compound, was the first successful treatment for syphilis, before the discovery of penicillin. Arsenic was also used to treat anaemia, rheumatism, asthma, cholera and malaria, and it may still be used in some Chinese herbal medicines. Intravenous administration of arsenic trioxide (As_2O_3) has been introduced in recent years for the treatment of leukaemia in patients who have not responded well to other medicines. Organic forms of arsenic are also still used to treat some tropical diseases, particularly African sleeping sickness, because of the high death rate from the disease and the lack of effective alternatives. Roxarsone, an organoarsenic compound, is used in poultry and pig food, partly to prevent bacterial gut infections but also because it is thought to promote growth and improve pigmentation. However, it is now restricted in many countries because it was found to convert in poultry to the more toxic, inorganic types of As.

Figure 6.4 Wooden decking is often coated with a wood preservative called chromated copper arsenate. This creates the potential for arsenic to leach into underlying soils or to be released into the atmosphere if the decking is later burned.

Credit: Kurtis Garbutt (Flickr).

Arsenates of calcium, copper, lead and zinc were historically used as insecticides for cotton, fruit and rice crops. These inorganic forms have been largely replaced by less toxic organoarsenic species such as monosodium methylarsenate (MSMA), which are still used as herbicides in agriculture and horticulture (e.g. lawns and golf courses) in some countries. The fungicidal properties of arsenic are employed in the widely used wood preservative, chromated copper arsenate (CCA; Fig. 6.4). Gallium arsenide (GaAs) and indium arsenide (InAs) are increasingly used as semiconductors in products such as photovoltaic cells (particularly in satellites and other 'high end' applications), light emitting diodes and integrated circuits (microchips), and these are likely to be the main uses of As in the coming decades. Arsenic is also alloyed with lead in car batteries and lead shot.

These uses of As, and others, create a continuing demand for the element, which is extracted around the world, often from the reprocessing of metal mine wastes. Approximately 4×10^4 t of As are produced every year, mostly in China; lesser amounts are extracted in South America and Morocco (USGS, 2014). Most As is processed to arsenic trioxide for use in product manufacture.

Pollution and environmental impacts

Arsenic pollution

Several countries have sites with significant areas of As wastes created by historic metal and As mining. In the UK, for example, the Devon Great Consols mining complex was the largest As works in the world in the 19th century, producing up to half the world's As, and today extensive areas of As-bearing wastes remain at such sites. Current mining operations in China and elsewhere are adding to the overall burden of As-contaminated wastes.

The presence of As-contaminated areas raises concerns about direct exposure and the potential for localised pollution of soils and rivers. The transport of As from mine wastes to rivers occurs in the form of fine particles carried in suspension during heavy rainfall, or as aqueous ions that have been leached out of the heaps and underground mine workings by percolating rainwater (Fig. 6.5). Particulate transport can lead to the accumulation of As in the sediments of rivers and estuaries downstream of metal mines. Aqueous As that has been leached into freshwaters may co-precipitate with iron and manganese oxyhydroxides, which are often abundant in such sediments; this will add to the sediment load and reduce As mobility, although, if reducing conditions develop, the bound As is likely to be re-released. There are limitations also on the amounts of aqueous As that may be leached in the first place; studies of mine wastes suggest that oxidation weathering of primary arsenic sulfide minerals forms secondary minerals such as the relatively insoluble scorodite, somewhat reducing the potential for further As leaching. Rivers are also polluted by As from non-mining sources (Box 6.1).

Atmospheric concentrations of As are typically in the range 20–100 ng m^{-3} in urban areas and 1–3 ng m^{-3} in rural areas, highlighting the impact of human activity. Estimates vary widely, but modern anthropogenic activities, chiefly fossil fuel combustion and non-ferrous metal smelting, are thought to release tens of thousands of tons of As to the atmosphere each year. The main species emitted by combustion sources is arsenic trioxide, As_2O_3 (equation 6.1), and, consequently, this is the most common inorganic As species in the atmosphere.

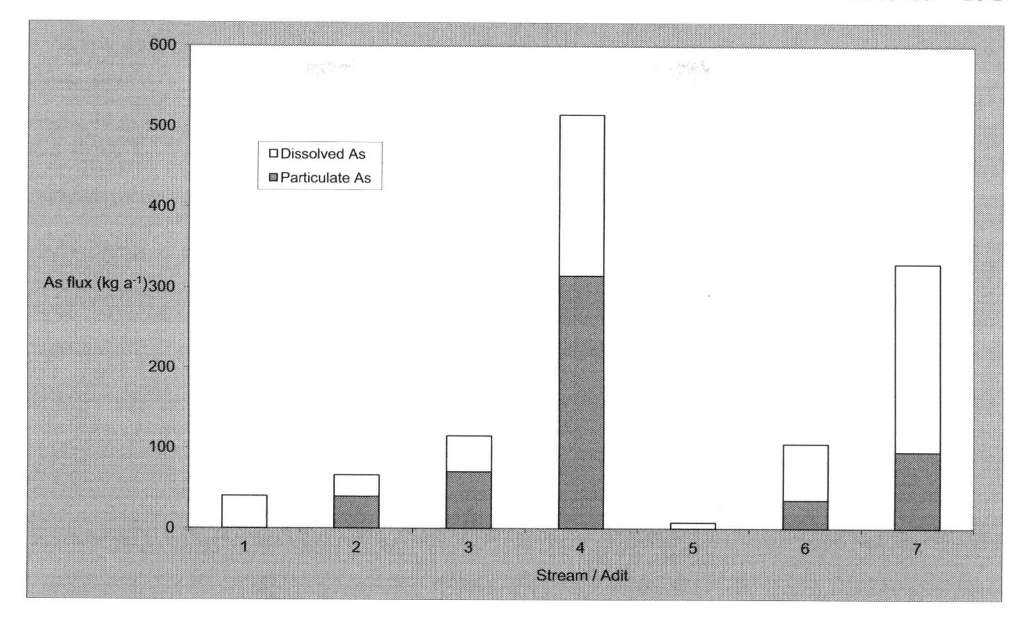

Figure 6.5 The annual fluxes of particulate and dissolved arsenic in selected streams and adits draining a former mining area in southwest England. Collectively these seven streams and adits alone add >1 t As to the River Tamar each year.

Source: Adapted from Mighanetara et al. (2009).

$$2FeAsS + 5O_2 \rightarrow Fe_2O_3 + 2SO_2 + As_2O_3 \quad [6.1]$$

Burning of CCA-treated wood by householders is another potential source of atmospheric As. Although this occurs on a relatively small scale, it is a potential hazard for those exposed to the fumes. In industry, there is the potential for toxic arsine gas (AsH_3) to be accidentally released during GaAs production, so strict safety measures are required. In the environment, GaAs undergoes slow dissolution, producing toxic arsenite, but GaAs is not considered a major pollutant.

Aerial deposition of industrially emitted As (and wind-blown mine waste) is a source of localised soil contamination and this can add to other sources of As in soils. Agricultural soils have been contaminated by pesticide applications and, to a lesser extent, by the use of rock phosphate and sewage sludge as fertilisers. In the Bengal Plain, the use of contaminated groundwaters for irrigation is another source of concern. Elsewhere, leaching of As from CCA-treated wood can also contaminate soils.

Human health effects

Human exposure to As mainly occurs via the ingestion of contaminated water, food, soil and dust. Inhalation and dermal contact are less important exposure routes, although As can be absorbed through the lungs and skin in addition to the gastrointestinal (GI) tract. The half-life of As in the human body is in the order of a

Box 6.1 Arsenic pollution in the USA.

In 2014, up to 8×10^4 t coal ash spilled into the Dan River in North Carolina, USA, contaminating the river water with As and other toxic elements, including chromium. In the weeks following the spill, coal ash deposits were found on the river bed up to 70 miles downstream and residents in affected areas of North Carolina and Virginia were told to avoid the river because concentrations of As were many times the recommended safe levels for human contact. Drinking water supplies were safeguarded, but concerns remained for aquatic organisms, including listed endangered species.

Figure 6.6 Coal ash spill on the Dan River, 2014. Image shows the ash basin and the upper stormwater drain.

Credit: United States Fish and Wildlife Service/Steven Alexander (www.fws.gov).

few days. Inorganic As is generally more toxic than organic forms, but the toxicity of As species is quite variable and there is general agreement that it follows the order: arsine (gas) > arsenite > trivalent organoarsenic (e.g. dimethylarsinous acid) > arsenate > pentavalent organoarsenic (e.g. MMA and DMA). Arsine is not generally an important environmental pollutant; ingestion of arsenites and arsenates via contaminated food or soil is of most concern.

Arsenite reacts with the sulfhydryl (–SH) groups of proteins and enzymes (Fig. 6.7), potentially disrupting the structure of the former and inactivating the latter. Because of the reaction with sulfhydryl groups, As concentrates in sulfurous keratin structures

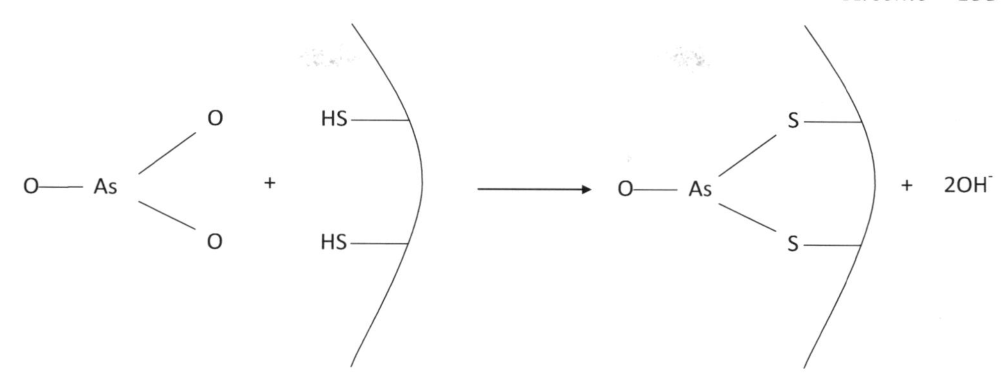

Figure 6.7 The inactivation of an enzyme by arsenite. Left: an arsenite ion (AsO_3^{3-}) approaches a pair of sulfhydryl (–SH) groups on an enzyme; Right: the arsenite ion has bonded with the S atoms of the –SH groups, possibly inhibiting the enzyme; two hydroxyl groups (–OH) are also produced.

like hair and nails. It also: (i) interacts with lipoic acid, inhibiting cellular respiration; (ii) coagulates proteins; (iii) complexes with coenzymes; and (iv) inhibits ATP (adenosine triphosphate) production. Arsenate is toxic in humans mainly because it is chemically similar to phosphate and can disrupt phosphate metabolism. There is some evidence that humans can become tolerant to As, as observed in some plants and animals.

Most inorganic As absorbed by humans is metabolised (methylated) to MMA and, more particularly, DMA; both compounds have relatively low toxicity and are excreted in urine. Unlike arsenite, arsenate is not methylated in the body and has first to be reduced to arsenite, in the kidneys for example, before methylation takes place. Proportions of As metabolites in the human body are, very approximately, 50% DMA, 25% MMA and 25% inorganic (unmethylated) As. Methylation is thought to make As less lipophilic, more hydrophilic and therefore more easily excreted, although methylated As may have a longer residence time in the thyroid. Although DMA and MMA are less toxic than arsenite, it is possible that the capacity of the body to methylate arsenite may be saturated at higher levels of inorganic As intake. Tolerance to As may be related to the rate at which an individual is able to methylate inorganic As and there is evidence that it may be lower in children and malnourished individuals. Unmethylated arsenite is transported from the bloodstream to the liver, kidneys and other major organs. In most humans, seafood is the main source of As, mainly in the form of the organoarsenic compound, arsenobetaine. It is absorbed from the gut, but is readily excreted in the urine; organoarsenic compounds, including MMA and DMA, do not appear to be metabolised (e.g. demethylated) to any great extent.

The main symptoms of acute exposure in humans include gastroenteritis, with severe inflammation of the GI tract, vomiting, abdominal pain and diarrhoea. Other common symptoms are muscle cramps, numbness and tingling in the hands and neurological effects, including delirium. Estimates of the lethal dose of inorganic As compounds lie in the range 50–300 mg. Most As toxicity is chronic and this type is, by definition, more difficult to diagnose. Chronic exposure results in increased risks of skin, bladder and lung cancer; this is known from epidemiological studies of occupational exposure in smelting and chemical industries, agricultural workers (pesticide exposure), studies of medicinal intake and observations of the high incidence of skin cancer (and other skin disorders) in areas with As-contaminated drinking water (Box 6.2). Other known symptoms of chronic exposure include 'blackfoot' disease (loss of circulation leading to gangrene of the feet), keratosis (thickening of the skin), GI tract effects, diabetes, kidney and liver dysfunction, and ischemic heart disease.

Box 6.2 Arsenic in Asian drinking waters.

In the 1970s a massive well-drilling programme was undertaken in the Bengal Delta of Bangladesh and India to provide clean drinking water and reduce the high numbers of deaths (250,000 per year in Bangladesh) caused by drinking river water containing pathogenic micro-organisms. In the 1980s and 1990s reports started to emerge of high As levels in many of the well waters (>50 $\mu g\ L^{-1}$ As, compared with the WHO recommended level of 10 $\mu g\ L^{-1}$) and high rates of hyperpigmentation (skin disorders) in local people. It is now thought that millions of people have been exposed to As-contaminated water and that thousands of deaths per year have occurred as a result. The cause of the disaster illustrates the importance of understanding environmental conditions and chemical speciation. The wells were drilled into Bengal Delta sediments that contain naturally high levels of both As – probably present in particles containing iron oxyhydroxides – and organic matter. The ongoing decomposition of the organic matter at depth depletes oxygen levels and in this anoxic environment the iron oxides dissolve, releasing the As as toxic arsenite, the prevalent form in reducing conditions.

Arsenic ecotoxicity

As in humans, the inorganic forms of As are generally accepted as being more toxic to wildlife than the organic forms and, of the dominant inorganic forms, arsenite is more toxic than arsenate. The effects of As in plants and animals are mainly related to its capacity to inactivate enzymes and disrupt phosphate-dependent processes. Arsenic toxicity varies widely among biological species and depends on the chemical forms of As present. Some species of animals, plants and particularly micro-organisms appear to be resistant to elevated levels of As.

Arsenic accumulates in the tissues of many organisms, particularly marine species, but it does not appear to biomagnify in terrestrial food webs to any great extent. A study of As concentrations in the sloughed skin of sperm whales (*Physeter macrocephalus*) from 17 regions around the world found relatively high levels in parts of the Indian Ocean (Fig. 6.8) and concentrations in the Maldives region were significantly higher than in all other regions. While acknowledging the possibility of seasonal effects, the authors attributed this to a mixture of natural As sources and the use of As-containing pesticides in areas to the north of the Maldives, particularly Sri Lanka, where relatively high As concentrations were also recorded.

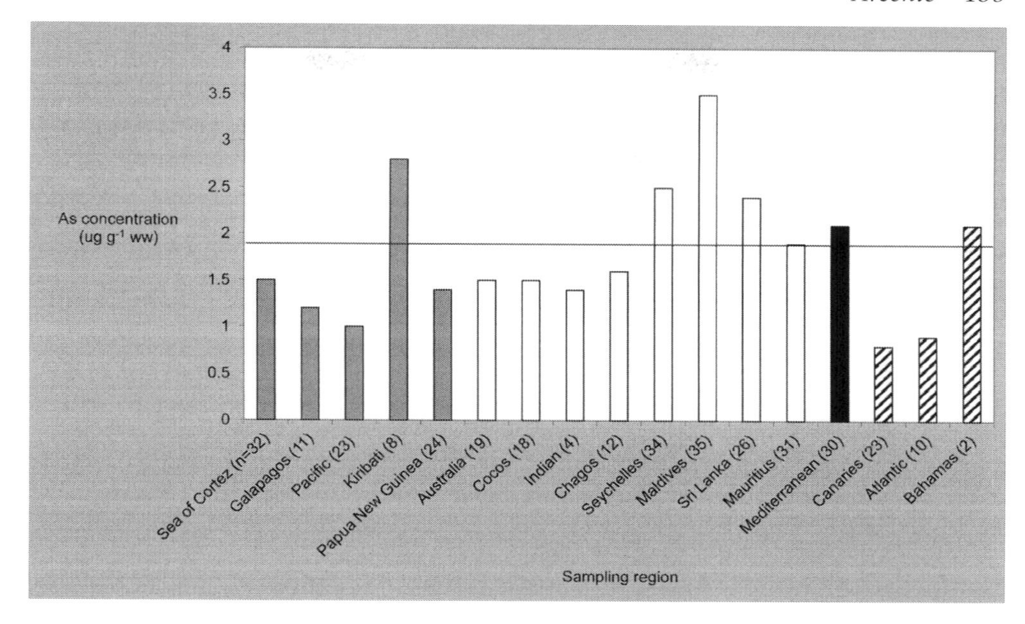

Figure 6.8 Mean arsenic concentrations in the sloughed skin of 342 sperm whales (*Physeter macrocephalus*) from 17 regions across the world (sample size in parentheses). Key to bar colours: grey – Pacific Ocean; white – Indian Ocean; black – Mediterranean Sea; hatched – Atlantic Ocean. Horizontal line denotes average concentration of all samples.

Adapted from Savery et al. (2014).

Figure 6.9 Sperm whale, New Zealand.

Credit: Strange One (Flickr).

Plant uptake of soil arsenite and arsenate ions (more typically the latter in well-drained soils) appears to be low in most plants. Typical uptake ratios are in the range 0.01–0.1, i.e. plant concentrations are between one-hundredth and one-tenth of those in corresponding soils. In contaminated soils, however, this can result in concentrations that are high enough to reduce crop yields and, in wild plants, affect growth or cause death. Inorganic As is particularly toxic in plants; arsenate is a phosphate analogue, so it is easily transported through the plasma membrane. Indeed, decreases in crop yield may be ameliorated in phosphate-rich soils because of preferential uptake of the phosphate ions by plants. Once in the cytoplasm, arsenate can replace phosphate in adenine triphosphate (ATP), affecting cell metabolism. Also, arsenate is rapidly reduced in cells (by glutathione, for example) to arsenite, which blocks sulfhydryl (–SH) groups in proteins and enzymes and interferes with light activation in plants.

Some plants have evolved tolerance to As, important examples being *Agrostis tenuis* (and some other *Agrostis* species), *Calluna vulgaris* and *Holcus lanatus*. In some of these plants, and others, inorganic As is complexed, and effectively detoxified, by organic molecules called phytochelatins. Some As-tolerant plants hyperaccumulate As, sequestering it away from sites where it might do harm. The phytochelatins contain cysteine, with constituent sulfhydryl groups that bond with arsenite. In other plants the mechanism of tolerance is exclusion, rather than hyperaccumulation and sequestration. Such plants appear to suppress phosphate uptake mechanisms (so that arsenate uptake is also suppressed), but they may also use phytochelatins as a secondary tolerance mechanism for any non-excluded As. Some of the As-hyperaccumulating plant species may be useful in the remediation of As-polluted soils. For example, the fast-growing Chinese brake fern (*Pteris vittata*) takes up and accumulates large quantities of As (Xie et al., 2009); detailed research into this plant and others continues and this includes the possibility of gene transfer to confer As-hyperaccumulation on plants that are not naturally tolerant to As.

Some animals are known to have evolved resistance to high levels of As, including earthworms that can tolerate highly contaminated soils at mine sites. Studies of the effects of As exposure on terrestrial wildlife are rare compared with laboratory studies using prescribed doses of As chemicals. However, exposure to As in polluted environments has been implicated in the deaths of wild mammals that had fed on plants treated with As-based herbicides; also, elevated levels of As have been recorded in birds preying on bark beetles in pesticide-treated trees. The organoarsenic species that accumulate in fish appear to be generally non-toxic to them.

References

Chilvers, D.C. and Peterson, P.J. 1987. Global cycling of arsenic. In: Hutchinson, T.C. and Meema, K.M. (Eds.). *Lead, Mercury, Cadmium and Arsenic in the Environment*. John Wiley & Sons, Chichester.

Cullen, W.R. and Bentley, R. 2005. The toxicity of trimethylarsine: an urban myth. *Journal of Environmental Monitoring* 7, 11–15.

Karim, M.M. 2000. Arsenic in groundwater and health problems in Bangladesh. *Water Research* 34(1), 304–310.

Mighanetara, K., Braungardt, C.B., Rieuwerts, J.S. and Azizi, R. 2009. Contaminant fluxes from point and diffuse sources from abandoned mines in the River Tamar catchment, UK. *Journal of Geochemical Exploration* 100(2–3), 116–124.

Savery, L.C., Wise, J.T.F., Wise, S.S., Falank, C., Gianios, C. Jr, Thompson, W.D., Perkins, C., Zheng, T., Zhu, C. and Pierce Wise, J. Sn. 2014. Global assessment of arsenic pollution using sperm whales (*Physeter macrocephalus*) as an emerging aquatic model organism. *Comparative Biochemistry and Physiology, Part C: Toxicology and Pharmacology* 163, 55–63.

Tamaki, S. and Frankenberger, W.T. Jr. 1992. Environmental biochemistry of arsenic. *Reviews of Environmental Contamination and Toxicology* 124, 79–110.

USGS (United States Geological Survey). 2014. Mineral commodity summaries: Arsenic. Available at: http://minerals.usgl.gov/minerals

Xie, Q.-E., Yan, X.-L., Liao, X.-Y. and Li, X. 2009. The arsenic hyperaccumulator fern, *Pteris vittata* L. *Environmental Science and Technology* 43(22), 8488–8495.

7 Bromine

Bromine in the natural environment

Bromine (Br) occurs in the natural environment as the bromide ion, Br^-. It is typically present at concentrations of <5 mg kg^{-1} in the Earth's crust, up to 10 mg kg^{-1} in some sedimentary rocks and higher still in soluble evaporite deposits. Crustal Br minerals are rare. Maximum concentrations in some soils may reach several hundred mg kg^{-1}, but average values are typically an order of magnitude lower. Bromide is incorporated into soils via weathering of minerals during pedogenesis and deposition of soluble atmospheric gases; some soil Br is retained by soil organic matter and clays, but much is readily leached from soils, its salts being very soluble. The high solubility of most Br compounds means it is present in higher quantities in natural waters than in the crust. Concentrations of Br in the oceans are approximately 60 mg L^{-1}, but it is more highly concentrated in salt lakes and inland seas; for example, the Dead Sea (Fig. 7.1) contains some of the highest known natural Br concentrations, some two orders of magnitude higher than in ocean waters.

Natural emissions of Br to the atmosphere are mostly in the forms of methyl bromide (also called bromomethane; CH_3Br) and bromoform ($CHBr_3$), both of which are produced by oceanic and terrestrial biomass and in natural forest fires; in total, >1 Mt Br a^{-1} is released to the atmosphere from natural sources. However, methyl bromide also has natural sinks that may exceed its sources; it dissolves in seawater, is oxidised in the atmosphere and is taken up by soil microbes and plant leaves. Despite this, a proportion reaches the stratosphere, where it can contribute to O_3 depletion (see below). Similarly, bromoform can contribute reactive Br to the stratosphere, despite a typical atmospheric lifetime of just a few weeks. Volcanoes emit relatively small amounts of the soluble gas hydrogen bromide, HBr, and this is readily deposited to the hydrosphere, either directly onto water surfaces or indirectly via leaching through soils.

Bromine has no known biological role in humans, but it is present in all organisms and is accumulated by marine plankton and terrestrial plants, with bioconcentration factors of $>10^3$.

Production and uses

The main uses of Br are as: brominated flame retardants (BFRs); dense drilling fluids in gas and oil extraction ($CaBr_2$, NaBr and $ZnBr_2$); soil fumigants (CH_3Br and other Br-containing gases); and water disinfectants, both in swimming pools (NaBr) and

Figure 7.1 Evaporite deposits, Dead Sea, Israel.
Credit: Israeltourism (Flickr).

some industrial processes. Polybrominated diphenyl ethers (PBDEs) are used as BFRs in plastic casings in electronic goods, polyurethane foam in furniture and in upholstery and textiles. There are more than 200 PBDE congeners, each with a different number and/or positions of Br atoms in the molecule; the general molecular structure is shown in Fig. 7.2. Because of concerns over their bioaccumulation, long-range transport and toxicity, production of four types of PBDE, namely tetraBDE, pentaBDE (both shown in Fig. 7.3), hexaBDE and heptaBDE, is restricted under the Stockholm Convention on Persistent Organic Pollutants (POPs), as is production of another organobromine compound, hexabromobiphenyl, which is a 'PBB' (polybrominated biphenyl), analogous to the more well-known PCBs (polychlorinated biphenyls) {→Cl}. Hexabromo-cyclododecane, a flame retardant used mainly in polystyrene, is also being considered for control under the Convention. Another toxic BFR, tetrabromobisphenol-A, is widely used in circuit boards but is considered less of an exposure risk to humans.

Methyl bromide (CH_3Br) is used as a gaseous fumigant against soil nematodes and other agricultural pests, particularly in fruit growing, and for pest control in buildings. While the use of CH_3Br is restricted internationally under the Montreal Protocol of 1987 (for protection of the ozone layer), annual exemptions are allowed for its use as

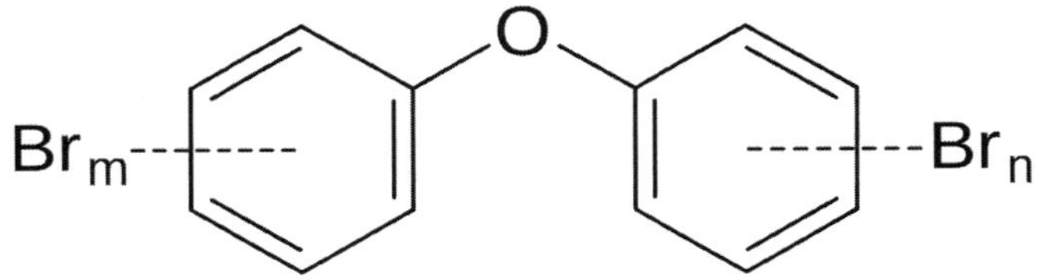

Figure 7.2 The general molecular structure of PBDEs, based on two phenyl rings. Various numbers (m, n) and positions of Br atoms in the molecule give rise to >200 congeners. Two examples are shown in Fig. 7.3.

Credit: Leyo (Wikimedia Commons).

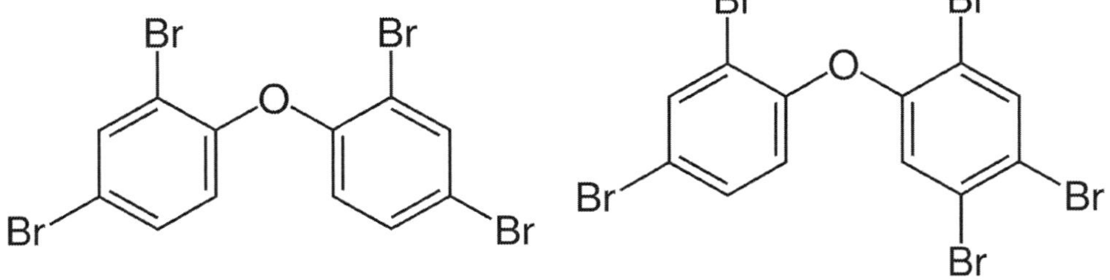

Figure 7.3 The molecular structures of a tetra-BDE isomer (BDE-47) and a penta-BDE isomer (BDE-99); these isomers are among the four PBDEs restricted under the Stockholm Convention. (a) 2,2',4,4'-tetrabromodiphenyl ether (BDE-47); (b) 2,2',4,4', 5-pentabromodiphenyl ether (BDE-99). 'Tetra' and 'penta' relate to the number of Br atoms in each isomer (4 and 5 respectively) and the prefix numbers for each isomer relate to the positions of the Br atoms in the molecule.

Credit: Leyo (Wikimedia Commons).

a soil fumigant because of a lack of suitable alternatives. Other organobromine (OB) compounds, including dibromochloropropane (DBCP) and ethylene dibromide, have also been used as soil fumigants but have been banned or restricted because of leaching risk and toxicity, particularly to workers. In addition to the main uses listed above, Br compounds are used in several other products and processes of note: to control Hg emissions in coal-burning power stations; as an emulsifier in some soft drinks (withdrawn in most countries, but still partly used in the USA); in paper manufacture; in photographic emulsion (declining) and in human and veterinary medicine as antihistamines and anticonvulsants. Global use of Br compounds requires the annual extraction of approximately 0.8 Mt Br, mostly from seawater and underground brines in Israel, Jordan, the USA, China and Japan (USGS, 2014). Extraction involves the oxidation of seawater bromide by Cl gas to bromine vapour, which is then chemically treated to produce an aqueous solution of bromine.

A former application of OB compounds, now strictly controlled under the Montreal Protocol, was the use of bromofluorocarbons ('halons') in fire extinguishers. These include: Halon 1211 (bromochlorodifluoromethane, CF_2ClBr); Halon 1301 (bromotrifluoromethane, CF_3Br); Halon 2402 (dibromotetrafluoroethane, $C_2F_4Br_2$); and Halon 1011 (bromochloromethane, CH_2BrCl). Thirty-four other Br compounds are listed in a separate group of the Protocol. Another OB that was commonly used in the past was ethylene dibromide $(CH_2Br)_2$; this compound was used as an additive in leaded petrol but phased out when the use of lead in petrol was banned.

Pollution and environmental impacts

Atmospheric Br pollution is mainly in the form of methyl bromide, CH_3Br. Anthropogenic sources, which add to the natural inputs of this gas, include vehicle emissions, deliberate biomass burning and soil fumigation; these sources appear to have increased the pre-industrial concentration of CH_3Br in the atmosphere by at least 50%. Wet and dry deposition of atmospheric Br onto soils and water surfaces, and dissolution in soil water of applied CH_3Br fumigant, will ultimately add to the Br loading of the hydrosphere. Direct sources of water pollution by OBs include sewage treatment plants, leachate from landfill sites, and waste effluents from power stations and gas and oil drilling (see Box 7.1). PBDEs and DBCP are detected in fresh and marine waters as well as in fish, seabirds and marine mammals in the Arctic and elsewhere. PBDEs are also found in household dusts (Fig. 7.5), raising particular concern for young infants, who have relatively high exposure to floor dust; PBDEs have also been detected in human milk (Table 7.1).

Concern over OB compounds is focused on their environmental persistence, bioaccumulation and toxicology to humans and animals. In particular, some PBDEs are lipophilic and very persistent, bioaccumulating in some organisms; intake is associated with neurological effects and endocrine disruption. In humans, research has indicated statistically significant relationships (adjusted for age and body mass index) between PBDE concentrations in household dust and hormone levels (e.g. Meeker et al., 2009). Similarly, suppressed levels of the thyroid hormone triiodothyronine were significantly associated ($p = 0.03$) with PBDE exposure in UK populations of the Eurasian dipper, *Cinclus cinclus* (Morrissey et al., 2014); a significant association ($p = 0.01$) was also observed between PBDE exposure and poor

Box 7.1 Bromide pollution associated with water from shale gas drilling.

In 2010, elevated concentrations of Br salts were detected in the Allegheny and Monongahela Rivers in the north-east of the USA. Concerns were raised because bromide can react with the Cl used in drinking water treatment, to produce brominated 'disinfection by-products' (DBPs), such as trihalomethane compounds which may be carcinogenic {→Cl}; further testing showed that brominated trihalomethanes were indeed present at elevated levels. Monitoring of the rivers indicated that a major source of bromide discharge was industrial wastewater treatment plants that were receiving wastewaters from shale gas drilling operations in the well-developed Marcellus Shale area. Bromide concentrations in the rivers decreased once the fracking wastewater was transferred to deep wells in a neighbouring state; however, power plant effluent and acid-mine drainage remain as local sources of bromide and this incident suggests the need for vigilance in the treatment of wastewaters from shale gas drilling operations.

Further reading: Ferrar et al. (2013), States et al. (2013) and Hladik et al. (2014).

Figure 7.4 Shale gas drilling site in the Marcellus shale, Lycoming County, Pennsylvania, USA.

Credit: Nicholas_T (Flickr).

body condition. One of the lower brominated PBDEs, detected in the hatchlings of European shag (*Phalacrocorax aristotelis*) in Norway, was significantly negatively correlated with levels of the antioxidant vitamin, tocopherol, in the liver (Fig. 7.6), indicating oxidative stress and potential health effects (Murvoll et al., 2006). The other toxic effects of Br pollution are mainly from occupational and accidental

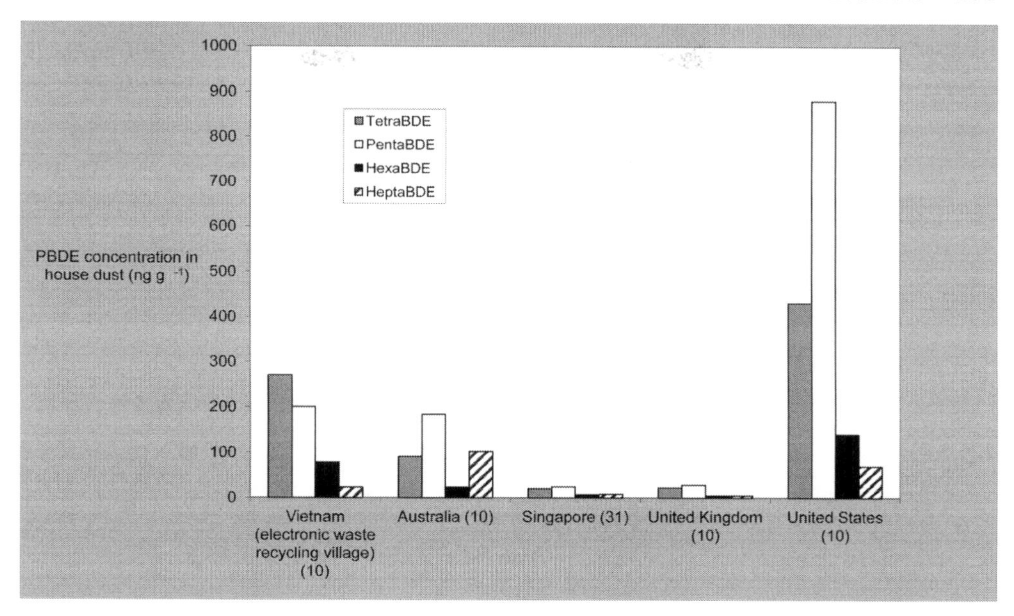

Figure 7.5 Median PBDE concentrations in house dusts of study areas in several countries (ng g^{-1}). In a Vietnamese village, the homes sampled were involved in informal recycling of electronic wastes in the back yard of the house. Note that domestic households in the USA have higher concentrations of all the PBDEs than in this village, despite domestic goods being the only likely source in these homes; this is likely to be linked to stringent flammability regulations in the USA.

Source: Adapted from Fromme et al. (2014) and Tue et al. (2013).

exposure of humans to high levels of CH_3Br (causing respiratory, neurological and kidney damage) and exposure of plants to bromide (which can replace chloride in some tissues) causing leaf chlorosis and necrosis.

Pollution of the atmosphere by OBs creates concern mainly because of their contribution to stratospheric ozone depletion. The production and use of halons are strictly controlled, but, like CFCs, they are still present in the stratosphere because of their long atmospheric lifetimes (65 years in the case of Halon-1301, CF_3Br). Methyl bromide, which reaches the stratosphere from both natural and anthropogenic sources, is also an ozone-depleting compound and is similarly controlled under the Montreal Protocol. These OB compounds are broken down in the stratosphere, releasing Br radicals (e.g. equation 7.1), which react with, and deplete, O_3 (equation 7.2). The BrO produced in the O_3 depletion reaction (equation 7.2) subsequently releases Br (equation 7.3) to react with more O_3. In the context of O_3 depletion, OBs, Br radicals and bromine monoxide (BrO) are direct analogues of CFCs, Cl radicals and chlorine monoxide, respectively {→Cl}.

$$CH_3Br + hv\ (\lambda < 260\ nm) \rightarrow CH_3^{\cdot} + Br^{\cdot} \quad [7.1]$$

$$O_3 + Br^{\cdot} \rightarrow BrO + O_2 \quad [7.2]$$

$$BrO + O \rightarrow Br^{\cdot} + O_2 \quad [7.3]$$

Table 7.1 Summary data of mean PBDE concentration (sum of several important congeners) in breast milk from various European countries and US states and cities. Note the much higher concentrations in the USA compared with Europe, following the same pattern noted in Fig. 7.5 for PBDE levels in house dusts.

Country	*n*	*PBDE concentration (ng g^{-1} lipid)*
Europe		
Belgium	14	2.9
England	54	8.9
France	77	2.5[a]
Germany	62	2.2
Norway	393	2.1
Russia	10	0.96
Slovakia	33	0.57
Spain	15	2.5
Sweden	93	4
USA		
Austin, Texas	47	74
Pacific Northwest	40	96
Central Massachusetts	38	76
Boston, Massachusetts	46	30[a]
Central North Carolina	301	89

a Median values.
Sources: Various sources reported and summarised in Chovancová et al. (2011) for European data and Daniels et al. (2010) for US data.

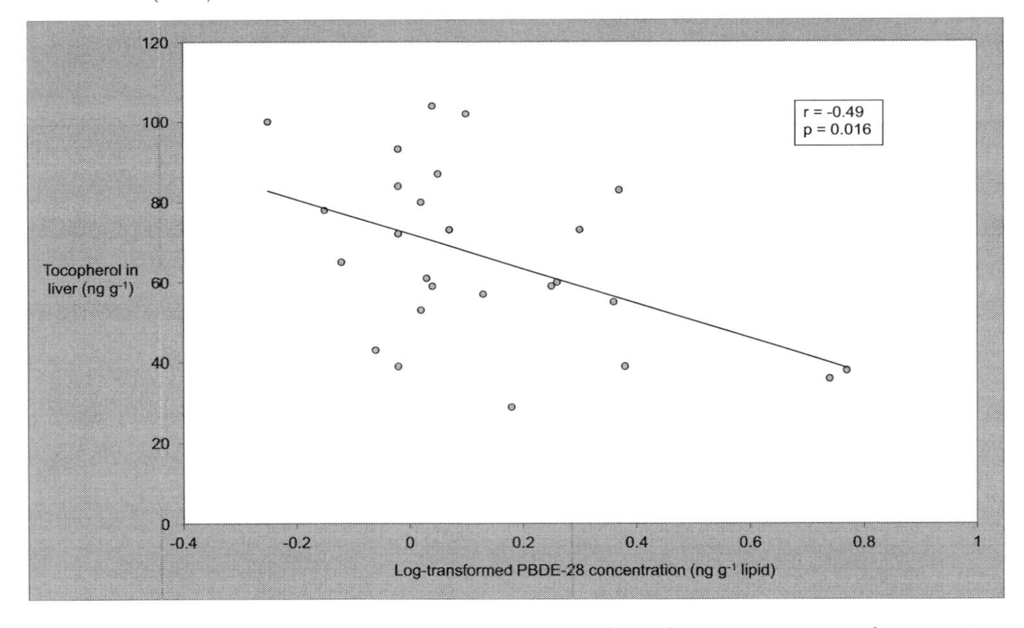

Figure 7.6 Significant negative correlation between lipid weight concentrations of PBDE-28, a tri-BDE, in European shag (*Phalacrocorax aristotelis*) hatchlings and levels of tocopherol in the liver.

Source: Adapted from Murvoll et al. (2006).

Overall, less Br has entered the stratosphere compared with Cl, but OBs cause concern because they are up to 16 times more potent than CFCs as ozone depleters (using CFC-11 as the benchmark). Furthermore, Br is removed from the atmosphere more slowly than Cl, by reaction with CH_4 and H_2.

References

Chovancová, J., Čonka, K., Kočan, A. and Stachova Sejáková, Z. 2011. PCDD, PCDF, PCB and PBDE concentrations in breast milk of mothers residing in selected areas of Slovakia. *Chemosphere* 83(10), 1383–1390.

Daniels, J.L., Pan, I.J., Jones, R., Anderson, S., Patterson, D.G. Jr, Needham, L.L. and Sjodin, A. 2010. Individual characteristics associated with PBDE levels in U.S. human milk samples. *Environmental Health Perspectives* 118(1), 155–160.

Ferrar, K.J., Michanowicz, D.R., Christen, C.L., Mulcahy, N., Malone, S.L. and Sharma, R.K. 2013. Assessment of effluent contaminants from three facilities discharging Marcellus Shale wastewater to surface waters in Pennsylvania. *Environmental Science and Technology* 47(7), 3472–3481.

Fromme, H., Hilger, B., Kopp, E., Miserok, M. and Volkel, W. 2014. Polybrominated diphenyl ethers (PBDEs), hexabromocyclododecane (HBCD) and 'novel' brominated flame retardants in house dust in Germany. *Environment International* 64, 61–68.

Hladik, M.L., Focazio, M.J. and Engle, M. 2014. Discharges of produced waters from oil and gas extraction via wastewater treatment plants are sources of disinfection by-products to receiving streams. *Science of the Total Environment* 466–467, 1085–1093.

Meeker, J.D., Johnson, P.I., Camann, D. and Hauser, R. 2009. Polybrominated diphenyl ether (PBDE) concentrations in house dust are related to hormone levels in men. *Science of the Total Environment* 407, 3425–3429.

Morrissey, C.A., Stanton, D.W.G., Tyler, C.R., Pereira, M.G., Newton, J., Durance, I. and Ormerod, S.J. 2014. Development impairment in Eurasian Dipper nestlings exposed to urban stream pollutants. *Environmental Toxicology* 33(6), 1315–1323.

Murvoll, K.M., Skaare, J.U., Anderssen, E. and Jenssen, B.M. 2006. Exposure and effects of persistent organic pollutants in European shag (*Phalacrocorax aristotelis*) hatchlings from the coast of Norway. *Environmental Toxicology and Chemistry* 25(1), 190–198.

States, S., Cyprych, G., Stoner, M., Wydra, F., Kuchta, J., Monnell, J. and Casson, L. 2013. Marcellus Shale drilling and brominated THMs in Pittsburgh, Pa., drinking water. *Journal of the American Water Works Association* 105(8), 432–448.

Tue, N.M., Takahashi, S., Suzuki, G., Isobe, T., Viet, P.H., Kobara, Y., Seike, N., Zhang, G., Sudaryanto, A. and Tanabe, S. 2013. Contamination of indoor dust and air by polychlorinated biphenyls and brominated flame retardants and relevance of non-dietary exposure in Vietnamese informal e-waste recycling sites. *Environment International* 51, 160–167.

USGS (United States Geological Survey). 2014. Mineral commodity summaries: Bromine. Available at: http://minerals.usgl.gov/minerals

8 Cadmium

Cadmium in the natural environment

The distribution of cadmium (Cd) in the natural environment reflects its nature as a lithogenic element and its association with Zn. The two elements are chemically similar, being found generally in the +2 oxidation state; Cd appears below Zn in the periodic table and they are found together in the Earth's crust. Cadmium is found particularly in deposits of zinc ore – mainly sphalerite (ZnS); it substitutes for Zn in the mineral lattice and is also present in Cd ores associated with the sphalerite – mainly greenockite (CdS), but also cadmoselite (CdSe) and otavite ($CdCO_3$). Zinc ores generally have Cd:Zn ratios of around 1:300, but can contain up to 5% Cd. Cadmium is also present in coals, black shales and, to a lesser extent, some phosphate rocks. During the formation of such sedimentary rocks in anoxic, organic-rich environments, Cd is sequestered by sulfides and organic complexes. The average crustal concentration is generally agreed to be ~0.1 mg kg^{-1}. Soil concentrations are generally low (Table 8.1), but can be elevated in mineralised areas. Except in alkaline soils, Cd is quite mobile in soils compared with many other trace elements; its retention in soil is mediated mainly by adsorption rather than precipitation and it is readily exchanged, especially in acidic soils. The leaching of soluble Cd, typically as free ionic Cd^{2+} or in aqueous chloro- and hydroxo-complexes, transfers the element to the hydrosphere, together with runoff of particulate Cd. The main mineral forms in soils are Cd hydroxide, $Cd(OH)_2$ and Cd carbonate, $CdCO_3$.

Table 8.1 Typical background concentrations of cadmium.

Environmental medium	*Typical background concentrations*[a]
Air	<5 ng m^{-3}
Soil	<2 mg kg^{-1}
Vegetation[b]	<0.5 mg kg^{-1}
Freshwater	<1 µg L^{-1}
Sea water	<50 ng L^{-1}
Sediment	<2 mg kg^{-1}

a These are *typical* values, encompassing the vast majority of reported concentrations, based on a large number of literature sources. Higher natural concentrations occur in some cases, for example in mineralised areas.

b Including crops. Not including metal-tolerant species.

Seawaters have higher Cd concentrations than freshwaters, with lower concentrations at the sea surface relative to deeper waters because of uptake by photosynthesising phytoplankton. The amount of Cd transferred to the atmosphere, mainly in mineral and soil dusts, may be higher than once thought, but there is very large uncertainty, with estimates spanning 3 orders of magnitude, from $\sim 1 \times 10^2$ to $\sim 1 \times 10^5$ t a^{-1}. Atmospheric Cd is wet- and dry-deposited and the atmospheric lifetime is fairly short (days to weeks).

Despite its toxicity to plants and animals, Cd is taken up by them because of its similarity to zinc, an essential element; it can substitute for Zn in enzymes. It has not generally been considered an essential element, but there is increasing evidence of essentiality in some organisms; Cd appears to act as a micronutrient in marine phytoplankton, facilitating photosynthesis. The element is strongly bound within proteins and amino acids, like cysteine, to sulfhydryl (thiol) groups. In animals, it is carried in the blood by red blood cells or by plasma proteins such as albumin. In the reproductive organs, the liver and, more especially, the kidneys, it is strongly bound by metallothionein, a cysteine-rich protein. It is stored mainly in the kidney, where it accumulates and has a half-life of some 10–35 years in humans. In plants, Cd is bound by cysteine-rich phytochelatins, which are somewhat analogous to metallothioneins in animals. Like Zn, Cd is readily translocated from the roots to the leaves and is therefore found particularly in leafy plants.

Production and uses

Cadmium has several important applications and is found in a variety of products (Table 8.2). It is also present as an impurity in some products that contain Zn; an example is tyres, which contain up to 90 µg kg^{-1} Cd because of impurities in the Zn oxide used in vulcanisation – hence, tyre wear can add to Cd in urban runoff (Fig. 8.1). The products listed in Table 8.2 can act as sources of Cd in the environment during their manufacture, use and/or disposal. Global reserves of Cd are 0.5 Mt and the main areas of primary production are North and South America, Australia and parts of Asia; particularly China, Russia, India and Kazakhstan (USGS, 2014).

Table 8.2 Major uses of cadmium.

Use	*Description*
Ni-Cd batteries	Main use of Cd; contain up to 20% Cd as CdO
Paints	CdS used as pigment
Electroplating of iron and steel	CdO and $CdSO_4$; form anti-corrosion $CdCl_2$ coating
Alloys	Cd metal in solders and manufacturing
Solar voltaic panels	Cd telluride is the main alternative to Si
Stabiliser in PVC plastics	Non-Cd alternatives now used in many countries
Nuclear reactor control rods	Often as silver-indium-Cd alloys (80:15:5)

Figure 8.1 Tyre wear is a source of cadmium in urban runoff.

Pollution and environmental impacts

Cadmium pollution

Anthropogenic emissions of Cd to the atmosphere decreased from around 8000 t a^{-1} in the 1980s to approximately 3000 t a^{-1} in the 1990s (UNEP, 2010); more recent estimates are not available and while emissions have decreased further in developed countries since the 1990s, they may have increased in newly industrialising countries. Cadmium is emitted into the atmosphere from smelters and other metal industries, partly because of its volatility above 400°C. The WHO guideline for Cd in air (5 ng m^{-3}, annual average) has been breached in the vicinity of smelters in particular; for example, 23 ng m^{-3} (annual average) was recorded ~1 km from a zinc smelter in Colorado (USDHHS, 2012). Despite this, most atmospheric Cd pollution arises from the more widespread combustion of coal, while other sources include mine waste heaps (wind-blown particles) and incineration of Cd-containing wastes (e.g. vinyl plastics and batteries, Fig. 8.2). Deposition of atmospheric emissions also contaminates soils, crops and waters, particularly in areas closest to emission sources; there is also evidence from ice and sediment cores of Cd deposition occurring in remote environments, including the Arctic (Fig. 8.3).

In addition to inputs from aerial deposition, agricultural soils are contaminated by the deliberate application of: (i) phosphate fertilisers, which contain Cd as a contaminant of phosphate rock, sometimes at concentrations in excess of 150 mg kg^{-1}

Figure 8.2 Old nickel-cadmium (NICAD) batteries are a potential source of cadmium pollution if they are landfilled or incinerated.

and (ii) sewage sludge, which is applied to soils as a convenient disposal method and to add nutrients and organic matter to agricultural fields. Phosphate fertiliser application can certainly increase the Cd concentration of soils, but tends also to decrease its mobility because of the strong affinity of Cd for phosphate groups. Sewage sludge incorporates several toxic metals, from industrial effluents and rural and urban runoff; however, Cd is typically of the most concern in relation to sewage sludge use. It is present in sludge at relatively low concentrations compared with other metals – typically <10 mg kg^{-1} (although much higher in individual cases) – but it is more readily taken up by crops, especially leafy vegetables such as lettuce and spinach. Cadmium is generally more mobile and bioavailable in soils than other toxic metals, being only weakly adsorbed; it is particularly mobile in acidic soils because of reduced cation exchange capacity and/or the dissolution of hydrous oxides. Despite this, in many countries the amount of Cd leaving soils is generally lower than inputs, leading to its gradual accumulation in the soil. The application of sewage sludge to agricultural land is an ongoing concern because the sludge is an important source of environmental Cd (and other metals) but not easily disposed of in other ways, meaning its continued use as a cheap fertiliser is likely. For these reasons, authorities are required to set maximum allowable concentration of metals in sewage sludges to be spread on land. Figure 8.4 shows an example of Cd uptake in sludge-amended soils.

Figure 8.3 The anthropogenic imprint of cadmium pollution is evident in Arctic lake sediments and ice cores.

Credit: NASA's Goddard Space Flight Center/Ludovic Brucker (www.nasa.gov).

Cadmium in contaminated soils and mine wastes is transferred into freshwaters via leaching and runoff. An additional source of Cd in waters, particularly groundwater, is landfill leachate, which can transport Cd from waste products such as Ni-Cd batteries; landfill is becoming less important as a waste management option in many countries, but landfill sites act as important reservoirs of Cd and other pollutants. Concentrations of Cd in mine-polluted waters can be 1–3 orders of magnitude higher than background. A portion of the Cd in freshwaters is inevitably carried to the oceans.

Human health effects

The effects of Cd on human health are known from occupational and environmental exposure. Chronic inhalation of particulate Cd in the workplace has been suspected of causing lung disease. Exposure of the general, non-smoking population is mainly (about 90%) via ingestion of foods from Cd-contaminated environments, particularly shellfish, liver, kidney and crops including leafy vegetables, root vegetables and rice. The remaining exposure is via ingestion of drinking water and inhalation of Cd in the air. Chronic exposure to high levels of Cd in the diet, particularly in malnourished subjects, can damage the kidneys and bone tissue. Kidney damage may disrupt vitamin D3 and Ca metabolism, decreasing Ca absorption. The element is stored in the kidney,

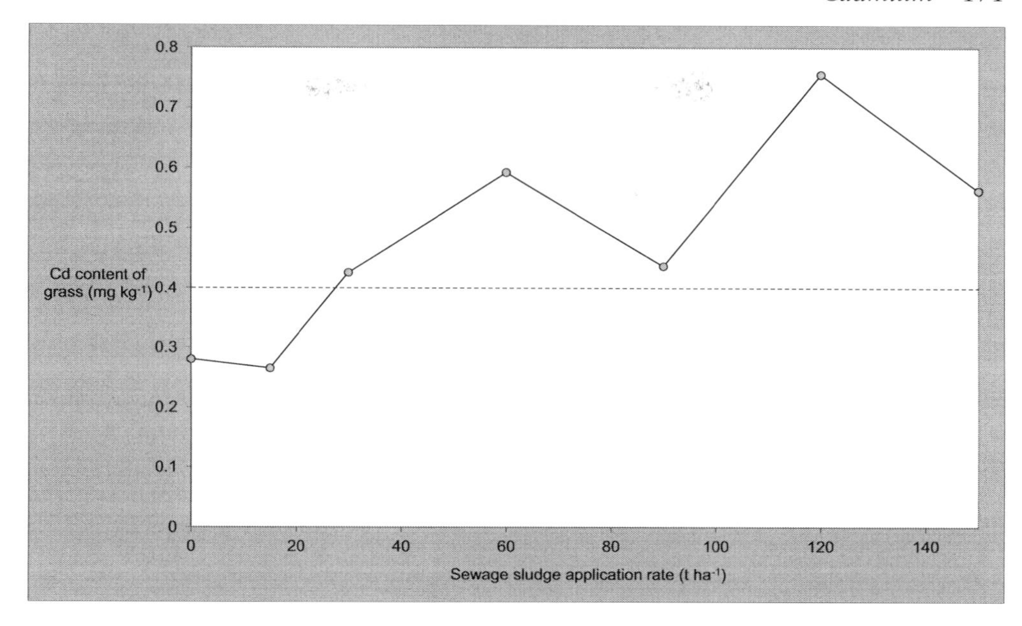

Figure 8.4 The effect of increasing sewage sludge application rates (to Chinese soils) on the uptake of cadmium by the grass species, *Poa annua*. The horizontal line shows the maximum environmental quality standard (EQS) for Cd in crops stated by the Food and Agriculture Organization of the United Nations (0.05–0.4 mg kg^{-1}). The high nutrient status of the sewage sludge increased grass yield but the authors state that it should not be used as a fertiliser because its Cd content (5 mg kg^{-1}) increased Cd concentrations in the amended soils above the Chinese EQS for farm soils (0.3 mg kg^{-1}); this was not the case for other heavy metals present in the sludge (Cu, Pb and Zn) and Cd contamination was therefore the limiting factor in the use of this sludge as a fertiliser. A second grass species, *Zoysia japonica*, did not show the same linear relationship, but the Cd content of the plant was again higher than in the control at most application rates (max. 1.4 mg kg^{-1}).

Source: Adapted from Wang et al. (2008).

accumulating there over time and Cd concentrations in the kidney are thought to become critical when they reach approximately 100 mg kg^{-1}. The WHO (2000) has expressed concern that Cd concentrations in the kidneys of the general population in Europe are only a few times lower than this critical level and states the need to prevent further increases of Cd in the diet by restricting soil inputs from fertilisers, sewage sludge and atmospheric deposition, particularly from waste incinerator emissions. With regards to carcinogenicity, studies of occupational exposure indicate that Cd inhalation contributes to incidence of lung cancer and there may be links with kidney and prostate cancers; however, ingestion of Cd is not a known cause of cancer. Boxes 8.1–8.3 show case studies from Japan, the UK and China.

Box 8.1 Itai-itai disease, Japan.

For a few decades, from 1910 until the mid-1940s, waste sludge containing Cd from the Kamioka Zn mines in the River Jinzu catchment of Japan gradually contaminated rice paddies and drinking water in the area. Cadmium uptake by rice was enhanced by the low pH of the contaminated paddy soils. Furthermore, the paddies had been drained before harvest, creating oxic conditions, which also increase Cd mobility (in reducing conditions, it tends to form relatively insoluble CdS). It is estimated that locals ingested 600 μg Cd per day, compared with the safe level of 60–70 μg per day. Ingestion of the contaminated rice over many years caused kidney dysfunction and osteomalacia (bone softening) leading to fragile and deformed bones, especially in elderly women. The associated joint and bone pain suffered by the victims led to the naming of the disease as 'itai-itai'; 'itai' means 'ouch' or 'it hurts'. The symptoms were most severe in malnourished women who suffered from Ca and vitamin D deficiency and who had been through multiple pregnancies. It is possible that bone damage was caused by Cd replacing Ca in the bones and/or because of deficient mineral metabolism caused by damage to the kidneys. Further investigations indicated that nearly 10% of Japan's paddy soils were contaminated by Cd from mines and aerial deposition from industry. Remediation of the Cd contamination in the Jinzu Valley was finally completed in 2012, but Itai-itai disease is still observed in elderly women, due to past exposure. Figure 8.5 shows the relationship between the level of Cd pollution in the River Jinzu catchment (Cd in rice) and prevalence of Itai-itai disease in the region.

Further reading: Ogawa et al. (2004), USDHHS (2012).

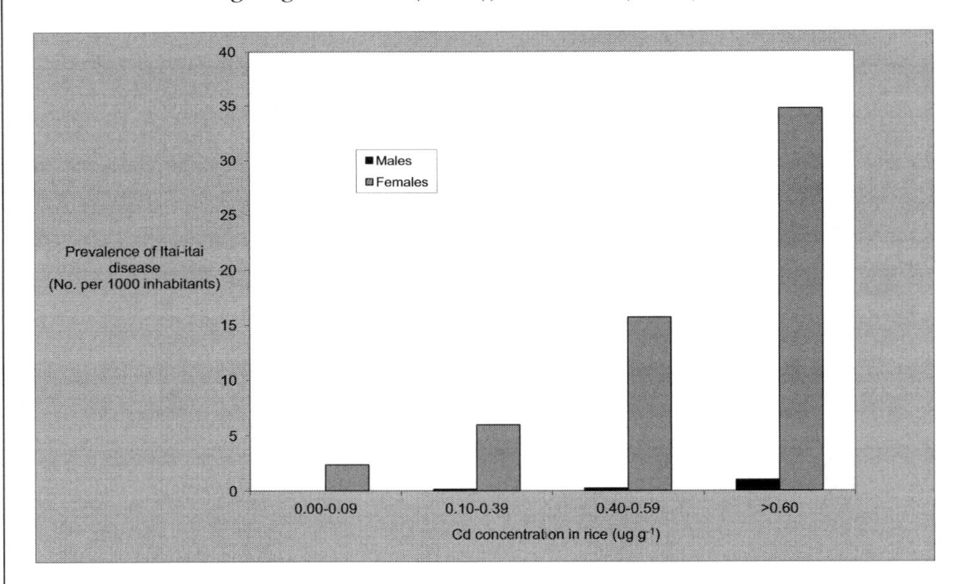

Figure 8.5 The relationship between concentrations of cadmium in rice and the prevalence of Itai-itai disease in inhabitants of 55 villages in the cadmium-polluted Jinzu River catchment area, Japan. Data from Ogawa et al. (2004), based on 2446 rice samples from the 55 villages, which were collected and analysed in the 1970s, and health statistics from the same villages, collected between 1967 and 1993; a highly significant correlation ($p < 0.001$) was reported between Cd in rice and prevalence of the disease. Note that the disease is much more prevalent in women (see text).

Box 8.2 Cadmium contamination in Shipham, UK.

In the 1980s, very high levels of Cd (up to 500 µg kg^{-1}) were recorded in soils in a former Zn mining area of the UK, centred on a village called Shipham. Locals ate vegetables grown in the contaminated soils, yet none of the symptoms observed in Japan (Box 8.1) were recorded. Protection may have been offered by the relatively high pH of the soils in what is a limestone area; other factors may have been better nutrition and shorter exposure times (housing in the affected area had been built mainly within the previous two decades).

Box 8.3 Rice contamination, Hunan Province, China.

In 2013, three rice mills in Hunan Province, China were shut down because of contamination of rice by Cd. The source appears to have been contamination of paddy soils by metal mining activities; the province is one of China's main rice producers but also has many metal mines. The problem first became apparent when Cd levels were found to breach allowable levels in six rice and two rice noodle samples (from a total sample size of 18) served at two restaurants and two university canteens in the city of Guangzhou, in neighbouring Guangdong province. One sample of rice contained double the permitted level of 0.2 mg kg^{-1}. For some years Cd contamination of rice has been problematic in China, with some researchers estimating that 10% of rice grown in the country may breach allowable levels. The contaminated rice findings came just a few weeks after reports of severe Cd pollution of a river, and associated fish kills, in China's Guangxi Province.

Cadmium ecotoxicity

The ecotoxicity of Cd is an additional cause for concern; for example, in some parts of Europe, environmental concentrations exceed estimated adverse effect thresholds for terrestrial and aquatic ecosystems. Animals exposed to Cd suffer from impairment of enzymes, changes in cell membrane permeability and effects on the kidney and Ca metabolism. Accumulation is known in many animals, including aquatic invertebrates, earthworms, seabirds and marine mammals. Fish suffer from inhibited Ca uptake and spinal deformation. In plants, Cd is relatively easily taken up and is extremely toxic, disrupting enzyme activities as a result of its affinity for the thiol group. Excess Cd uptake can affect respiration, transpiration and photosynthesis, with visible symptoms of leaf chlorosis and retarded growth. Some algal species bioaccumulate Cd. Microbial biomass and enzyme activities can be inhibited at soil concentrations of 10 mg kg^{-1} (e.g. Table 8.3); at higher levels rates of nitrification and N-mineralisation also decrease. Some species of fungi are particularly affected, becoming absent in Cd-contaminated soils. As with other toxic elements, the presence of Cd in wastewaters can impair sewage treatment because of its toxicity to microbes.

Table 8.3 Effect of soil cadmium concentration on microbial biomass and enzyme activity.

	Soil microbial biomass µg C g soil^{-1}			*Dehydrogenase activity[a,b] µg TPF g soil^{-1} 24 h^{-1}*		
Soil Cd[c] mg kg^{-1}	*Soil 1*	*Soil 2*	*Soil 3*	*Soil 1*	*Soil 2*	*Soil 3*
0	142	280	188	50	233	180
10	135	271	186	46.5	227	178
25	122	247	172	43	215	168.5
50	101	218	149	33	175	138

a Other enzyme activities (alkaline phosphatase and arginine ammonification) were measured and found to be similarly affected by Cd.
b Dehydrogenase activity expressed as triphenyl formazone (TPF) formation.
c Soil Cd concentrations attained by addition of cadmium chloride, $CdCl_2$; microbial parameters were measured 60 days after the Cd addition. Soils 1, 2 and 3 were, respectively, a sandy loam, a loam and a clay loam.

Source: Dar (1996).

References

Dar, G.H. 1996. Effects of cadmium and sewage sludge on soil microbial biomass and enzyme activities. *Bioresource Technology* 56, 141–145.

Ogawa, T., Kobayashi, E., Okubo, Y., Suwazono, Y., Kido, T. and Nogawa, K. 2004. Relationship among prevalence of patients with Itai-itai disease, prevalence of abnormal urinary findings, and cadmium concentrations in rice of individual hamlets in the Jinzu River basin, Toyama prefecture of Japan. *International Journal of Environmental Health Research* 14(4), 243–252.

UNEP (United Nations Environment Program). 2010. *Final Review of Scientific Information on Cadmium*. Available at: www.unep.org

USDHHS (US Department of Health and Human Services: Public Health Service, Agency for Toxic Substances and Disease Registry, ATSDR). 2012. *Toxicological Profile for Cadmium*. ATSDR, Atlanta.

USGS (United States Geological Survey). 2014. *Mineral Commodity Summaries: Cadmium*. Available at: http://minerals.usgl.gov/minerals

Wang, X., Chen, T., Ge, Y. and Jia, Y. 2008. Studies on land application of sewage sludge and its limiting factors. *Journal of Hazardous Materials* 160, 554–558.

WHO (World Health Organization). 2000. Air Quality Guidelines for Europe, 2nd Edition. WHO Regional Office for Europe, Copenhagen.

9 Chlorine

Chlorine in the natural environment

Lithosphere

Chlorine (Cl) occurs in igneous, metamorphic and sedimentary rocks, mainly as chloride, Cl^-. The main minerals are evaporites, particularly halite (NaCl) and various potassium chlorides, including sylvite (KCl) and carnallite ($KMgCl_3 \cdot 6H_2O$). Evaporites form in arid basins with intermittent water inputs (salt pans or flats; Fig. 9.1). The chloride ion is not bound strongly to soils; a typical mean soil concentration is ~100 mg kg^{-1}, but this approximation masks wide variations depending on proximity to coastal areas, where sea salt deposition increases Cl^- concentrations.

Figure 9.1 The Salar de Uyuni in southwest Bolivia, the largest salt flat in the world. It covers >1 × 10^4 square kilometres and is composed of chlorides of sodium, potassium, magnesium and lithium.

Credit: Pedro Szekely (Flickr).

Atmosphere

The atmosphere contains many natural Cl-bearing gases, but only one, chloromethane, is present in significant quantities. This gas, chemical formula CH_3Cl (also known as methyl chloride), has a longer atmospheric residence time (between 1 and 2 years) than other naturally derived Cl gases and therefore has time to accumulate to some degree. It is present at a concentration of 0.55–0.65 parts per billion volume (ppbv) (Khalil and Rasmussen, 1999). The other main naturally produced Cl gases in the atmosphere are chloroform ($CHCl_3$), dichloromethane (CH_2Cl_2), tetrachloroethylene (C_2Cl_4) and trichloroethylene (C_2HCl_3). Hydrogen chloride (HCl) is also present but has a short residence time in the atmosphere because it is soluble in rainwater (equation 9.1). The naturally produced Cl-bearing gases reside mainly in the troposphere, although a small proportion of tropospheric chloromethane reaches the stratosphere each year. The atmosphere also contains particulate Cl in the form of soil and desert dust.

$$HCl + H_2O \rightarrow H_3O^+ + Cl^- \quad [9.1]$$

Hydrosphere

Chlorine in the hydrosphere exists almost entirely as Cl^-, salts of which are highly soluble. It is the most abundant dissolved ion in the oceans and estimations of seawater concentrations range from 1.8 to 1.94×10^4 mg L^{-1}. Concentrations of Cl^- in unpolluted freshwaters are generally accepted as being <20 mg L^{-1}.

Biosphere

Chlorine is one of life's essential elements. It is found mainly in body fluids (as Cl^-), where it provides osmotic potential and ionic balance for cells. However, organisms also use Cl in other vital processes. White blood cells manufacture hypochlorous acid (HOCl) from Cl^- and hydrogen peroxide (H_2O_2) and use this reactive oxidant against invading pathogens. Hydrochloric acid (HCl) is the main component of gastric fluid in the mammalian stomach, aiding food digestion. Chloride plays a key role in photosynthesis; it is present in chloroplasts, where it activates enzymes involved in the splitting of water molecules to produce oxygen.

Chlorine cycling: inputs to the atmosphere

The largest natural input of Cl into the atmosphere is the chloride contained within minute droplets of airborne sea spray. Evaporation of water from the droplets leaves behind suspended sea salt aerosols. The process is thought to transfer around 6×10^3 Mt Cl^- to the atmosphere each year, but nearly all is rapidly re-deposited and the atmospheric Cl reservoir is not large (Fig. 9.2). Volcanoes are another important source of atmospheric Cl, mainly in the form of hydrogen chloride (HCl). The eruption of Mount Pinatubo in the Philippines in 1991, for example, released an estimated 4.5 Mt of HCl (Tabazadeh and Turco, 1993). Additionally, particulates containing Cl^- are carried into the atmosphere from wind-blown surface dusts.

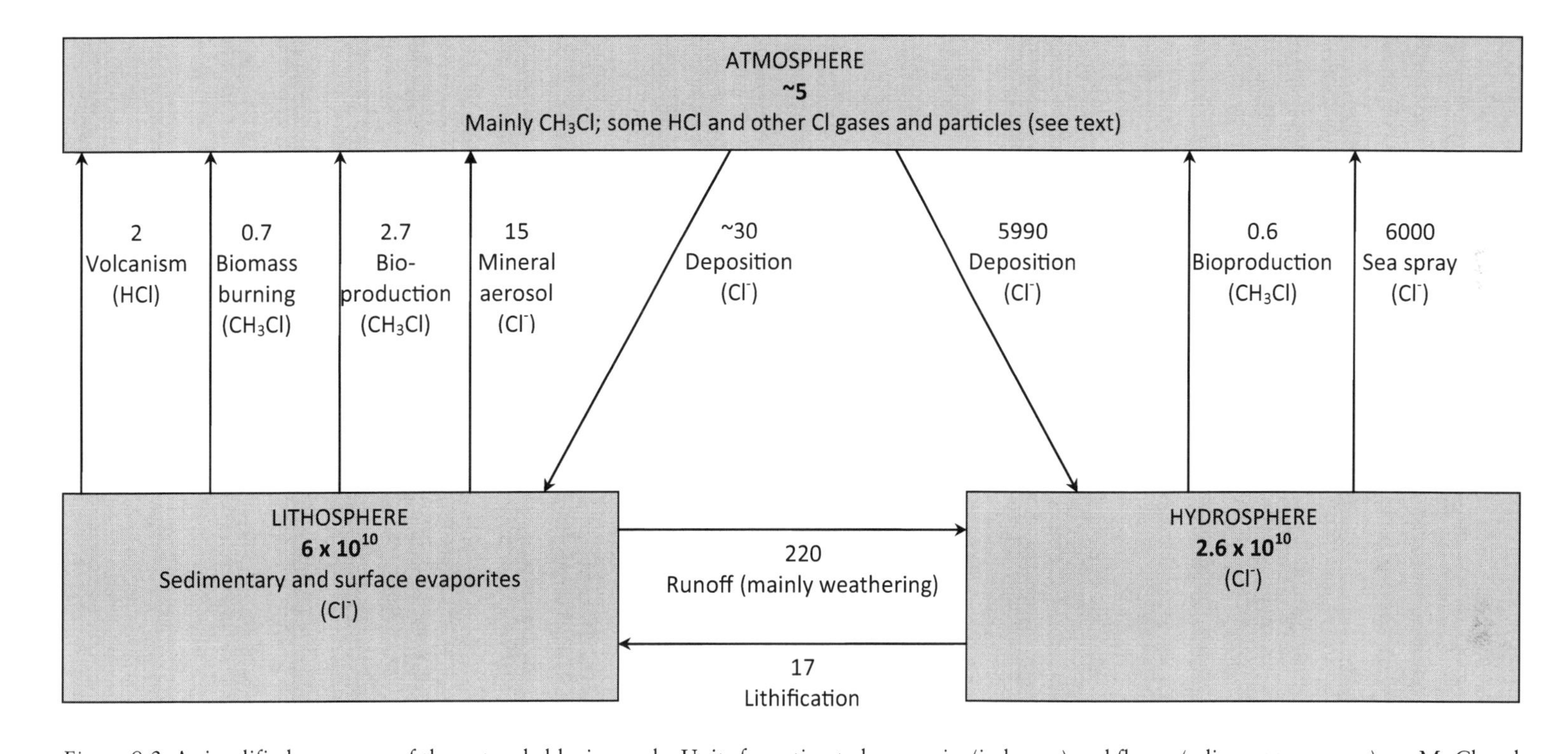

Figure 9.2 A simplified summary of the natural chlorine cycle. Units for estimated reservoirs (in boxes) and fluxes (adjacent to arrows) are Mt Cl and Mt Cl a^{-1}, respectively. The net input to the hydrosphere is consistent with increasing Cl^- concentrations in the oceans, resulting from a relative global scarcity of extensive salt pans.

Source: Adapted from Graedel and Keene (1996) and Schlesinger and Bernhardt (2013).

Various biotic processes result in the transfer of Cl to the atmosphere, mainly in the form of chloromethane (CH_3Cl). Marine algae produce methyl iodide, which reacts with Cl^- in seawater to produce chloromethane, which, in turn, diffuses into the atmosphere. Chloromethane is also produced in the terrestrial environment, particularly in tropical regions, by ferns and saltmarsh plants and by microbial decomposition of soil organic matter and dead wood. It is also released by biomass burning and possibly by the methylation of Cl^- in decomposing plant leaves by pectin, a polysaccharide also present in leaves (Hamilton et al., 2003). Other gases are produced in smaller amounts, including aromatic and phenolic compounds such as 1,2,3,4-tetrachlorobenzene.

Chlorine cycling: inputs to the hydrosphere

Chloride is released into freshwaters by weathering of Cl-bearing rocks and runoff or leaching of sea salt Cl^- that has been deposited on land. Chloride is a conservative ion, meaning most land-deposited Cl^- enters stream flow and is returned to the oceans (Fig. 9.3); i.e. little Cl^- is held long-term in soils (via ion exchange, for example) or by terrestrial organisms. The main input to the oceans, however, is from the rapid re-deposition of sea salt aerosol (Fig. 9.2).

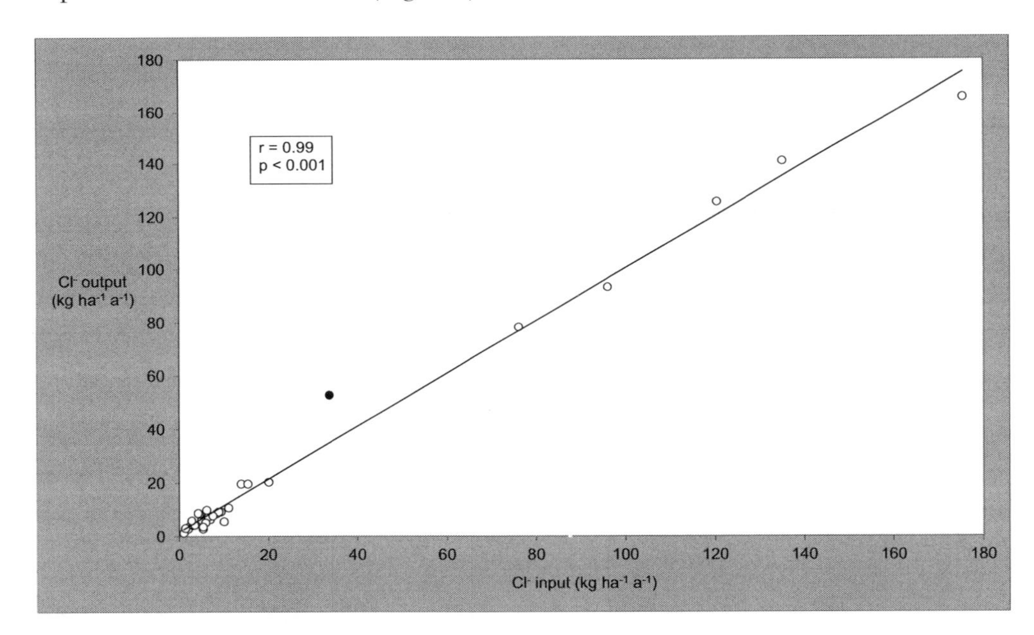

Figure 9.3 Inputs and outputs of chloride in 32 temperate forest catchments in Europe and North America. In all but one case (black data point) the input amount, based on measurements of throughfall, bulk deposition or wet deposition, is close to the outflow from the catchment; i.e. close to net balance, as shown by the close 1:1 ratio.

Source: Svensson et al. (2012).

A small fraction of the total deposited Cl^- shown in Fig. 9.2 is derived from the reaction of chloromethane with hydroxyl radicals (equation 9.2) and dissolution of the HCl gas (equation 9.1) emitted from volcanoes.

$$CH_3Cl + OH^{\cdot} \rightarrow CH_3OH + Cl^- \qquad [9.2]$$

Chlorine cycling: inputs to the lithosphere

Precipitation of halite and other chloride salts in shallow saltwater bodies may ultimately become geological deposits in the Earth's crust, where subsequent conditions allow (e.g. burial of the evaporites by overlying sedimentary layers). Chlorine-bearing rocks (although with much lower Cl concentrations) can also arise where seawater, with its Cl^- load, fills the pores of suspended particles which then settle to the sea bed and subsequently become lithified, along with the Cl^-, as sedimentary rocks.

Inputs of Cl^- to soils are large, particularly in coastal areas; some catchment studies have reported terrestrial deposition of >150 kg Cl^- ha^{-1} a^{-1} (Svensson et al., 2012). The Cl^- ultimately flows out of catchments (see above), but because of microbial and plant uptake, fresh soil organic matter (SOM) tends to act as a short-term sink of Cl^- before it is released again as the SOM is slowly decomposed to residual humus. In laboratory experiments, up to 24% of Cl^- added to soil has been taken up by micro-organisms for short periods of up to one month (Bastviken et al., 2007). Despite the conservative nature of Cl^-, in arid parts of the world with insufficient throughflow of water, some soils can be highly saline, particularly in low-lying areas or natural depressions; in such cases, as water evaporates, salts dissolved from the soil form a deposit on the soil surface (Fig. 9.4).

Production and uses

The main raw material in the production of chlorine (for manufacturing processes, see below) and chloride (for de-icing salts and the food industry) is sodium chloride (NaCl). The mineral form, halite, is mined in many parts of the world, including North America, Europe and parts of Asia; it is physically extracted from underground and surface (dry salt lake) deposits and is dissolved and pumped from underground deposits using hot water followed by evaporation in vacuum plants. Sodium chloride is also extracted from seawater, by precipitation in solar ponds or concentration into brine.

The chlorine required for manufacturing is produced by the chlor-alkali industry. The chlor-alkali process refers to the production of chlorine gas (Cl_2) as well as sodium hydroxide, or 'caustic soda' (NaOH, an alkali), which is itself an important raw material for the chemical industry. The process centres on the electrolysis of brine, a concentrated solution of NaCl and/or KCl. The salts required to make up the concentrated brine solutions are derived from mineral deposits (NaCl and KCl) and/ or seawater (NaCl). A current is passed through the brine in a reaction cell. This is typically a membrane-type cell, now that traditional mercury cells are being phased out because of mercury pollution concerns; diaphragm cells are also used, but to a lesser degree. Electrolysis of NaCl brine produces chlorine gas, hydrogen gas and caustic soda:

$$2NaCl + 2H_2O \rightarrow Cl_2 + H_2 + 2NaOH \qquad [9.3]$$

Figure 9.4 Salt-encrusted agricultural soil in Colorado, USA.
Credit: Tim McCable, USDA Natural Resources Conservation Service (Wikimedia Commons).

The resulting chlorine is stored as either hydrochloric acid (HCl) or pressurised liquid chlorine (Cl_2), which are the feedstocks for the production of chlorine compounds. Great care has to be taken in the storage and transport of Cl_2 because it is highly toxic.

The main direct and indirect uses of Cl are in the manufacture of polyvinyl chloride (PVC) plastics, polyurethane foams, pesticides, disinfectants and solvents (Fig. 9.5). The organochlorine compound required for PVC production (Fig. 9.6) is chloroethene (vinyl chloride, C_2H_3Cl), a toxic gas. For polyurethane production, another organochlorine, carbonyl dichloride ('phosgene', $COCl_2$), is reacted with amines (general formula RNH_2, where R is an alkyl or aryl group) to form isocyanates (RN=C=O), which are the essential precursors for polyurethane production:

$$COCl_2 + RNH_2 \rightarrow RN{=}C{=}O + 2HCl \qquad [9.4]$$

Polyurethanes are mainly used in the manufacture of rigid and flexible foams for use in products ranging from upholstery and refrigerators to footwear and carpet underlay. Other uses include varnishes, adhesives and sealants.

Other categories of organochlorine (OC) compounds include those that are well known for causing environmental pollution concerns, such as pesticides and polychlorinated biphenyls (PCBs). The main OC pesticides in use, past and present, include DDT (dichlorodiphenyltrichloroethane), aldrin, dieldrin, endrin, chlordane and heptachlor. The use of all of these pesticides is banned or severely restricted under

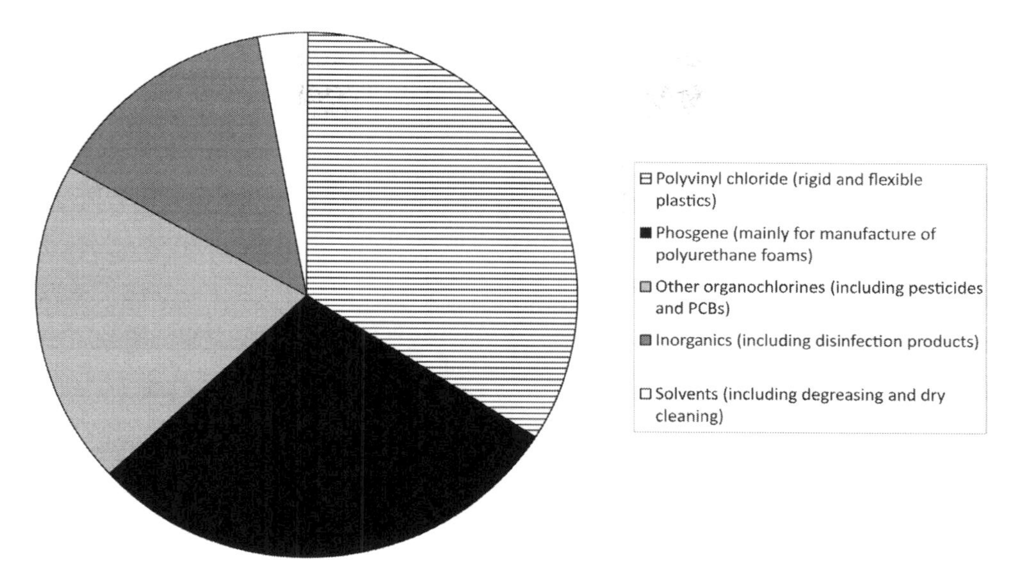

Figure 9.5 The main uses of chlorine; approximate proportions based on various sources.

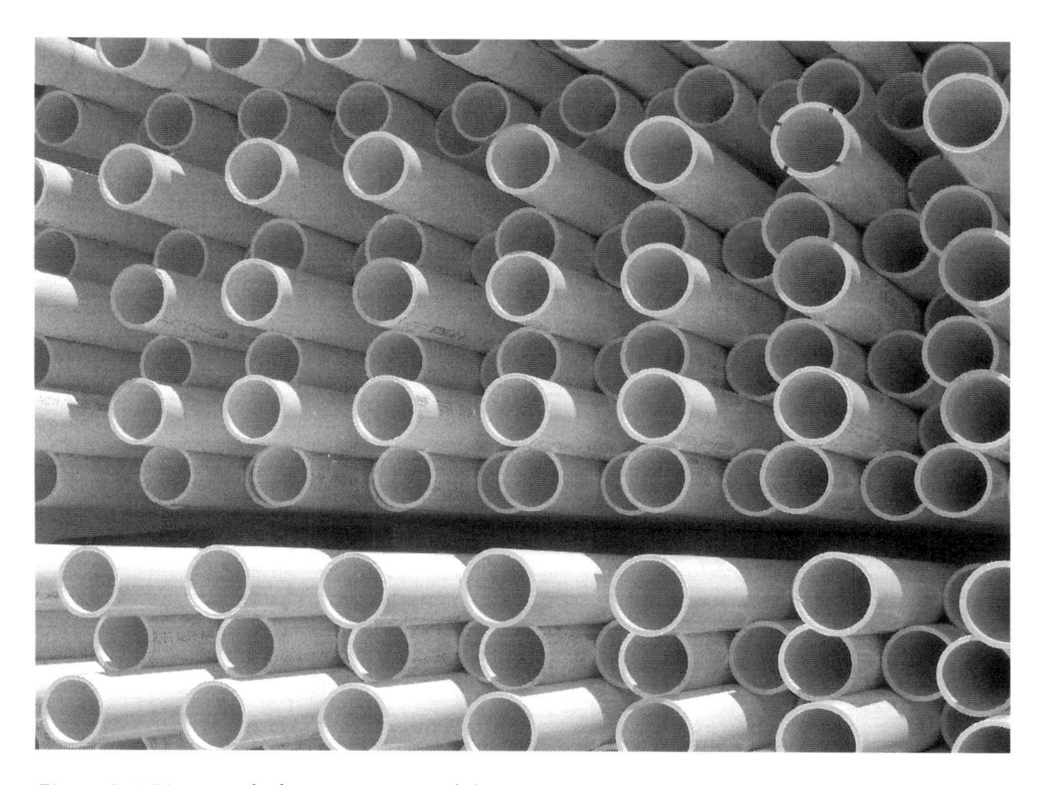

Figure 9.6 Pipes made from PVC, one of the main commercial uses of chlorine.
Credit: Dennis Hill (Flickr).

the Stockholm Convention of 2001, an international treaty under which signatories are to take measures to eliminate or reduce release of these pesticides into the environment. Of the signatory countries, China, India and some African countries still produce DDT for its use in mosquito control in malaria-prone areas; it is recommended by the World Health Organization (WHO) for internal wall-spraying in affected homes. Other OC pesticides include 2,4-dichlorophenoxyacetate (2,4-D), which is still used in most countries, and 2,4,5-trichloropenoxyacetate (2,4,5-T), also known as Agent Orange, which was famously used as a defoliant by the USA in the Vietnam War.

The OCs known as polychlorinated biphenyls (PCBs) are a group of 209 compounds with between 1 and 10 Cl atoms attached to a biphenyl molecule; the chemical formula is $C_{12}H_{10-X}Cl_X$ (Fig. 9.7). They were first manufactured in the 1920s and valued as stable, fire-resistant compounds with low vapour pressures and high electrical resistance. This made them very useful as insulators and coolants in electrical transformers and as dielectric fluids in capacitors. They were also used as additives in paints and engineering oils, as plasticisers (to reduce brittleness in plastics) and as flame retardants. However, by the 1970s there were concerns over their environmental effects (see below) and their production was banned by most countries in the 1980s. By 2008 there were still an estimated 3 Mt of PCBs and associated equipment to be safely disposed of. An estimated 0.6 Mt are thought to have been disposed already, to add to the unknown amounts that have escaped into the environment. The Stockholm Convention states that all use of PCBs should be phased out by 2025 and that all known stocks should be safely disposed of by 2028.

Historically, an important use for OCs has been as solvents for dry cleaning and degreasing, but this has been declining for some time because of their toxicity. For example, carbon tetrachloride (CCl_4), also known as tetrachloromethane, can affect the central nervous system when inhaled. Other common OC solvents include dichloromethane (CH_2Cl_2), trichloroethane ($C_2H_3Cl_3$), trichloroethene (C_2HCl_3) and tetrachloroethene (C_2Cl_4). Carbon tetrachloride is also known to aid in the destruction of stratospheric ozone and, together with the more well-known ozone-depleting compounds, chlorofluorocarbons (CFCs), it is now banned under the Montreal Protocol, an international treaty formulated in 1987. The main two CFC compounds produced, CFC-11 [$CFCl_3$] and CFC-12 [CF_2Cl_2], were used in refrigerators, air conditioners, aerosols and foam blowing. The initial CFC replacements, hydrochlorofluorocarbons (HCFCs), which are also ozone depleters (but less potent), can still be used under the Protocol until 2030.

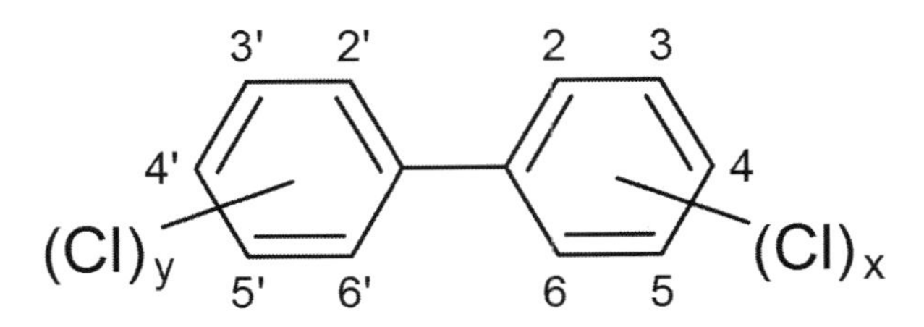

Figure 9.7 The general molecular structure of PCBs.

Credit: Dschanz (Wikimedia Commons).

Chlorine gas is used widely in the disinfection of bathing and drinking waters. It was first used to disinfect tap water in Britain in the late 19th century and is still preferred worldwide over the other main disinfectants, primarily ozone, because it is effective at very low concentrations and over relatively long time periods. The process typically involves the dissolution of chlorine gas in water to produce hypochlorous acid (HClO):

$$Cl_2 + H_2O \rightarrow H^+ + Cl^- + HClO \qquad [9.5]$$

The hypochlorous acid dissociates to yield ions of hydrogen and hypochlorite (ClO^-), which is very effective at killing pathogenic bacteria and some viruses:

$$HClO \leftrightarrow H^+ + ClO^- \qquad [9.6]$$

Hypochlorite salts are sometimes used instead of Cl_2 in the first stage of this process because of their safer handling and transportation. Chlorine dioxide (ClO_2) and chloramines (NH_2Cl) are also used in water disinfection, but on a smaller scale. Chlorine is also used as the disinfectant agent in sodium hypochlorite bleach, which is produced by bubbling chlorine gas through sodium hydroxide. Sodium hypochlorite has further uses in the removal of ink from recycled paper and the bleaching of cotton.

Pollution and environmental impacts

Salinity

In contrast with the major elements (C, N, P, S) and many trace elements, most pollution from Cl is not associated with disturbances of natural reservoirs or cycles but with the introduction of man-made, synthetic compounds such as CFCs into the environment. However, there is one important example of a perturbation of the natural cycle; the extraction of halite from the Earth's crust followed by its distribution onto road surfaces in winter as road salt. Sodium chloride is the main salt used for road de-icing, but smaller amounts of other chloride salts are also used, including those of calcium and magnesium. When rain or melting snow carries the salt away from treated roads, the Cl^- loading of local rivers is significantly raised. Roadside soils are also affected, by salt-spreading activities and by the spray from passing traffic. This can cause environmental problems in higher latitudes, such as northern parts of North America and Eurasia, where road salting is routine in winter-time.

Chloride concentrations in road runoff can reach levels of 1–2 g L^{-1} (Environment Canada and Health Canada, 2001); such elevated levels are diluted on entry into bodies of freshwater, but concentrations are still high enough to cause concern. For example, in North American lakes, Cl^- concentrations of up to 5000 mg L^{-1} have been recorded, while 10% of aquatic species are estimated to be affected (LC_{50}) at exposure to 240 mg L^{-1} (Environment Canada and Health Canada, 2001). Published field surveys show that salt-intolerant amphibians such as spotted salamanders and wood frogs are excluded from affected waterways as more tolerant species such as American toads become dominant (Figs. 9.8 and 9.9).

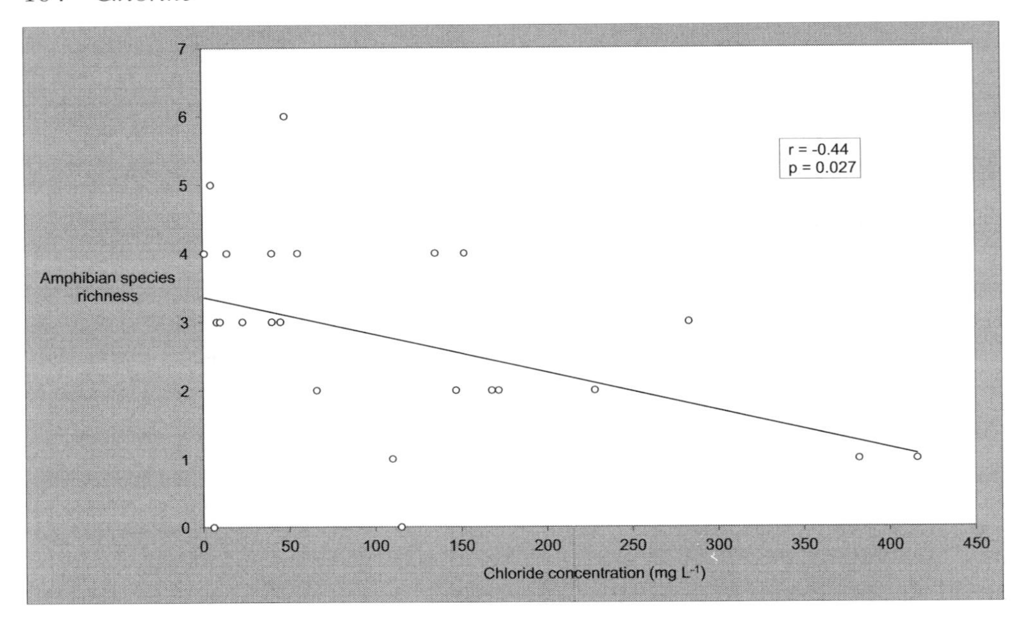

Figure 9.8 Effect of chloride concentration in pond waters on amphibian species richness in Nova Scotia, Canada. Field results from 26 ponds within 60 m of a secondary road.

Source: Adapted from Collins and Russell (2009).

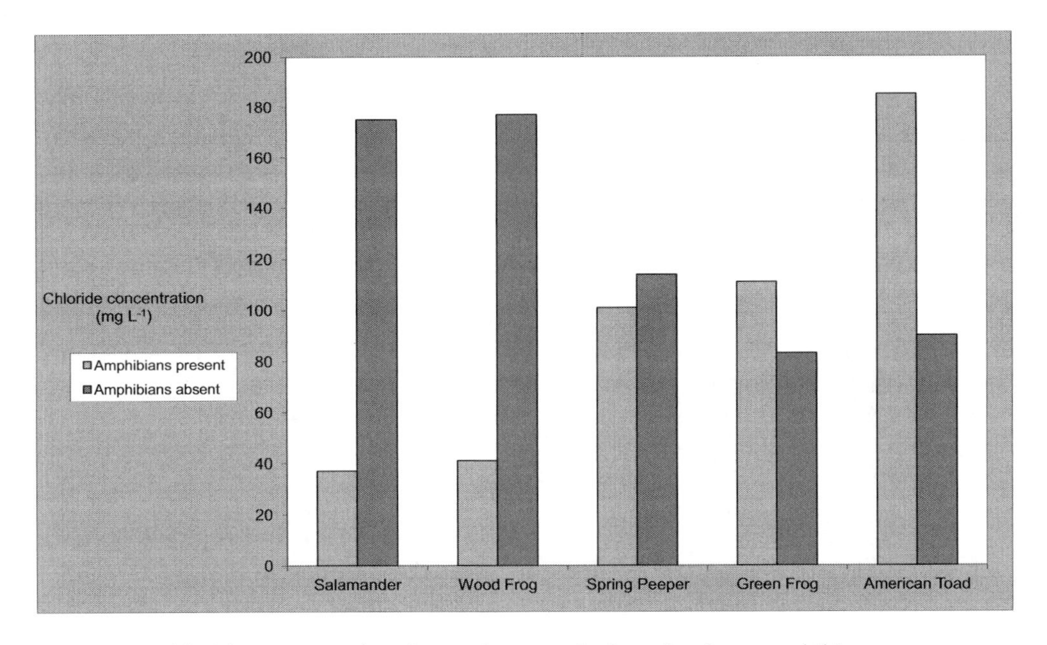

Figure 9.9 Chloride concentrations in ponds near salted roads where amphibians were present or absent, Nova Scotia, Canada. High salt levels appear to influence community structure, favouring salt-tolerant species at the expense of non-tolerant species.

Source: Adapted from Collins and Russell (2009).

Chloride concentrations in roadside soils can reach several hundred parts per million and, although the worst contamination occurs within 10 m, levels of >200 mg kg^{-1} have been recorded in Canada at up to 200 m distance; soil bacteria inhibition has been observed at 90 mg kg^{-1} (Environment Canada and Health Canada, 2001). Elevated Cl^- inhibits plant uptake of water and nutrients because of osmotic imbalance. Grass and wildflower seed germination is inhibited and contaminated soils can limit plant growth. This has ecological effects on affected areas; for example, coastal halophytes (salt-tolerant plants) have spread inland alongside treated roads, becoming dominant over the original plant populations. Chloride-free road salts, such as calcium magnesium acetate (CMA) are available, but are generally more expensive and/or less effective than NaCl at very low temperatures. Authorities in affected areas have tended to focus more of their efforts on the suitable storage of NaCl and minimisation of its overuse.

Disinfection by-products

The hypochlorite ion used in water chlorination can react with the dissolved organic matter that is naturally present in the water and form low concentrations of organochlorine compounds, examples of 'disinfection by-products' (DBPs), that have been linked with cancer. Trichloromethane (chloroform) is the most commonly formed compound, along with other 'trihalomethanes', particularly dibromochloromethane ($CHBr_2Cl$) and bromodichloromethane (BDCM; $CHBrCl_2$). Long-term ingestion of chlorinated drinking water with elevated DBP levels has been linked with small but consistent increases in incidence of colon and bladder cancer and with early-term miscarriage; the latter is especially linked with BDCM. Drinking water consumers who are located furthest away in the water distribution network from the point of chlorination appear to be at most risk because time is required for DBPs to form. DBPs can be removed by tertiary treatment (activated charcoal) but this is resource intensive. The WHO drinking water quality guideline for trichloromethane (acceptable concentration) is 0.3 mg L^{-1}. The International Agency for Research on Cancer (a WHO body) states that health risks from DPBs are low and should be balanced against the risks of not disinfecting water.

Persistent organic pollutants – an introduction to organochlorines

The occurrence of organochlorine compounds in the environment is of serious concern because of their persistence, bioaccumulation and toxicity. Organochlorine, or chlorinated hydrocarbon, compounds form when one or more Cl atoms appear in place of, or in addition to, the hydrogen atoms in hydrocarbon molecules. The precursor hydrocarbon structures of most relevance to organochlorine chemistry are described in Table 9.1.

Table 9.1 Hydrocarbon structures relevant to organochlorine chemistry.

Hydrocarbon type	*General structure*
Alkanes	Single-bonded C atoms in chains or rings with side bonds to H
Alkenes	Double-bonded C atoms in chains or rings with side bonds to H
Alkynes	Triple-bonded C atoms in chains or rings with side bonds to H
Aryls	Aromatic rings of 6 C atoms with alternate single and double bonds, with side bonds to H

In alkanes, Cl replacement gives alkyl halides such as chloromethane (CH_3Cl), which can form naturally, initially by the production of Cl free radicals resulting from the photolysis of molecular chlorine:

$$Cl_2 + uv \rightarrow 2Cl^{\cdot} \qquad [9.7]$$

The Cl free radical needs only one electron for stability and is highly reactive. Reaction with methane, present naturally in the atmosphere, yields hydrogen chloride gas and a methyl radical (equation 9.8), which goes on to react with more Cl_2 to yield chloromethane and another Cl radical (and thus a potential chain reaction as long as sufficient reactants are present):

$$Cl^{\cdot} + CH_4 \rightarrow HCl + CH_3^{\cdot} \qquad [9.8]$$

$$CH_3^{\cdot} + Cl_2 \rightarrow CH_3Cl + Cl^{\cdot} \qquad [9.9]$$

CFCs and HCFCs are compounds of this type, with the H atoms surrounding the central C being replaced by Cl and F; for example, the chemical formula of CFC-11 is CCl_3F.

The multiple bonds of alkenes and alkynes are unsaturated and elements like Cl do not replace H in these molecules but form additional bonds. In aryl compounds, H can be replaced by Cl in mono-aromatic compounds such as monochloro- to hexachloro-benzene and in molecules like trichlorophenol and pentachlorophenol, which have been used as pesticides. Good examples of the chlorination products of polyaromatic aryl compounds are the 209 different congeners of PCB.

Organochlorines are very stable, which has traditionally made them attractive to manufacturers and users of a wide range of products, from pesticides to refrigerators. However, it also means they are persistent in the environment, being relatively impervious to normal breakdown by micro-organisms or chemistry. Being non-polar, they are relatively water-insoluble (hydrophobic) and fat-soluble (lipophilic) and are therefore bioaccumulative. Some organochlorine compounds, mainly polychlorinated solvents like chloroform and carbon tetrachloride, are classed as 'DNAPLs' (dense non-aqueous phase liquids), which can cause deep contamination of groundwaters if they leak into the ground surface. OCs that have one or more C-H bonds can react with atmospheric OH radicals and are therefore relatively less resistant, while those with no C-H bonds (e.g. CFCs) can persist in the atmosphere. This is because C-Cl bonds such as those in CFCs have higher enthalpy (325 kJ mol^{-1}) than the O-Cl bond (214 kJ mol^{-1}) that would form in a hypothetical reaction of CFC and OH. In contrast, C-H bonds in OCs have *lower* enthalpy (416 kJ mol^{-1}) than the O-H bond (~464 kJ mol^{-1}) so are relatively less stable.

Because of their benefits, OCs have been widely manufactured for many decades and have become an important category of pollutants. They have polluted soils and waters, particularly those OCs that have been spread on the land as pesticides, landfilled as hazardous wastes or illegally dumped. The volatility of OCs contributes to their transfer to the atmosphere, as does emission during manufacture (e.g. hexachlorobenzene and others in pesticide production and chloroethene in PVC production). At least 1500 OCs are known to occur naturally, but only one of these, chloromethane, is present in relatively high concentrations in the atmosphere. The OCs with the highest atmospheric concentrations are mainly human-made gases (Table 9.2).

Table 9.2 Atmospheric concentrations of some anthropogenic chlorine gases (pptv).

Gas	*Peak concentration (and year of peak)*	*2014 concentration*	*Natural concentration*
CCl_4 (Carbon tetrachloride)	105 (1991)	81	38
CFC-11	270 (1994)	234	0
CFC-12	547 (2003)	524	0

Source: Data from Bullister (2014).

International concern about the environmental effects of OCs led to the Stockholm Convention of 2001. This treaty, signed by more than 90 countries, sought to reduce or eliminate the production, use and release of 12 'persistent organic pollutants' (POPs), all of which were OCs. More POPs, including some non-chlorinated compounds, have been included in subsequent updates to the Convention (all are listed in Table 9.3). Action was taken because of concerns over the possible human health impacts of POPs (including carcinogenicity, suppression of the immune system, neurological disorders, infertility and endocrine disruption), their ecological impacts (including oestrogen-mimicking by OCs such as endosulfan) and their persistence and long-range transport in the environment; they are commonly detected in the polar regions, for example.

Organochlorine pesticides

Organochlorines were used extensively as pesticides from the 1940s onwards, both in agriculture and for malaria control. DDT (dichlorodiphenyltrichlorethane) was the first widespread OC pesticide. Such chemicals were valued because they killed a wide variety of insects but did not appear to have effects on mammals. Furthermore, their persistence and water-insolubility meant they did not have to be frequently reapplied. A notable landmark in the history of OC pesticide use, and indeed of environmentalism, was the publication of Rachel Carson's book *Silent Spring* in 1962. This drew attention to the effects of DDT and other OC pesticides on fish, birds and other wild animals and research in the following years showed the tendency of OCs to bioaccumulate in organisms and to biomagnify in food chains (Fig. 9.10). In particular, it was found to bioaccumulate in fish and cause population declines in fish-eating birds such as bald eagle, osprey and brown pelicans. An accepted effect on some bird species was the production of abnormally thin egg shells and their breakage before full development of the embryo. Egg-shell thinning was caused by disruption of calcium metabolism in the birds by DDE (dichlorodiphenyldichloroethylene), the main breakdown product of DDT.

The acute toxicity of DDT to humans has traditionally been stated as being low, but concerns remain about chronic exposure. Exposure to DDT was linked to premature birth and low birth weight by a study of blood samples from 2613 pregnant women in the USA in early 1960s, the main period of DDT use there (Longnecker et al., 2001). Levels of DDE were higher in mothers whose children were premature or had low birth weight and hormone disruption by DDE was postulated as the mechanism. In 2014, an association was reported between blood levels of DDE and Alzheimer's Disease (AD); the mean level was 3.8 times higher in 86 AD patients than

Table 9.3 Persistent organic pollutants of the Stockholm Convention.

Name	*Type*[a]	*Molecular formula*
Aldrin	P	$C_{12}H_8Cl_6$
Chlordane	P	$C_{10}H_6Cl_8$
Chlordecone	P	$C_{10}Cl_{10}O$
DDT[b]	P	$C_{14}H_9Cl_5$
Dieldrin	P	$C_{12}H_8Cl_6O$
Endosulfan and related isomers	P	$C_9H_6Cl_6O_3S$
Endrin	P	$C_{12}H_8Cl_6O$
Heptachlor	P	$C_{10}H_5Cl_7$
Hexachlorobenzene	P, I	C_6Cl_6
Hexachlorocyclohexane (α and β)	P	$C_6H_6Cl_6$
Lindane (γ-hexachlorocyclohexane)	P	$C_6H_6Cl_6$
Mirex	P	$C_{10}Cl_{12}$
Pentachlorobenzene	P, I	C_6HCl_5
Polychlorinated biphenyls	I	$C_{12}H_{10\text{-}X}Cl_X$
Polychlorinated dibenzo-p-dioxins[c]	U	$C_{12}H_4Cl_4O_2$[d]
Polychlorinated dibenzofurans[c]	U	$C_{12}H_4Cl_4O$[e]
Toxaphene	P	$C_{10}H_{10}Cl_8$[f]
Non-chlorinated POPs:		
Hexabromobiphenyl and some polybrominated diphenyl ethers (PBDEs) {→Br}		
Perfluorooctane sulfonic acid[b], its salts[b] and perfluorooctane sulfonyl fluoride[b] {→F}		

a P = pesticide; I = industrial chemical; U = unintentional production.
b Production and use to be restricted, not eliminated.
c Unintentional production to be reduced and, where feasible, eliminated.
d Molecular formula for 2,3,7,8-tetrachlorodibenzodioxin.
e Molecular formula for 2,3,7,8-tetrachlorodibenzofuran.
f This is an example formula, because more than 670 compounds comprise toxaphene.

in 79 control subjects (Richardson et al., 2014). DDT is classed as a possible carcinogen by the International Agency for Research on Cancer and its breakdown product DDE is known from animal studies to be antiandrogenic, meaning it is able to block male hormone receptors. Because of its ecological effects and concerns over human health impacts, DDT was first banned in the USA in 1973 and, under the Stockholm Convention of 2001, its global production and use is currently restricted to malaria control. DDT is the most well-known of the OC pesticides but the others listed in Table 9.3 are also classed as POPs for similar reasons: i.e. their environmental persistence, bioaccumulation and biomagnification, their ecological impacts and their suspected chronic health effects, including carcinogenicity.

Polychlorinated biphenyls

PCB pollution of soils and water bodies (surface and sub-surface) may arise from leakage at their point of use, from irresponsible disposal or from aerial deposition. PCBs are not particularly volatile but can enter the atmosphere from high-temperature

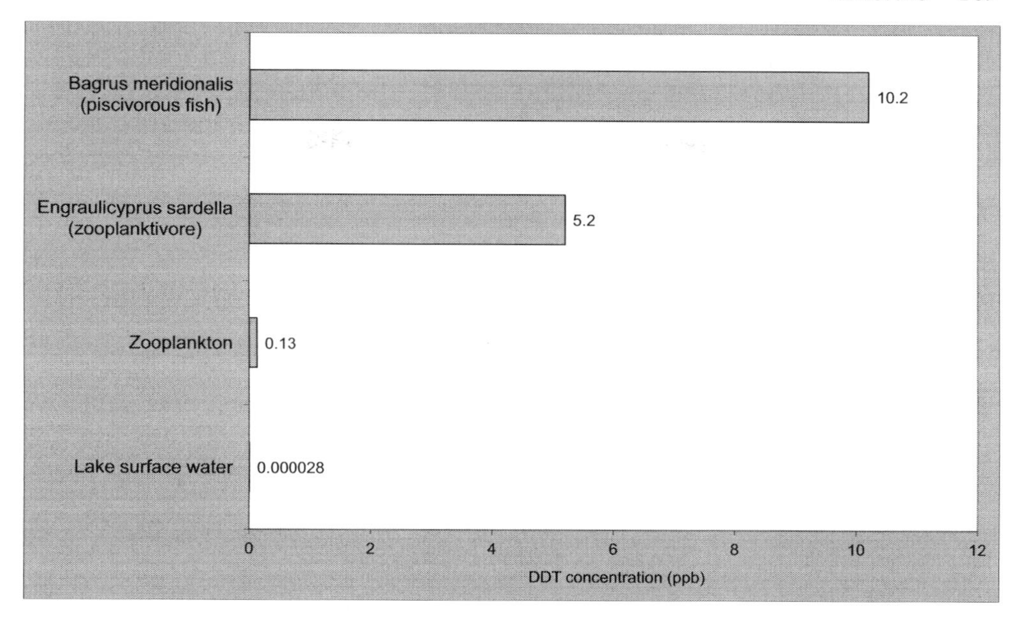

Figure 9.10 DDT biomagnification in Lake Malawi, East Africa. Units are shown as ppb (μg L^{-1} for water; μg kg^{-1} for zooplankton and fish). The biomagnification is 5–6 orders of magnitude (i.e. a 1×10^5 to 1×10^6 times increase in concentration), but the authors state that the DDT concentrations measured in the piscivorous fish were lower than in similar species measured in other parts of East Africa.

Source: Data from Kidd et al. (2001).

processes, particularly waste incineration. In most countries (signatories to the Stockholm Convention), PCB wastes and PCB-contaminated soil must be disposed of by high-temperature incineration or in licensed hazardous waste landfill sites. Incineration temperatures of >1200°C are required to ensure their complete destruction (conversion to HCl) and this may be prohibitively expensive in some cases, with landfilling then being the preferred option. Landfilling does not preclude the long-term risk of PCB release to groundwater and the atmosphere. PCBs are dense and water-immiscible so if they pollute waters they tend to sink and accumulate in sediments. Because of their persistence they may remain at harmful levels in soils and sediments for many years or decades after pollution has occurred.

PCBs are bioaccumulated due to their lipophilicity and are biomagnified in food webs. They are found in marine mammals and fish predators such as seabirds and otters, including in remote areas such as the polar regions. The extent of biomagnification varies among the 209 PCB congeners and is typically found to be highest among the heavier and more chlorinated congeners. This is probably because the latter undergo less biotransformation (i.e. they are less easily metabolised); they are also more liphophilic, so have longer biological half-lives and therefore can be transferred to more trophic levels. The pattern of chlorination of the congener is also important – PCB 138 and PCB 153 are the most recalcitrant and this is thought to be due partly to their 2,4,5 Cl-substitution pattern. Effects of PCB bioaccumulation include egg-shell thinning, as observed for DDT, and disruption of reproduction in aquatic organisms.

Human health effects include neurotoxicity, chloracne (skin lesions), liver disorders and stillbirths, and PCB is linked to birth defects, cancers and reproduction problems.

Production of PCBs is banned under the Stockholm Convention, but their continued use is permitted until 2025 to allow for safe disposal; however, disposal has to be carefully managed and monitored. Furthermore, because of the recalcitrance of PCBs, pollution incidents that occurred before the production ban have left a legacy of contamination that still requires attention (see Box 9.1).

Box 9.1 Ongoing remediation of PCB contamination, New York State, USA.

An estimated 600 tonnes of PCBs were released into the Hudson River in New York State, USA from two General Electric Company (GE) capacitor manufacturing plants between the 1940s and 1970s. Subsequently, extensive, long-term studies of the river showed that PCBs contaminated the river's sediments for many miles downstream. This left concentrations that were still hazardous to ecology and human health several decades later. The US Environmental Protection Agency (USEPA) declared a 197-mile stretch of the river as a Superfund site (i.e. a hazardous waste site, requiring remediation), with dredging to occur on a 40-mile stretch of the Upper Hudson River. Remediation by GE, overseen by USEPA, started in 2009; 2.8×10^5 m^3 of contaminated sediment (containing 18 t of PCBs) were dredged and transported by rail to a permitted out-of-state landfill. After an evaluation of this activity, a second phase of the remediation project began in 2011, to dredge and treat nearly 2×10^6 m^3 of sediment and remove 87 t of PCBs.

Information from USEPA (personal communication).

Figure 9.11 PCB-contaminated sediments are dredged from the Hudson River, New York State, USA.

Credit: United States Environmental Protection Agency (www.epa.gov).

Dioxins and furans

The terms 'dioxins' and 'furans' refer to two categories of polychlorinated hydrocarbons; dioxins are a group of 75 polychlorinated dibenzo-p-dioxins (PCDDs) and furans are 135 polychlorinated dibenzofurans (PCDFs). The term 'dioxin' is often also applied to the most toxic and most studied of all the dioxins, which is 2,3,7,8-tetrachlorodibenzo-p-dioxin (2,3,7,8-TCDD). Dioxins and furans are not produced deliberately, but are by-products of the manufacture of other OCs and are sometimes present within them, good examples being PCBs and herbicides like 2,4-D. Therefore, there is the potential for PCDD/PCDF contamination to arise from the use of such herbicides and from the unauthorised release or dumping of OCs such as PCBs. The toxicity of the various dioxins and furans is usually expressed in units of 'toxic equivalent' (TEQ), based on a TEQ of 1 for 2,3,7,8-TCDD.

Dioxins and furans are released to the atmosphere from incineration of Cl-containing wastes, particularly plastics in domestic waste streams. However, emission can be minimised to very low levels in modern incinerators or 'energy-from-waste' (EfW) plants as long as the operating temperature remains above 850°C and sufficient oxygen is present during combustion, conditions that facilitate dioxin and furan destruction by 'cracking' of the molecular structure. In older mass-burn incinerators, this was not always the case, resulting in emission of relatively high levels. Control of particulate matter to minimise dioxins and furans is also vital, as is careful control of the flue gas's cooling rate because they can reform at 200–450°C. As a result of improved technology and practices, atmospheric emissions of dioxins and furans in Europe have decreased by up to 98% between 1990 and 2011 (Fig. 9.12). In the UK, for example, annual emissions have declined from 1037 g to 186 g TEQ, a drop of 83%; however, it is still the third highest emitter of the countries shown (behind Poland, 269 g, and Italy, 242 g). Concentrations decreased particularly after 1997 as a result of improved modelling to ensure appropriate temperature and residence time of the waste and rapid cooling of combustion gases.

Emitted dioxins and furans are ultimately deposited to the Earth's surface, but contamination of soils and waters can also occur more directly. Paper mills have been known to pollute watercourses with dioxins contained in paper bleach effluent. Soils have been contaminated by unregulated waste disposal; for example, in Newcastle, UK incinerator fly ash, containing dioxins, was spread onto paths on vegetable allotments during the 1990s, contaminating the adjacent soil. Follow-up studies found no measurable transfer of dioxins to vegetables, but it was estimated that regular consumption of eggs laid by chickens on the allotments would have measurably increased the body burden of dioxins and furans.

Some of the best known and well-studied examples of dioxin pollution are related to industrial accidents, particularly an incident that occurred at Seveso, Italy (Box 9.2). Other notable incidents of dioxin pollution include those at Times Beach, Missouri and Love Canal, New York, both in the USA. More recently, there have been several incidents in Europe relating to dioxin contamination of food via animal feed that has been contaminated, either by inappropriate mixing of industrial fatty acids (used in paper processing, for example) with vegetable fats, or by the drying of foodstuffs using PCB-contaminated fuel oil.

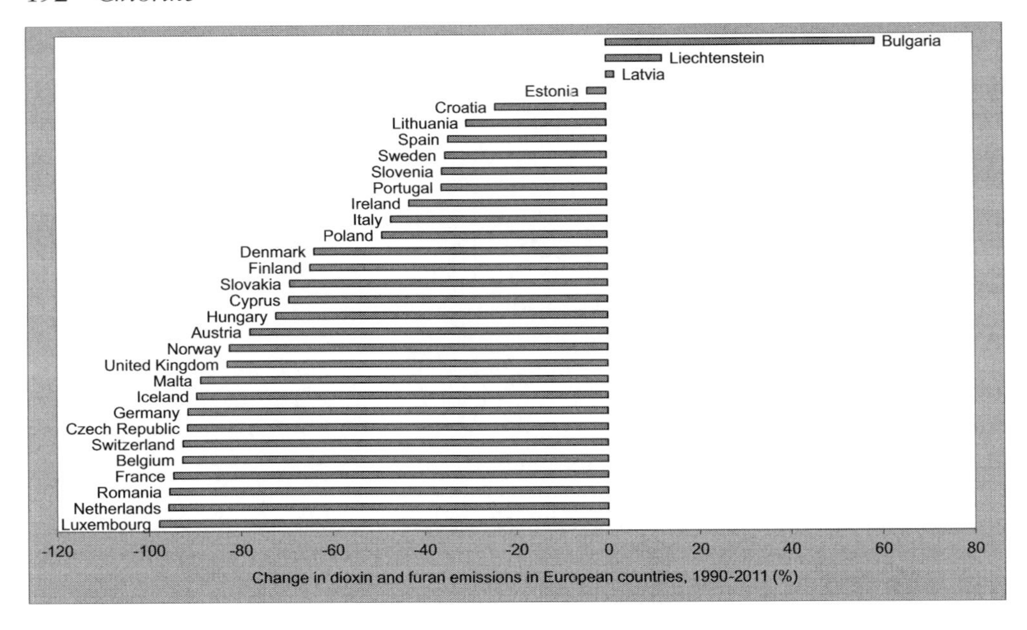

Figure 9.12 Percentage change in dioxin/furan emissions in European countries, 1990–2011.
Source: Data from EEA (2013).

Box 9.2 Dioxin pollution incident, Seveso, Italy.

In 1976, an explosion occurred in a 2,4,5-trichlorophenol reactor in a pesticide manufacturing plant in the town of Meda, Italy and the resulting leak of dioxin affected a 15 km^2 area inhabited by 37,000 people in 11 villages, Seveso being the village most badly affected by the pollution. Animals died and more than 600 residents were evacuated from their homes. Approximately 2000 people were treated for dioxin poisoning, with around 200 suffering chloracne, a clearly visible symptom affecting the skin. Long-term health studies at Seveso continue to this day and, despite difficulties caused by a lack of appropriate exposure assessments, increases in cancer rates and reproductive effects are reported.

The human effects of short-term exposure to dioxins and furans are unclear apart from chloracne and, possibly, altered liver function. The World Health Organization states that long-term chronic exposure has been linked to disorders of the immune system, central nervous system and endocrine system and to reproductive function. Very low LC_{50} values have been reported in animal tests; i.e. values that are much lower than non-lethal concentrations in humans. There is also evidence of carcinogenicity to animals.

CFCs and stratospheric ozone depletion

Normal stratospheric ozone concentrations vary widely, both spatially, on a global scale, and temporally, over the seasons of the year. Therefore, only large-scale ozone depletion is particularly noticeable. However, in the southern hemisphere spring of 1982, following several decades of CFC use, ozone depletion was first recorded over Antarctica. This was so unexpected at the time that the monitoring scientists initially thought their measurements were incorrect. Monitoring of atmospheric ozone concentrations had begun at the British Antarctic Survey's Halley Research Station in the late 1950s, about 10 years after CFCs had first been released to the atmosphere. At this time, the concentration in springtime (September to November in the southern hemisphere), when concentrations above Antarctica are at their minimum, for reasons we will come to, was generally between 280 and 300 Dobson Units (DU[1]). By the 1990s the springtime minimum had decreased to a little over 100 DU (Fig. 9.13), and the term 'ozone hole' was born; the ozone hole is actually defined by a concentration of <220 DU, the lowest concentration recorded before 1979, rather than a complete absence of ozone. For reasons to be explained below, ozone depletion has stabilised since the mid-1990s, and minima over Antarctica since then have been typically 100–140 DU. Stratospheric ozone depletion is important because of the protection O_3 provides to humans and other organisms from damaging ultraviolet (UV) radiation; the O_3 molecule absorbs UV of wavelengths between approximately 200 nm and 310 nm, approximating to parts of the UV-B and UV-C spectra.

Anthropogenic ozone depletion has been caused mainly by human emissions of CFCs, although it should be noted that there are other important ozone-depleting compounds, including brominated hydrocarbons and nitrous oxide (N_2O). In common with many other organochlorine compounds, CFCs are very stable in the environment, including the atmosphere. The C-Cl bonds in these molecules are too strong to be broken down by OH radicals in the troposphere, so CFCs have a long atmospheric

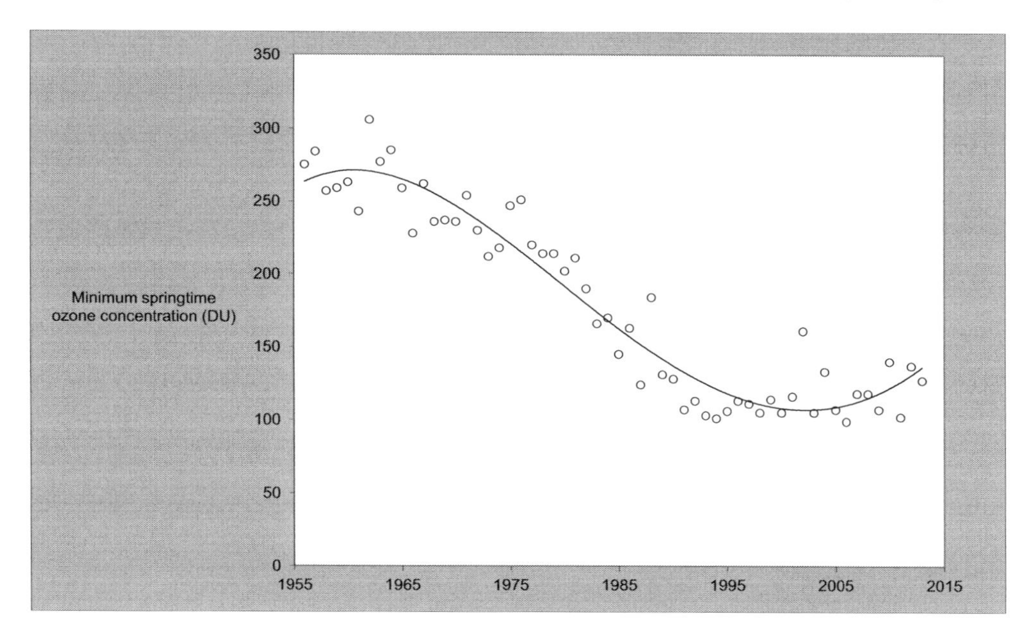

Figure 9.13 Minimum springtime ozone concentrations recorded at Halley Research Station, Antarctica.

Source: Data from British Antarctic Survey (www.antarctica.ac.uk).

lifetime (45 years for CFC-11 and 111 years for CFC-12) and ultimately are able to reach the stratosphere. Once in the stratosphere, the C-Cl bonds can be broken because of the strong UV wavelengths that reach the upper atmosphere. This allows the release of Cl atoms, which then react with stratospheric ozone to form chlorine monoxide, ClO:

$$Cl + O_3 \rightarrow ClO + O_2 \qquad [9.10]$$

The ClO produced can react with monatomic oxygen, which is present in the stratosphere because of the breakdown of O_2 by the strong UV. This releases the Cl atom again (equation 9.11), which can then go on to deplete more ozone.

$$ClO + O \rightarrow Cl + O_2 \qquad [9.11]$$

This means that a small amount of CFC in the stratosphere can deplete an appreciable amount of O_3, as one Cl atom can go on to destroy an estimated 1×10^5 molecules of O_3 before it is removed, for example by reaction with methane, CH_4. The chemistry of O_3 depletion is actually a little more complicated than the above reactions would suggest. Nitrogen dioxide (NO_2) in the stratosphere is normally able to react with ClO and therefore reduce the rate of ClO-induced O_3 depletion. However, in the very low temperatures of the dark polar winter, some of the NO_2 is taken out of the atmosphere's gaseous phase by its condensation to solid $HNO_3 \cdot 3H_2O$ particles which, together with ice crystals, form polar stratospheric clouds. These cloud particles are also the location for a reaction between two other Cl-containing compounds, hydrogen chloride (HCl) and chlorine nitrate ($ClONO_2$), and this reaction, replicated many times, produces a store of Cl_2 molecules:

$$HCl + ClONO_2 \rightarrow Cl_2 + HNO_3 \qquad [9.12]$$

It is worth noting that the two reactants in the above equation are formed from the following precursor reactions:

$$Cl + CH_4 \rightarrow HCl + CH_3 \qquad [9.13]$$

$$ClO + NO_2 \leftrightarrow ClONO_2 \qquad [9.14]$$

Very low atmospheric temperatures are required for polar stratospheric clouds (and thus for equation 9.12), which is why the main regions of anthropogenic O_3 depletion are over the polar regions. The cold conditions are particularly pronounced over the cold land mass of Antarctica, which gives rise to a stratospheric vortex that is isolated from warmer air at higher latitudes. In comparison, the Arctic has no land mass at the pole and has a less stable stratospheric vortex that gives more potential for mixing with air from higher latitudes.

When the polar spring arrives and UV reappears above the horizon, the accumulated Cl_2 molecules on the stratospheric cloud particles (equation 9.12) are split into free Cl atoms.

$$Cl_2 + h\nu \rightarrow 2Cl \qquad [9.15]$$

This sudden liberation of Cl atoms means that O_3 depletion above the poles (equation 9.10) can be rapid and significant. The depletion is fairly short-lived however; the gradual warming of the stratosphere also leads to the breakdown of the vortex so that warmer air arrives from lower latitudes, carrying with it air that is relatively rich in O_3, together with fresh NO_2 to react with any remaining ClO (equation 9.14).

The main concern relating to stratospheric ozone depletion is the decreased capacity of the ozone layer to protect us from the high-frequency, short-wavelength forms of UV which are capable of damaging DNA and causing malignant melanomas (skin cancer). The 'ozone hole' is limited to the largely uninhabited Antarctic (Fig. 9.14), but the prospect of ozone depletion spreading to inhabited parts of the globe (particularly southernmost parts of South America and Australasia) has been the source of particular concern (Fig. 9.15). There are also important effects of ozone depletion on global climate

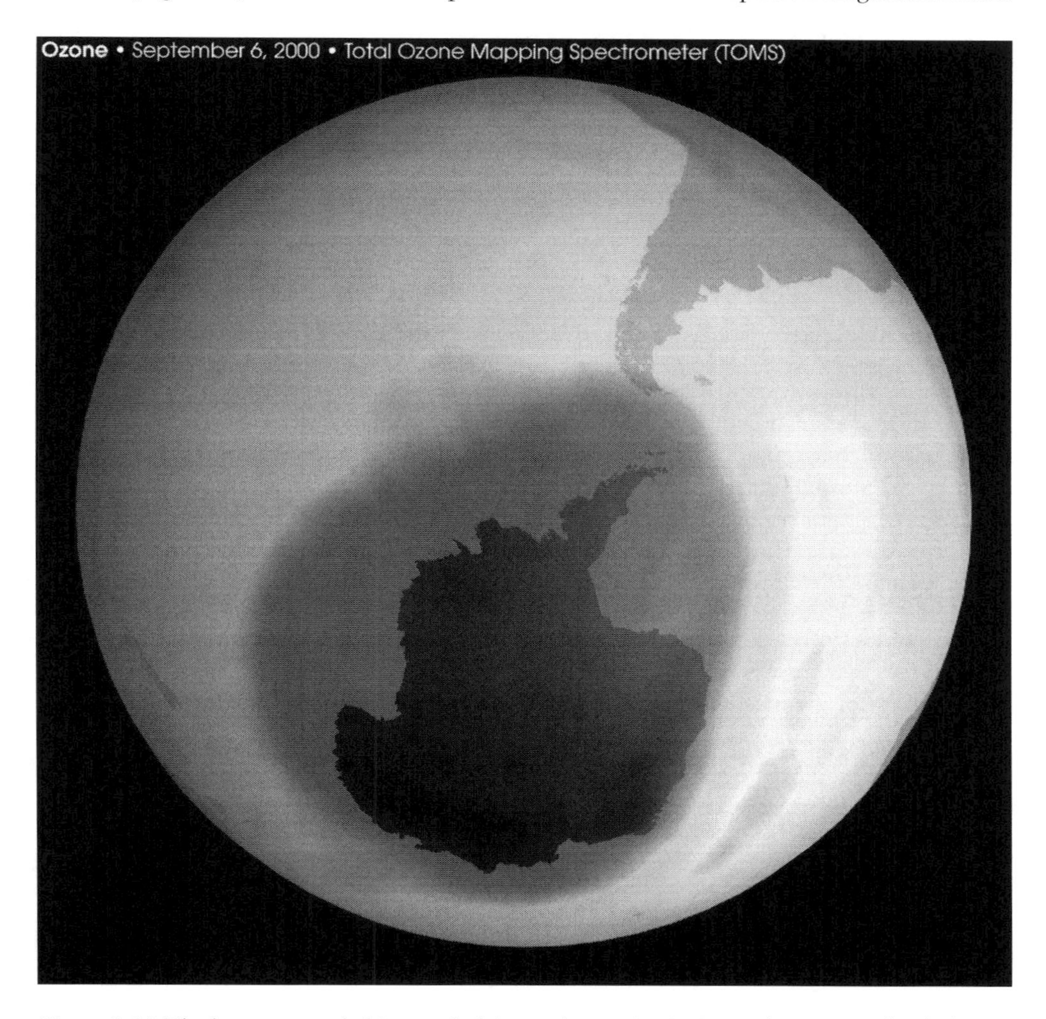

Figure 9.14 The largest recorded 'ozone hole' over Antarctica in September 2000. The dark grey area covering the whole continent corresponds to ozone concentrations of <220 DU. Most of Antarctica is still covered by such an area in springtime each year.

Credit: NASA (Wikimedia Commons).

Figure 9.15 Campaigns to educate the public against the dangers of sunburn became more prevalent with the onset of stratospheric ozone depletion.

Credit: US Government (Wikimedia Commons).

because of the changes in overall UV absorption by the atmosphere; furthermore, CFCs and their replacements (see below) are potent greenhouse gases.

International agreements, particularly the first Montreal Protocol of 1987, led to the replacement of ozone-depleting CFCs. The first replacements were hydrochlorofluorocarbons (HCFCs), which have less Cl and more of the weaker C-H bonds, thus decreasing their atmospheric lifetime as they are more readily broken down in the troposphere. HCFCs have only 1–12% of the ozone-depleting potential of CFC-11. Subsequently, hydrofluorocarbons (HFCs) were introduced, with no Cl and therefore no ozone-depleting potential. Under the Protocol, production and consumption of CFCs (and that of the OC solvent, CCl_4) were banned in developed countries in 1996 and developing countries in 2010. However, the long residence times of CFCs, and the potential for further releases from existing products (e.g. refrigerators) have to be borne in mind. All countries can currently use HCFCs until 2030, when they will be phased out. HFCs are allowed under the Montreal Protocol (as they have no ozone-depleting potential), but they have significant global warming potentials, particularly as they absorb infra-red radiation (i.e. thermal radiation from the Earth's surface) at wavelengths that are not absorbed by CO_2 and other greenhouse gases. For example, HFC-23 (CHF_3), which has an atmospheric lifetime of 260 years, is 12,400 times more potent as a greenhouse gas, molecule-for-molecule, than CO_2 (over a 100-year timescale).

Since the phaseout of CFCs, the Antarctic ozone hole has stopped expanding and is expected to shrink in the coming decades. This success story shows that major pollution problems can be tackled and is sometimes proclaimed as proof that international action could overcome other major environmental problems (particularly global warming). A note of caution is required, however; while it is indeed an encouraging sign, it should be remembered that the phasing out of CFCs was viable because the chemical industry was able to produce (and market) ready replacements and we, as consumers, were not particularly inconvenienced by the switch. Transition to a low carbon economy will be much more difficult.

Note

1 Atmospheric O_3 concentrations are measured by a Dobson spectrophotometer, which measures incoming solar radiation at the wavelengths absorbed by O_3. Concentrations of O_3 are expressed in terms of O_3 abundance in a column of air from the bottom to the 'top' of the atmosphere. 100 Dobson Units equates to a 1-mm thick layer of O_3 if all the O_3 molecules in the measured air column were accumulated at ground level (at standard atmospheric pressure). Although O_3 concentrations are highest in the stratosphere, the term 'ozone layer' is something of a misnomer because the gas is thinly distributed throughout the troposphere and stratosphere.

References

Bastviken, D., Thomsen, F., Svensson, T., Karlsson, S., Sanden, P., Shaw, G., Matucha, M. and Oberg, G. 2007. Chloride retention in forest soil by microbial uptake and by natural chlorination of organic matter. *Geochimica et Cosmochimica Acta* 71, 3182–3192.

Bullister, J. 2014. Atmospheric CFC-11, CFC-12, CFC-113, CCl_4 and SF_6 histories (1910–2014). Carbon Dioxide Information Analysis Centre, Oak Ridge National Laboratory. Available at: www.cdiac.ornl.gov

Collins, S.J. and Russell, R.W. 2009. Toxicity of road salt to Nova Scotia amphibians. *Environmental Pollution* 157, 320–324.

EEA (European Environment Agency). 2013. *European Union Emission Inventory Report 1990–2011 Under the UNECE Convention on Long-Range Transboundary Air Pollution (LRTAP)*. EEA Technical Report 10/2013. Publications Office of the European Union, Luxembourg.

Environment Canada and Health Canada. 2001. *Canadian Environmental Protection Act Priority Substances List Assessment Report: Road Salts*. Minister of Public Works and Government Services, Canada.

Graedel, T.E. and Keene, W.C. 1996. The budget and cycle of the Earth's natural chlorine. *Pure and Applied Chemistry* 68(9), 1689–1697.

Hamilton, J.T., McRoberts, W.C., Keppler, F., Kalin, R.M. and Harper, D.B. 2003. Chloride methylation by plant pectin: an efficient environmentally significant process. *Science* 301, 206–209.

Khalil, M.A.K. and Rasmussen, R.A. 1999. Atmospheric methyl chloride. *Atmospheric Environment* 33(8), 1305–1321.

Kidd, K.A., Bootsma, H.A., Hesslein, R.H., Muir, D.C.G. and Hecky, R.E. 2001. Biomagnification of DDT through the benthic and pelagic food webs of Malawi, East Africa: importance of trophic level and carbon source. *Environmental Science and Technology* 35, 14–20.

Longnecker, M.P., Klebanoff, M.A., Zhou, H. and Brock, J.W. 2001. Association between maternal serum concentration of the DDT metabolite DDE and preterm and small-for-gestational-age babies at birth. *The Lancet* 358, 110–114.

Richardson, J.R., Roy, A., Shalat, S.L., von Stein, R.T., Hossain, M.M., Buckley, B., Gearing, M., Levey, A.I. and German, D.C. 2014. Elevated serum pesticide levels and risk for Alzheimer's Disease. *JAMA Neurology* 71(3), 284–290.

Schlesinger, W.H. and Bernhardt, E.S. 2013. *Biogeochemistry: An Analysis of Global Change*. Academic Press, Waltham, USA.

Svensson, T., Lovett, G. and Likens, G.E. 2012. Is chloride a conservative ion in forest ecosystems? *Biogeochemistry* 107(1–3), 125–134.

Tabazadeh, A. and Turco, R.P. 1993. Stratospheric chlorine injection by volcanic eruptions: HCl scavenging and implications for ozone. *Science* 260, 1082–1086.

10 Chromium

Chromium in the natural environment

There are several oxidation states of chromium (Cr); the commonest form in the natural environment is the trivalent Cr^{3+}, which forms the most stable compounds, but hexavalent Cr^{6+} is of importance because of its relative mobility, bioavailability, oxidising potential and toxicity. The metallic form, Cr^0, is not found in nature. Concentrations of Cr in acidic-igneous, acidic-metamorphic and sedimentary rocks are <100 mg Cr kg^{-1}. The main Cr mineral is chromite ($FeCr_2O_4$), in which Fe can be replaced by Mg. Chromite has a high melting point and crystallises directly from magma; therefore concentrations of Cr in basic and ultrabasic rocks such as serpentinite can reach 2000–3000 mg kg^{-1}. The other mineral of note is crocoite (lead chromate; $PbCrO_4$), which was once the main source of the element but is quite rare. Chromium is present as a minor constituent of other minerals, where it can replace Fe^{3+} and Al^{3+}; in this context it gives rubies and emeralds their colours and, in fact, the element gets its name from the Greek 'chroma' (colour).

A typical mean concentration in soils is 50 mg kg^{-1} but levels in excess of 500 mg kg^{-1} may be observed in serpentinite areas, where the range of plants is restricted to those that can tolerate the elevated levels of Cr (Fig. 10.1 and Table 10.1). Most soil Cr is present as Cr^{3+} compounds in mineral particles and insoluble oxyhydroxides or as strongly adsorbed Cr^{3+}; in the trivalent form it is insoluble and essentially immobile in soils, except in very acid conditions or where complexed with dissolved organic matter. The hexavalent form, Cr^{6+}, exists mainly within the chromate anion, CrO_4^{2-}; this anion is mobile in acid and alkaline soils and waters but is typically reduced to insoluble Cr^{3+} forms by organic matter and also, in anoxic conditions, by S and Fe. The stable Cr^{3+} forms are not typically oxidised to Cr^{6+}. For these reasons, concentrations of dissolved Cr (mainly free Cr^{3+} and aqueous Cr hydroxide species) in unpolluted freshwaters are typically low and lower still in unpolluted seawaters (Table 10.2). Most Cr in water bodies is contained in suspended and bottom sediments at concentrations approximating those of soils. Windblown soils, and to a lesser extent volcanic emissions and forest fires, transfer particulate Cr into the air, but the atmosphere is not a significant reservoir of the element.

Most organisms, including bacteria, plants and animals, take up Cr mainly in the relatively rare but bioavailable Cr^{6+} form. It is typically reduced within organisms to the less toxic, and less mobile, Cr^{3+} form, which can accumulate in tissues. The element is essential for glucose, triglyceride and/or cholesterol metabolism in some micro-organisms and animals, including mammals; in the context of glucose metabolism, Cr

Figure 10.1 Kynance Cove, Lizard Peninsula, southwest England. The unusual geology of this area is based on Cr-rich serpentinite and is home to rare plants that are tolerant to the elevated levels of the metal in the soils (see Table 10.1).

Credit: zooK2 (Flickr).

Table 10.1 Naturally elevated chromium concentrations in serpentinite soils of the Lizard Peninsula, southwest England. Note how the concentrations increase in the deeper, mineral layers of the soil profile, underneath the upper, organic horizons.

Soil depth (cm)	*Cr concentration (mg kg^{-1})*
0–9	50
9–17	78
17–28	45
28–45	136
45–75	354
75–100	380
100–130	988
130–160	1056

Source: Unpublished data belonging to the author.

Table 10.2 Typical background concentrations of chromium.

Environmental medium	*Typical background concentrations*[a]
Air	<10 ng m^{-3}
Soil	<500 mg kg^{-1}
Vegetation[b]	<0.5 mg kg^{-1}
Freshwater	<10 µg L^{-1}
Sea water	<1 µg L^{-1}
Sediment	<80 mg kg^{-1}

a These are *typical* values, encompassing the vast majority of reported concentrations, based on a large number of literature sources. Higher natural concentrations occur in some cases, for example, in mineralised areas.

b Including crops. Not including metal-tolerant species.

facilitates insulin action and deficiency may cause mild diabetes. There is no evidence of essentiality in plants.

Production and uses

Production of Cr has increased significantly during the last century and continues to do so. Most Cr resources are located in South Africa and Kazakhstan and of the 20–25 Mt of Cr extracted annually, roughly half is mined in South Africa; recycled Cr, mainly from scrap steel, is also used (USGS, 2014). The major use of Cr, a hard and corrosion-resistant metal, is in steel. Stainless steel and superalloy, which is used in high-powered engine turbines, both contain up to 25% Cr. Chromium is electroplated onto metal surfaces for corrosion resistance and decoration (Fig. 10.2), using chromate or dichromate ($Cr_2O_7^{2-}$), solutions containing hexavalent Cr. Other uses include: (i) the production of pigments for paints, plastics, glass and cosmetics (mainly green chromium(III) oxide, Cr_2O_3 and yellow lead chromate, $PbCrO_4$); (ii) tanning, where Cr(III) sulfate, $Cr_2(SO_4)_3 \cdot 12(H_2O)$, is added to provide stronger, more resistant leather; (iii) production of refractory bricks and foundry sand, utilising the heat resistance of Cr(III) oxide; (iv) catalysis in the chemical industry; and (v) dyeing of textiles. Chromium is also present in miscellaneous products such as chromated copper arsenate (CCA) wood preservative.

Pollution and environmental impacts

Chromium pollution

Industrial effluent discharged to rivers can contain high levels of Cr. Tanneries in particular discharge wastewater containing >1 mg Cr L^{-1}, and much more than this in some countries; for example, up to 5000 mg L^{-1} in some parts of India (Mandal et al., 2011). This can result in Cr concentrations in rivers of >100 µg L^{-1} (Santonen et al., 2009), at least 10–100 times higher than in unpolluted rivers. In comparison, the WHO guideline for total Cr in drinking water is 50 µg L^{-1}. Groundwaters can also be polluted; for example, in 2013, Cr^{6+} contamination of drinking water wells was reported in the USA, although the source of the Cr was not officially verified (Box 10.1).

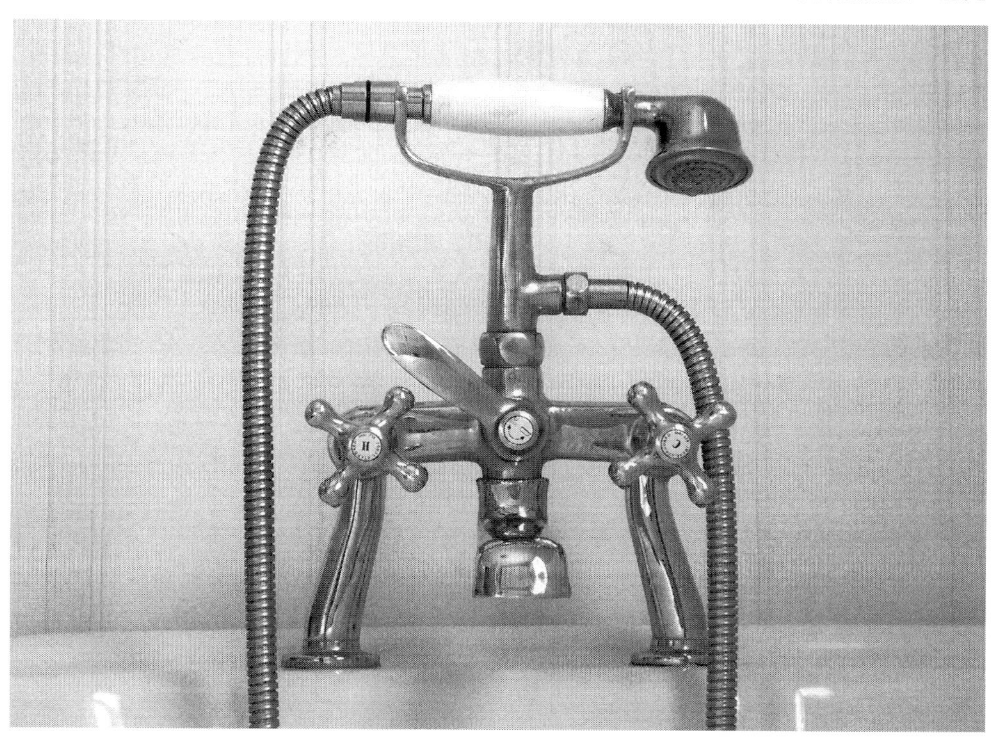

Figure 10.2 A major use of chrome is for decorative plating of objects like bathroom fittings.

Box 10.1 Chromium pollution of drinking waters, USA.

In 2013, state authorities in Texas, USA undertook a detailed investigation of a groundwater plume of Cr, including hexavalent Cr, in the town of Midland (USEPA, 2014). The pollution came to light after a resident informed the authorities that the water from his private well was yellow. Testing showed a *hexavalent* Cr concentration in the water of >5 mg L^{-1}, more than 100 times the WHO recommended guideline for *total* Cr in drinking water (50 µg L^{-1}). Subsequent testing of >200 wells across the neighbourhood revealed a 3.2-km-long contaminant plume, with total Cr concentrations of >100 µg L^{-1} recorded over a 2-km length and in 46 private wells. Anion-exchange filtration systems were installed because the wells are the only source of water supply for the residents. The pollutant source was not officially identified but an Environmental Protection Agency manager was quoted in a local newspaper as saying that a former plastics factory appeared to be the source.

Soils are contaminated by the deposition of atmospheric Cr emitted by smelters; for example, soil concentrations of up to 3000 mg Cr kg^{-1} have been recorded in the vicinity of nickel smelters at Sudbury, Canada. Application to agricultural soils of sewage sludge and phosphate fertilisers, which can contain up to 3000 mg Cr kg^{-1}, also adds Cr to soils; insoluble Cr^{3+} is considered to be the main form present, diminishing the risk of leaching or plant uptake. Other sources of soil contamination include the leaching of Cr from CCA-treated wood and leaks or spills of liquid wastes containing Cr^{6+}. In the latter case, some or all of the Cr^{6+} is reduced to Cr^{3+} forms over time, but usually only if the soil contains sufficient organic matter. An important additional source of soil contamination is the disposal of tannery sludges (Box 10.2).

Box 10.2 Contamination of soils by tannery sludge.

In the tanning process, Cr sulfate is not wholly absorbed by treated hides, leaving a large amount in waste tanning solution that has to undergo wastewater treatment, generating Cr-rich sludge. One kilogram of tannery sludge can hold as much as 30 g Cr, at least some of which will be present as toxic hexavalent Cr. Land application of tannery sludge is practised in the USA, Mexico, India, Italy and other countries, and in some areas it is a widespread practice, being seen as a convenient disposal method that adds organic matter and fertility to soils at low cost; the conventional wisdom is that the Cr is present as immobile and less toxic Cr^{3+} because of reduction of any Cr^{6+} present by the organic matter making up the sludge. An investigation in Missouri, USA found that ongoing sludge disposal resulted in contamination of farm soils and private wells by Cr^{6+}, but at concentrations below US Environmental Protection Agency guideline levels (Missouri Department for Natural Resources, 2010). Other studies have reported that up to two-thirds of Cr in sludge-treated soils is mobilisable.

Atmospheric concentrations of Cr in urban areas are often 1–3 orders of magnitude higher than background levels and are higher still in some industrial settings. Anthropogenic sources include refractory brick production, coal combustion, electroplating facilities and electric arc furnaces in steel works. In the USA, Europe and elsewhere, modern pollution control technologies have significantly reduced atmospheric levels of Cr compared with previous times. In such areas, burning of CCA-treated wood is an increasingly important source. Additional sources in urban areas are the brake linings and catalytic converters of petrol vehicles, which add Cr to road dusts, kerbside soils and urban runoff.

Human exposure and health impacts

For most human populations, total exposure to Cr from air, food and drinking water is <0.1 mg per day and exposure to the bioavailable and toxic Cr^{6+} form is a fraction of this. Dermal exposure can occur via contact with contaminated bathing water, CCA-treated wood or Cr-tanned leather; Cr-induced dermatitis has been attributed to skin contact with leather (e.g. sandals) that contains approximately 3% Cr. Occupational exposure to airborne Cr^{6+} has been found to cause lung cancer, associated with accumulation of Cr-containing particles in the bronchioles. Chromosomal abnormalities have also been recorded in workers exposed to Cr^{6+}, as have allergic reactions and the appearance of small pits in the skin ('chrome ulcers').

In polluted areas the risks of harmful exposure to Cr are increased, via ingestion of contaminated drinking water for example (Boxes 10.1 and 10.3). Ingested Cr^{6+} is readily reduced to the less toxic Cr^{3+} in the stomach but, prior to reduction, is easily transported through the cellular membrane allowing the possibility of damage within cells. For example, significantly higher levels ($p < 0.001$) of DNA-protein crosslinks (which can block DNA replication) were found in residents of New Jersey, USA, who were at risk of exposure to Cr mine wastes and who had relatively high urinary Cr compared with a control population (Taioli et al., 1995).

Ecotoxicity

In micro-organisms, Cr exposure can have population effects and, at the individual level, impacts on cell metabolism; in algal cells, growth and photosynthesis can be affected. In plants, Cr causes oxidative stress, affects photosynthesis and impairs enzymatic processes, including nitrate reduction. Symptoms of Cr toxicity in plants include decreased uptake of major nutrients, chlorosis, wilting and root damage. A study by López-Luna et al. (2009) showed that seed germination and root growth in wheat, oat and sorghum plants were affected by increasing additions of tannery sludge to soils (Fig. 10.3).

Chromium is bioaccumulative in some animals. Bioconcentration factors of 10–1000 are generally accepted in some marine species and factors of several thousand have been recorded in *Daphnia* spp. Fish exposed to Cr in polluted effluents have suffered gill damage, with possible effects on respiration and osmoregulation; exposure also affects fish behaviour and growth (e.g. Fig. 10.4). Certain invertebrate species,

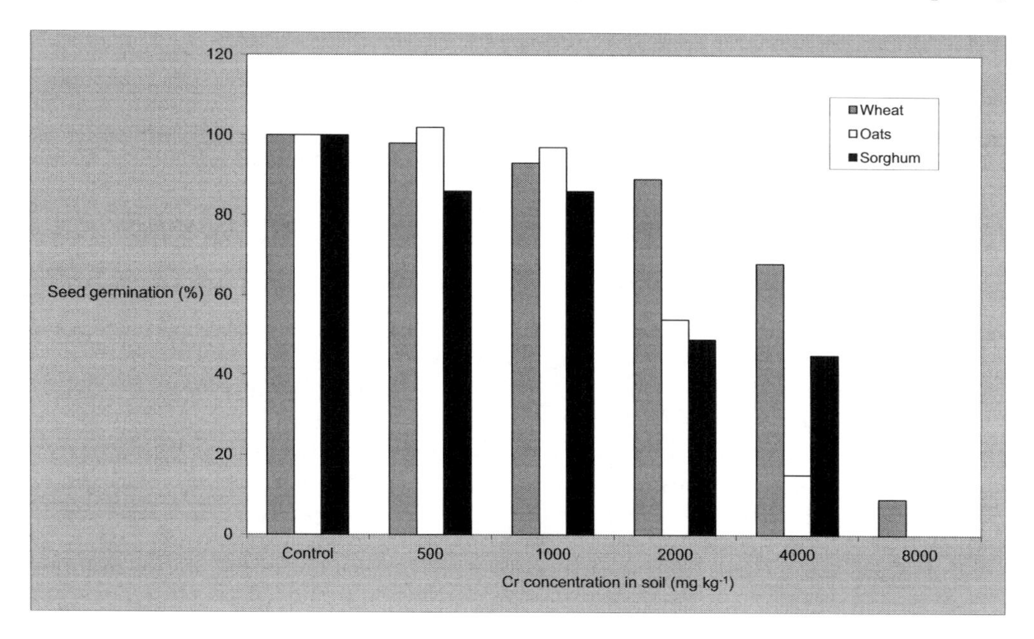

Figure 10.3 Effects on seed germination of increasing chromium concentrations in soils via additions of tannery sludge.

Source: Adapted from López-Luna et al. (2009).

including some snails, larvae and crustaceans, are similarly sensitive to Cr pollution; possible effects include reduced reproductive capacity. Livestock have been affected by grazing in fields amended with Cr-rich sewage sludge and by drinking contaminated water (see Box 10.3).

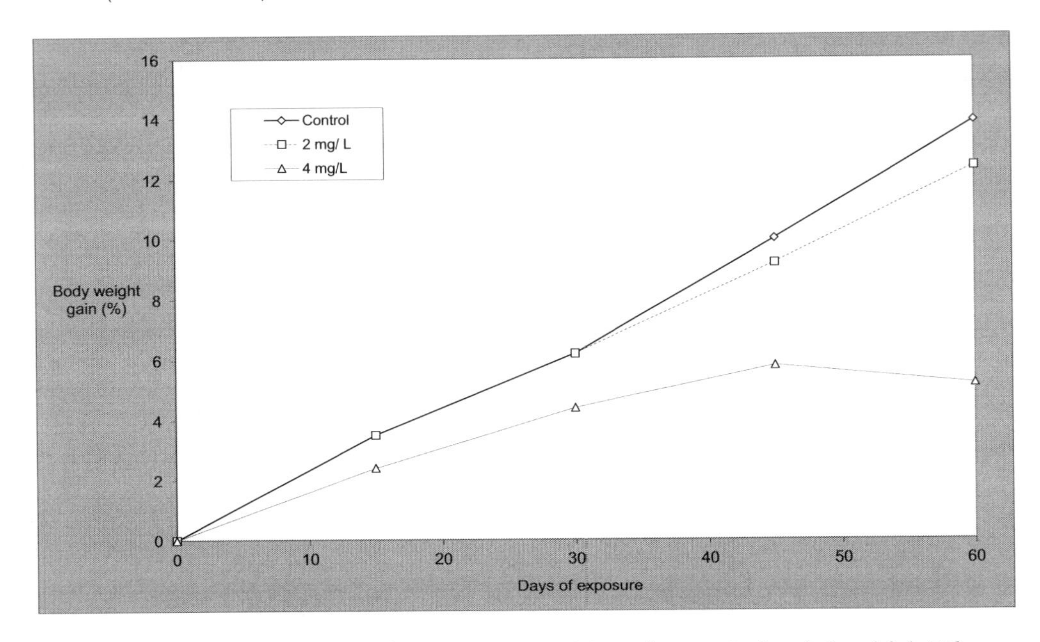

Figure 10.4 Body weight gain in three groups (n = 10 in each group) of snakehead fish (*Channa punctata*) exposed to different concentrations of Cr^{6+} (potassium dichromate, $K_2Cr_2O_7$) for 60 days. After 60 days, there was a significant ($p < 0.05$) reduction in growth in the group exposed to 4 mg L^{-1} compared with the control group. Behavioural abnormalities in the Cr-exposed groups included lethargy and convulsions. Damage to gills, kidneys and livers was also reported.

Source: Adapted from Mishra and Mohanty (2009).

Box 10.3 Chromium contamination of drinking water, China.

In 2011, Chinese news outlets reported hexavalent Cr contamination of public water supplies in Yunnan province. The source of the pollution was several thousand tonnes of waste tailings from the production of the tanning chemical chromium sulfate, which had been illegally dumped on wasteland. Rainwater percolated through the waste and drained into the nearby water supply reservoir, resulting in concentrations of Cr^{6+} in drinking water that were 200 times higher than the official quality standards. More than 70 goats and pigs died after drinking from a local pond that contained higher levels still. The growth of locally planted rice and other crops was impaired. Damage to human health was more difficult to ascertain. The authorities first covered the waste to prevent further rainwater infiltration, installed a ditch to collect remaining drainage water and then treated contaminated waters by converting the toxic Cr^{6+} to

Cr^{3+}, using the reducing agent, sodium metabisulfite ($Na_2S_2O_5$). The chemical company was ordered to remove the waste and contaminated soil to a controlled disposal facility and production at the plant was halted. In 2012, the company was fined 0.3 million yuan and seven employees and sub-contractors of the company were given prison sentences; subsequently two non-governmental organisations sued for compensation of 10 million yuan.

References

López-Luna, J., González-Chávez, M.C., Esparza-Garcia, F.J. and Rodríguez-Vázquez, R. 2009. Toxicity assessment of soil amended with tannery sludge, trivalent chromium and hexavalent chromium, using wheat, oat and sorghum plants. *Journal of Hazardous Materials* 163, 829–834.

Mandal, B.K., Vankayala, R. and Kumar, L.U. 2011. Speciation of chromium in soil and sludge in the surrounding tannery region, Ranipet, Tamil Nadu. *ISRN Toxicology* 2011, 1–10. International Scholarly Research Network.

Mishra, A.K. and Mohanty, B. 2009. Chronic exposure to sublethal hexavalent chromium affects organ histopathology and serum cortisol profile of a teleost, *Channa punctatus* (Bloch). *Science of the Total Environment* 407, 5031–5038.

Missouri Department for Natural Resources. 2010. *Soil and Water Sampling for Hexavalent Chromium in Northwest Missouri*. Hazardous Waste Program Factsheet 10/2010. Available at: www.dnr.mo.gov

Santonen, T., Zitting, A. and Riihimaki, V. 2009. *Concise International Chemical Assessment Document 76: Inorganic Chromium(III) Compounds*. World Health Organization. Available at: http://www.inchem.org/documents/cicads/cicads/cicad76.pdf

Taioli, E., Zhitkovich, A., Kinney, P., Udasin, I., Toniolo, P. and Costa, M. 1995. Increased DNA-protein crosslinks in lymphocytes of residents living in chromium-contaminated areas. *Biological Trace Element Research* 50(3), 175–180.

USEPA (United States Environmental Protection Agency). 2014. Site update. Available at: www.epa.gov/region6/6sf/pdffiles/west-county-road-112-tx.pdf

USGS (United States Geological Survey). 2014. *Mineral Commodity Summaries: Cadmium*. Available at: http://minerals.usgl.gov/minerals

11 Copper

Copper in the natural environment

The main oxidation state of copper (Cu) in the environment is Cu^{2+} (cupric Cu); Cu^0 (native metal) and Cu^+ (cuprous Cu) are relatively rare and Cu^{3+} is much rarer, existing only in a few compounds. The average crustal concentration is approximately 25–50 mg kg^{-1}; concentrations in most major rock types are generally <50 mg kg^{-1}, but are higher (>100 mg kg^{-1}) in some basic igneous rocks. The commonest mineral is chalcopyrite ($CuFeS_2$); other important minerals include cuprite (Cu_2O), a secondary mineral, and chalcocite (Cu_2S), which contains up to 80% Cu by mass but is rarer than chalcopyrite, which contains approximately 30% Cu. Some deep ocean ferromanganese nodules contain 4000 mg kg^{-1} Cu.

Copper minerals are quite soluble, particularly under acidic weathering conditions, releasing Cu^{2+} and $CuOH^+$ ions into soil solution, the latter dominating in neutral and alkaline soils. Free ionic Cu binds strongly to soil particles, sometimes by specific adsorption, and therefore becomes less mobile in soils than many other toxic metals. Most Cu in soils is associated with soil organic matter (OM), particularly with the carboxylate functional group, COO^-, and most Cu in soil solutions, both acid and alkaline, exists in the form of dissolved Cu–organic complexes (chelates); organic complexes with a low molecular weight are likely to be taken up readily by plants along with the aqueous inorganic forms. Most dissolved Cu in freshwaters exists in organic complexes, but rivers also carry Cu in the particulate phase and, in fact, most Cu in water bodies is present in bottom sediments. In saline waters, the main inorganic species of Cu include hydroxides and carbonates. Volcanoes and mineral dust contribute some particulate Cu to the atmosphere. Typical background concentrations of Cu are shown in Table 11.1

Copper is an essential element in micro-organisms, plants and animals, and Cu deficiency impairs organism health. In plants, it activates enzymes, confers disease resistance and has roles in photosynthesis, respiration and nutrient metabolism. In animals and humans, it is required for growth, iron transport, neurological development, metabolism of cholesterol and glucose, and various other vital functions. In some arthropods and molluscs, Cu atoms carry oxygen around the body in haemocyanin, in the same way as iron in haemoglobin in other animals (Fig. 11.1).

Production and uses

Global reserves of Cu are 680 Mt and approximately 17 Mt are mined annually; most Cu is extracted in Chile, but considerable amounts are also mined in China, Peru and

Table 11.1 Typical background concentrations of copper.

Environmental medium	*Typical background concentrations*[a]
Air	<1 ng m^{-3}
Soil	<150 mg kg^{-1}
Vegetation[b]	<30 mg kg^{-1}
Freshwater	<2 μg L^{-1}
Sea water	<0.5 μg L^{-1}
Sediment	<60 mg kg^{-1}

a These are *typical* values, encompassing the vast majority of reported concentrations, based on a large number of literature sources. Higher natural concentrations occur in some cases; for example in mineralised areas.
b Including crops. Not including metal-tolerant species.

Figure 11.1 Horseshoe crabs (Fam. *Limulidae*), in common with some molluscs and crustaceans, have copper-containing haemocyanin in their blood giving it a blue colour.

Credit: Bill Perry, US Fish and Wildlife Service (Wikimedia Commons).

the USA (USGS, 2014). Recycled Cu is also utilised. Copper is used in the production of alloys such as bronze, brass and Cu-Ni alloys, and the main end uses are electrical wiring (Fig. 11.2), plumbing, roofing, machinery and electronics. Copper-bottoming of ships was the original antifouling technology and Cu is still used in antifouling paints. It has also long been used as a fungicide in orchards, vineyards and gardens, particularly as Bordeaux Mixture (containing Cu sulfate) and as an algicide in reservoirs and ponds. Copper is added to animal feed to counteract deficiency, typically where soils are low in Cu or high in molybdenum, which is antagonistic to

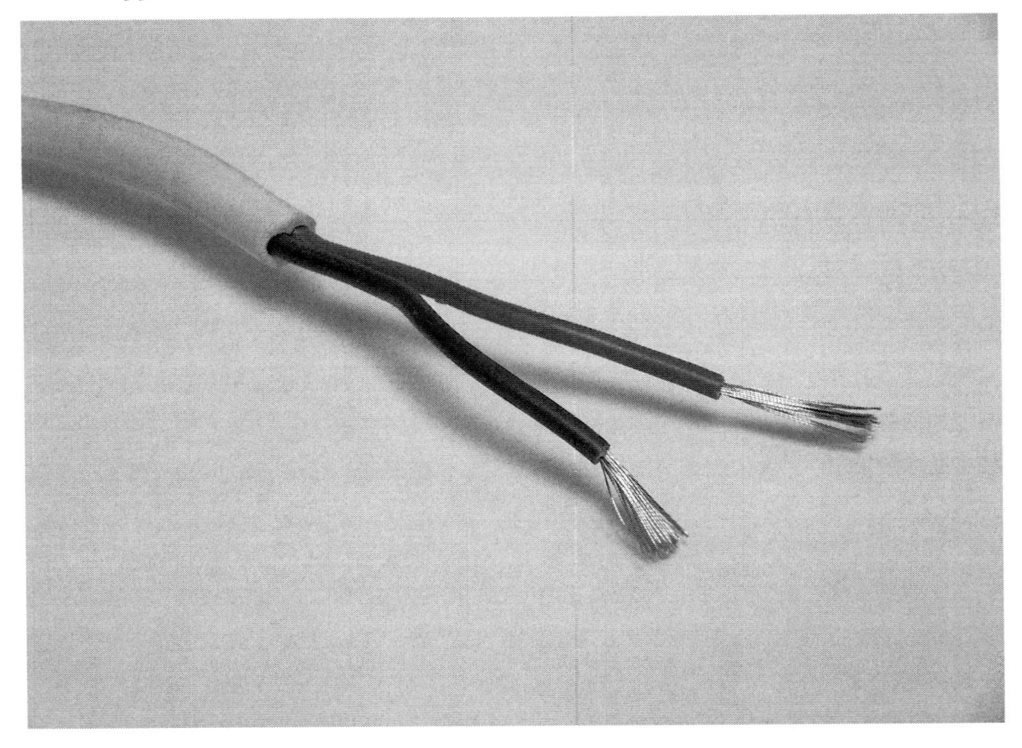

Figure 11.2 Copper wires in electrical flex.

Cu and reduces its uptake in the body; however, over-supplementation may lead to toxicosis in cattle.

Pollution and environmental impacts

Copper pollution

Emissions of Cu to the atmosphere are mainly from non-ferrous metal smelting and refining; concentrations well in excess of 1 μg m^{-3} have frequently been recorded near Cu smelters. Other sources include fossil fuel combustion, steel works, waste incineration and, on urban streets, vehicles. As with other metals, contamination of polar ice indicates global transport of anthropogenic emissions to the atmosphere.

Atmospheric deposition is an important source of Cu contamination of soils and waters in polluted areas. Elevated soil concentrations of up to 6.9 g kg^{-1} have been recorded near metal smelters (Table 11.2). In areas of excessive deposition, Cu is usually elevated in the organic surface horizons, indicative of its presence in decaying plant materials and its association with soil OM. Horticultural soils receive excess Cu from the use of pesticides containing Cu sulfate, particularly in vineyards, orchards and tea plantations. Sewage sludges containing >1.5% Cu have been reported (Baker, 1990) and application of Cu-rich pig and poultry manure to agricultural soils is also practised in some areas. Despite specific adsorption of Cu in soils and its close association with

Table 11.2 Copper concentrations in contaminated soils.

Contamination source	*Location*	*Concentration mean or range (mg kg^{-1})*
Former mines	United Kingdom	13–2000
Metal smelting/processing	Belgium	16–1089
	Canada	1400–3700
	Canada	6912
	Russia	121–4622
Gardens, orchards, parks	Japan	31–300
	Poland	12–240
Agriculture	Poland	80–1600
	Germany	273–522

Source: Data from Kabata-Pendias (2001) and USDHHS (2004).

solid OM, sufficient mobile Cu may be present in polluted soils to leach into surface and ground waters and be taken up in excess by plants; specific adsorption occurs quite slowly and excess Cu may be leached or taken up from the soil solution before it has time to occur.

A major anthropogenic source of Cu in rivers and estuaries is the wastewater treatment plant, which takes in domestic and, sometimes, industrial wastewater as well as urban runoff; the latter can carry Cu abraded from roofs and emitted by vehicles, for example. Another important source in some areas is mining effluent, both from underground drainage and rainwater percolating through waste heaps. The scale of Cu mining operations is important in this context; chalcopyrite and other Cu ores are finely distributed through host rocks and their extraction generates huge mining operations like those at Chuquicamata Cu mine in Chile and Bingham Canyon Cu mine in Utah, USA (Fig. 11.3), which are among the largest metal mines in the world. As well as significant landscape impacts, Cu mines also generate massive waste piles; however, modern operations may, if tightly regulated, be less significant as sources of Cu in rivers than abandoned and/or unregulated mine wastes. Mine drainage adits, or their receiving streams, carrying excess Cu ions are identifiable by the blue and green discolorations that occur as Cu compounds precipitate as solids on the stream bed. Other sources of Cu in waters include the algicides (mainly copper sulfate pentahydrate) that are applied to ponds and reservoirs and anti-fouling paints, which can pollute ocean and estuarine waters. Concentrations of Cu in polluted river waters can reach well in excess of 1 mg L^{-1}, 2–3 orders of magnitude higher than the lowest observable effect levels for some aquatic species. Similarly, sediments in some polluted areas exhibit gross contamination with concentrations of up to several thousand mg kg^{-1} (USDHHS, 2004).

Human exposure and toxicity

In most human populations, exposure to excess Cu is mainly via ingestion of water contaminated by Cu plumbing; however, Cu levels in drinking waters are rarely above the WHO guideline value of 2 mg L^{-1}, which is primarily set for taste considerations

Figure 11.3 Bingham Copper Mine, Utah, USA; one of the largest mines in the world. The dark specks near the bottom of the pit are huge trucks.

Image source: Spencer Musick (Wikimedia Commons).

rather than health concerns. The use of Cu algicide in waters has been linked with human health effects, however (Box 11.1). In areas with Cu-contaminated soils, exposure may occur via the ingestion of soil and dust particles, particularly by young infants who tend to play on the floor and engage in hand-to-mouth activity. Other potential exposure pathways, for both children and adults in such areas, are the ingestion of home-grown fruits and vegetables and the inhalation of Cu in fine particulates. The human body is able to regulate Cu levels very efficiently, but toxicity can arise as a result of excess Cu intake. Toxicity is caused partly by Cu's catalysis of highly oxidative hydroxyl (OH^-) radicals, which can cause oxidative stress and tissue damage, and partly by its binding to thiol groups, causing protein cross-links that can reduce enzyme activity. Symptoms of Cu toxicity include gastrointestinal, respiratory and liver disorders. A link with neurodegenerative disorders has been posited but not confirmed; in fact, as with some of the other disorders listed above, Cu deficiency is likely to be of more significance. Copper is not thought to be carcinogenic.

Ecotoxicity

Copper is bioaccumulative in some organisms: it bioconcentrates in some plankton, oyster and squid species by factors of up to 10^7, but does not biomagnify in the food chain. Impacts in micro-organisms are manifested by decreases in the mineralisation

Box 11.1 Copper algicide and human health.

The use of Cu algicide was at the centre of a case of human poisoning in Australia in 1979, although the exact cause of illness is still disputed. Copper sulfate was applied to a domestic water supply reservoir in Palm Island, Queensland, to control a bloom of the cyanobacteria, *Cylindrospermopsis raciborskii*. This caused the hospitalisation of 138 consumers, all children, with gastroenteritis and kidney damage. These illnesses were attributed to Cu sulfate splitting the algal cells, causing the release of cylindrospermopsin, a cyanotoxin. However, in more recent times scientists have hypothesised that the illnesses may have been a direct result of Cu sulfate consumption by the children.

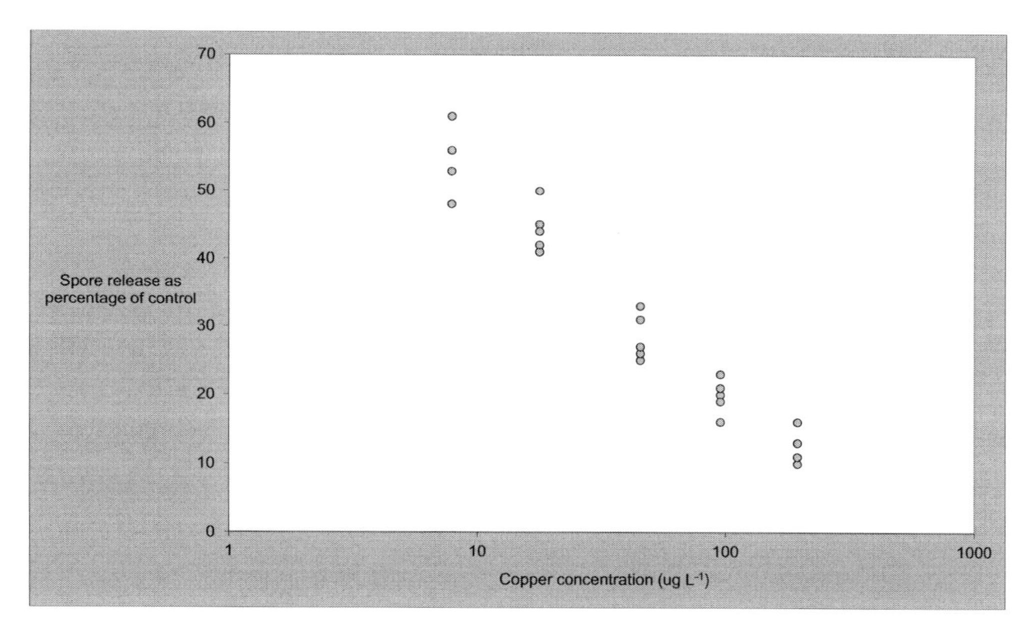

Figure 11.4 The effect of copper in seawater on spore release from the macroalga, *Lessonia nigrescens*, a kelp organism. Copper ($CuCl_2$) was added to seawater samples from northern Chile in the laboratory to yield a series of increasing concentrations; 4 or 5 replicate samples of kelp were exposed to each concentration. Spore release and subsequent spore development (not shown) were both affected by the lowest added Cu level of approximately 8 µg L^{-1}, which is lower than concentrations recorded in the seawater from this area over the previous decade. The authors suggest that the relative scarcity of this species from the study area, and of other organisms that are dependent on it, results from the high sensitivity to Cu of the kelp's early life stages.

Source: Adapted from Contreras et al. (2007).

of OM, P and N; this is observed in sewage effluents with high Cu concentrations, for example. Impacts of Cu contamination have also been observed in microalgae and macroalgae (Fig. 11.4). In plants there is a narrow range of Cu concentrations between deficiency, typically occurring where soil levels are <5 mg kg^{-1}, and toxicity, which may occur at 60 mg kg^{-1} for some sensitive species. Some tolerant plants in contaminated soils accumulate Cu in the roots (away from the aerial parts), a tolerance response called exclusion, while others sequester excess Cu in the oldest leaves, a form of hyperaccumulation. Symptoms of toxicity are leaf chlorosis, following an initial darkening of the leaves, damaged roots and restricted growth. Leaf necrosis in trees has been observed in metal smelting areas. As well as higher plants, species of bryophyte and moss can be adversely affected in Cu-polluted areas.

Effects of excess Cu exposure have been noted in various animal categories, particularly in aquatic organisms. Molluscs, especially bivalves, are sensitive to Cu and are scarce in some polluted estuaries; they can accumulate large quantities of Cu (and Zn) via their ingestion of particulate matter and common species are sometimes absent in mine-polluted catchments, for example. Fish and some aquatic invertebrates are also known to be affected by Cu pollution (Boxes 11.2 and 11.3); fish suffer gill damage, affecting osmoregulation and respiration. Damage to the liver and olfactory organs is also reported in fish, the latter affecting predation avoidance and reproductive success. Copper can also cause endocrine disruption in aquatic organisms. In amphibians, effects include reduced growth and embryonic abnormalities.

Box 11.2 Ecotoxicity of copper revealed by palaeoecology.

Palaeoecologicial techniques show the significant damage that can be inflicted on aquatic organisms by Cu pollution. Lake Orta, Italy received Cu effluents from a rayon factory from approximately 1930 to 1960. Lake sediments dating from this time show high Cu levels compared with older sediments, and fossil evidence in the same sediments shows that several common diatom and invertebrate species were quickly killed by the Cu pollution and that others were deformed or more limited in size. A detailed account of these studies appears in Smol (2008).

Box 11.3 Fish kill resulting from copper pollution, China.

In July 2010, a chemical spill at a Cu mining-smelting complex in southeast China entered a major river, resulting in a significant pollution event. It was reported that ~9000 m^3 of Cu-contaminated acidic waste spilled into the Tingjiang River, killing a large number of fish; it was estimated that 1900 tons of dead fish collected on the river surface. Local fishermen suffered economic losses and the mine-owners were fined 22 million yuan; prison sentences were given to managerial staff.

Effects on animals in the terrestrial environment are generally less pronounced than in waters, but Cu toxicity is observed. The cocoon production rate of earthworms was reduced in soils containing >50 mg Cu kg^{-1} (Eisler, 1998). In some birds, Cu has been found to concentrate in the feathers, although Cu appears to be less of an avian toxicant than some other toxic metals. Copper toxicity is observed in livestock, especially in sheep; the key symptoms displayed are jaundice, reduced reproduction and liver damage. As in plants, Cu tolerance may be displayed by organisms in polluted environments, in some species of algae and polychaete worms, for example.

References

Baker, D.E. 1990. Copper. In: Alloway, B.J. (Ed.). *Heavy Metals in Soils*. Blackie, Glasgow.

Contreras, L., Medina, M.H., Andrade, S., Oppliger, V. and Correa, J.A. 2007. Effects of copper on early developmental stages of *Lessonia nigrescens* Bory (Phaeophyceae). *Environmental Pollution* 145, 75–83.

Eisler, R. 1998. *Copper Hazards to Fish, Wildlife, and Invertebrates: A Synoptic Review*. U.S. Geological Survey, Biological Resources Division, Biological Science Report USGS/BRD/BSR-1998-0002.

Kabata-Pendias, A. 2001. *Trace Elements in Soils and Plants*, 3rd Edition. CRC Press, Boca Raton.

Smol, J.P. 2008. *Pollution of Lakes and Rivers: A Paleoenvironmental Perspective*, 2nd Edition. Blackwell, Malden.

USDHHS (US Department of Health and Human Services: Public Health Service, Agency for Toxic Substances and Disease Registry). 2004. *Toxicological Profile for Copper*. ATSDR, Atlanta.

USGS (United States Geological Survey). 2014. *Mineral Commodity Summaries: Copper*. Available at: http://minerals.usgl.gov/minerals

12 Fluorine

Fluorine in the natural environment

Fluorine (F) is present in the fluoride form (F^-) in all major rock types, typically at concentrations of 50–1000 mg kg^{-1}. The most common minerals are fluoroapatite ($Ca_5F(PO_4)_3$), a component of phosphate rock deposits, and fluorite (or 'fluorspar'; CaF_2), which is most often found in mineral veins (especially of Pb and Ag) but also in limestone beds and around hot springs. Another well-known mineral, cryolite (Na_3AlF_6), is relatively rare and is now made artificially for industrial use. Fluorite and fluoroapatite are quite insoluble and most F is retained in the Earth's crust. Soil concentrations vary quite widely; typically, they are less than 500 mg kg^{-1}, but in areas of mineralisation they can be an order of magnitude higher than this. Most of the F in soils is present in weathered mineral particles, particularly the clay fraction, where F^- substitutes for hydroxyl groups. Fluoride ions released by dissolution may be leached from the soil or, alternatively, held by adsorption and by complexation with Al. Solubility of F in soils is generally limited, but it is higher in alkaline soils because of F^- repulsion by anionic surface charge.

Background concentrations of F in the atmosphere are <2 µg m^{-3}. Fluorine is almost never present in its pure, gaseous form (F_2). Most F present naturally in the atmosphere is likely to be associated with suspended mineral particles but several Mt of fluoride are emitted by volcanoes each year (Fig. 12.1). Volcanoes emit F in the ash and in gases, mainly hydrogen fluoride (HF), and smaller amounts of sulfur tetrafluoride (SF_4). Volcanic HF is either dry or wet deposited, because the gas dissolves readily in water, forming hydrofluoric acid. Ingestion of the deposited fluorides can poison grazing livestock and other animals, as occurred in New Zealand in 1995, when 2000 sheep died in the weeks following the eruption of Mount Ruapehu; fluoride toxicity was thought to be one of the main causes of death.

Freshwaters contain some F from weathering and dissolution. Typical groundwater concentrations are <1 mg L^{-1}, but higher levels are often recorded, particularly in arid areas of the world that also have fluoride-containing rocks (Table 12.1). Freshwater F concentrations tend to be lower in hard water areas because of precipitation with Ca and Mg. The concentration of F in seawater is approximately 1 mg L^{-1}.

Fluorine is essential in many animals for formation of dental enamel and mineralisation in bones; a small amount is present in soft tissues. Highly toxic organofluorine compounds, such as fluoracetic acid, fluoropalmitic acid and fluoroleic acid, are formed naturally in the leaves and seeds of several plant species. Naturally elevated fluoroacetate concentrations in shrubs such as *Dichapetalum*

Figure 12.1 Eruption of Eyjafjallajökull, Iceland in May 2010. Volcanic gases and ash contain fluorine, giving rise to concerns over toxicity.

Credit: David Karnå (Wikimedia Commons).

Table 12.1 Naturally elevated fluoride concentrations in groundwaters (mg L^{-1}).

Country	*Mean/maximum fluoride concentration*	*No. of samples*	*Reference*
China, Datong Basin	3.2/22	353	Li et al. (2012)
Ethiopia	9.4/68	112	Rango et al. (2012)
India, Andhra Pradesh	3.6/7.6	433	Reddy et al. (2010)
Malawi	NS/7	61	Msonda et al. (2007)

NS = not stated.

toxicarium ('ratsbane', west Africa) and *Gastrolobium bilobum* ('heart-leaved poison', Australia) are sufficient to kill grazing livestock (Fig. 12.2).

Production and uses

Hydrogen fluoride (HF), which is manufactured by heating fluorspar with sulfuric acid, is the main F compound used in the chemical industry. Important compounds used in manufacturing include: fluorocarbons, used in industry and as refrigerants, solvents and for non-stick surfaces in cookware and packaging; uranium hexafluoride (UF_6), used in uranium enrichment; and sulfur hexafluoride (SF_6), used mainly in

Figure 12.2 A 19th-century lithograph of *Gastrolobium bilobum* ('heart-leaved poison'), an Australian shrub that contains naturally elevated levels of fluoroacetate. Native grazers are tolerant to it, but it is toxic to introduced animals and livestock.

Credit: Anon (Wikimedia Commons).

circuit breakers in the electrical industry. An important and well-known fluorocarbon is the non-stick surface compound polytetrafluoroethylene (PTFE, trade name Teflon™); similar fluoropolymers are produced for other applications, including expanded-PTFE (ePTFE, trade name Gore-tex™). Another important group of fluorocarbons, the perfluorooctane sulfonates (PFOS) and perfluorooctanoic acid (PFOA), are used in fire-fighting, surface treatments and manufacturing industry. Natural and synthesised minerals such as cryolite, fluorite and Al fluoride are used in the manufacture of Al (Fig. 12.3), steel, toughened glass and enamelware. Fluorine compounds such as sodium fluoroacetate are used as pesticides against mammalian pests and other F-containing compounds are used as household insecticides (e.g. Transfluthrin).

Another important use of F in many countries is the deliberate fluoridation of public water supplies for medical purposes; fluoride in water at levels up to 1 mg L^{-1} increases the formation of fluoroapatite in bones and teeth. Water fluoridation was first undertaken in Michigan, USA in 1945 and now covers two-thirds of US citizens. Some other countries have similar or higher proportions, while others have lower rates (Table 12.2). Fluoridated toothpaste and salt are commonly available in most countries as additional sources.

Figure 12.3 Molten aluminium. Synthetic cryolite (sodium hexafluoroaluminate) is used as a flux in the production of Al. The mineral fluorite was traditionally used in metal smelting and the name of the mineral, and ultimately the element fluorine, is derived from the Latin 'fluo' (flow).

Credit: US Marine Corps (Wikimedia Commons).

Table 12.2 Estimates of the scale of fluoridation of water in selected countries.

Country	*Population with deliberately fluoridated water (%)*	*Population with naturally fluoridated water (%)*
Argentina	8	11
Australia	80	<1
Brazil	41[(1)]	Unknown
Canada	44	<1
China	0	15
France	0	3
India	0	5
Israel	68	2
Libya	6	16
Malaysia	75[a]	Unknown
New Zealand	61[a]	Unknown
Peru	2	<1
Republic of Ireland	69	4
Spain	11	<1
Sweden	0	8
United Kingdom	10	<1
USA	63	3
Zambia	0	7

a May include a proportion of naturally fluoridated water.

Source: Data adapted from British Fluoridation Society (2012).

Pollution and environmental impacts

Fluoride pollution

Industrial processes are important local sources of atmospheric F. For example, the use of fluorite as a flux in smelting and steel-making produces silicon tetrafluoride, SiF_4, which is a pollutant gas. The toxic gas HF is generated by Al smelting, brick manufacture (which also produces SiF_4) and by the production of P fertiliser from fluoroapatite (rock phosphate; $Ca_5F(PO_4)_3$):

$$Ca_5F(PO_4)_3 + H_2SO_4 + 2H_2O \rightarrow CaSO_4.2H_2O + HF + 3H_3PO_4^{3-} \quad [12.1]$$

Concentrations of total atmospheric F in industrialised areas are typically 1–2 $\mu g\ m^{-3}$, but much higher levels have been recorded nearer to sources. Coal combustion is an additional source of atmospheric F; some samples of coal contain >1 g F kg^{-1} (Table 12.3). Contamination of waters is caused by the leaching of F compounds from mine waste heaps and from soils contaminated by deposition of industrial emissions or applications of sewage sludge.

Table 12.3 Fluorine concentration in coals.

Country	n	*Mean F concentration[a] (mg kg^{-1})*	*Reference*
Australia	NS	98 (340)	Dale (2006)
Bulgaria	142	74–459 (1742)	Greta et al. (2013)
China	305	147–229 (1230)	Wu et al. (2004)
USA	799	30–160	Swanson et al. (1976)

a Except for Australia, ranges of mean values are shown (i.e. for several types of coal); figures in parentheses show the maximum concentration recorded.
NS = not stated

Global warming

An important environmental impact of F pollution is global warming {→C}. The greenhouse gases listed by the United Nations Framework Convention on Climate Change (UNFCCC) are, after carbon dioxide, methane and nitrous oxide, all fluorine compounds; one of these is sulfur hexafluoride (SF_6), and the remainder are classified as seven separate perfluorocarbons and 13 hydrofluorocarbons (some of these are listed in Table 12.4). The perfluorocarbons are essentially fluoroalkanes ranging from perfluoromethane (CF_4) to perfluorohexane (C_6F_{14}). These gases are very persistent and have large global warming potentials (GWPs), retaining thousands of times more heat in the atmosphere than an equivalent mass of CO_2 (Table 12.4). Sulfur hexafluoride has the highest GWP of the main greenhouse gases. The HFCs have smaller atmospheric lifetimes but are potent greenhouse gases with high GWPs. Another F compound with a long lifetime (800 years) and large GWP (17,400) is trifluoromethyl sulfur pentafluoride (SF_5CF_3), but its atmospheric concentration is relatively low at <1 ng m^{-3}.

Table 12.4 Atmospheric lifetimes and global warming potentials of selected fluorine compounds listed by the UNFCCC.

Compound	*Atmospheric lifetime (years)*	*100-year global warming potential[a]*
Sulfur hexafluoride	3,200	23,500
	Hydrofluorocarbons	
HFC-23	222	12,400
HFC-125	28.2	3,170
HFC-134a	13.4	1,300
HFC-143a	47.1	4,800
HFC-236fa	242	8,060
	Perfluorocarbons	
Perfluoromethane, PFC-14	50,000	6,630
Perfluoroethane, PFC-116	10,000	11,100
Perfluorohexane, PFC-51-14	3,100	7,910

a Based on a 100-yr GWP of 1 for CO_2.

Source: Data from IPCC (2013).

Human exposure and toxicity

Emissions of gaseous and particulate fluorides (F^-) by polluting industries, and their subsequent deposition to soils and waters, are of particular concern for human health; as a result, the WHO has set an air quality guideline for F^- of 1 μg m^{-3}. The main health impact of elevated F^- intake, via drinking water and food, is the development of skeletal and dental fluorosis. Skeletal fluorosis is characterised by increased density and brittleness of bones and calcification of ligaments. Symptoms of dental fluorosis are the staining and pitting of tooth enamel (Fig. 12.4). The cause of fluorosis is accumulation of F^- and depletion of Ca in bones and teeth (Ca deficiency can be an additional factor in fluorosis). In the vicinity of industrial emissions and exposed mine waste, ingestion of F^- can be significant. Chronic exposure to >2 mg F^- per litre of water during early childhood can cause dental fluorosis, while longer-term exposure to levels of >3 mg L^{-1} can also result in skeletal fluorosis. The WHO guideline value for F^- in drinking water is 1.5 mg L^{-1} (WHO, 2011). High F^- intake in humans can also affect the neurological and gastrointestinal systems and disrupt food metabolism, with symptoms of weight loss, weakness and anaemia. High intakes can also cause thyroid and kidney disruption and cardiac arrhythmia.

Cases of fluorosis in polluted areas have been recorded in many areas, particularly Europe, North Africa and parts of Asia, including China. High rates of fluorosis in some parts of central and western China are attributed, in part, to the combustion of high-F coal. Another major cause seems to be the use of coal-clay briquettes; the clay, which is used to bind coal dust into the briquette, also contributes to F^- exposure. Exposure in these cases seems to be via inhalation of combustion fumes and/or

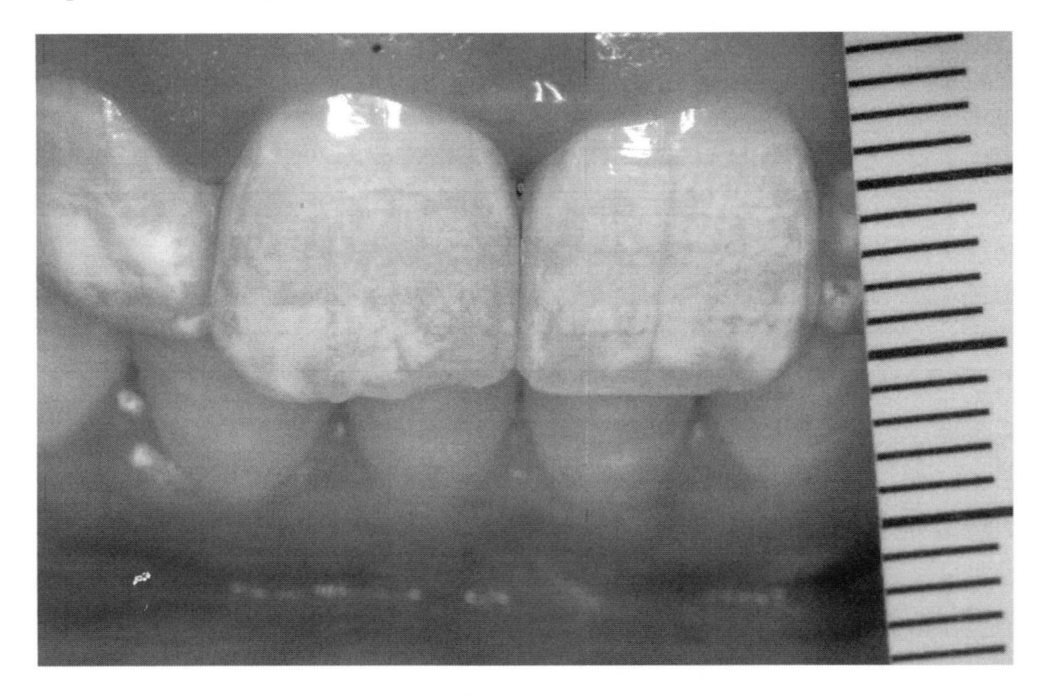

Figure 12.4 Mottling of teeth caused by fluorosis.

Credit: Castro et al. (2014), courtesy of Elsevier.

ingestion of food that has been contaminated by the fumes. Another ongoing source of concern with regards to F^- pollution is associated with brick kilns; a study in India showed elevated levels in soils and crops in brick fields and concludes, from estimations of daily F^- intakes, that children are at heightened risk of fluorosis from ingestion of vegetables and cereals in such areas (Fig. 12.5). The use of F in smelting is another source of human exposure; for example, an ongoing international dispute between the Asian republics of Tajikistan and Uzbekistan centres on emissions of HF from one of the world's largest Al smelters and has led to claims that subsequent F^- exposure is having serious health impacts. It should be remembered that fluorosis is particularly problematic in countries with *naturally* contaminated drinking water; the main affected areas are in China, India, Central Africa and South America. Data collected by the Chinese Ministry of Health indicated the possibility of 18×10^6 cases of dental fluorosis and 1.6×10^6 cases of skeletal fluorosis across the country (Wu et al., 2004); many of these are likely to be caused by naturally elevated F levels in parts of the country in addition to exposure in polluted areas.

The fluorocarbons perfluorooctane sulfonate (PFOS) and perfluorooctanoic acid (PFOA) have long half-lives of several years in humans; exposure is primarily via fish ingestion and accumulation appears to be mainly in the blood, kidney and liver. Occupational exposure may be associated with incidence of bladder cancer, based on findings in a small number of studies. Inhalation of fluorine gas and HF can be fatal, but environmental concentrations are generally too low to be of concern.

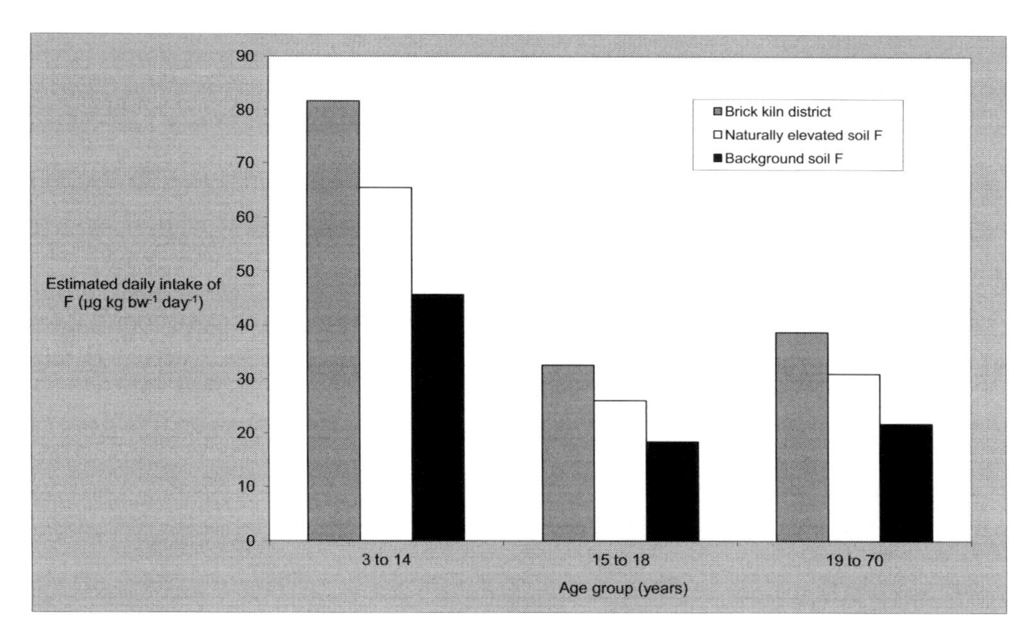

Figure 12.5 Estimated daily intakes of fluorine (µg kg body weight^{-1} day^{-1}) for three age groups (shown on horizontal axis) in three areas of Uttar Pradesh, India. The authors calculated estimated daily intakes from measured fluorine concentrations in replicate samples of 10 commonly eaten vegetables and two staple cereals (rice and wheat) that were sampled from the study areas, combined with data on amount and frequency of consumption, estimated absorption rates and body weight.

Data adapted from Jha et al. (2011).

Ecotoxicity

Plants suffer from chronic exposure to gaseous fluorides at concentrations of up to three orders of magnitude lower than observed for other phytotoxic gases such as O_3 and SO_2. Fluorides are readily taken in by the leaves and phytotoxicity is most likely related to the interaction of F^- with Ca^{2+} and Mg^{2+} within plant tissues. Elevated levels lead to chlorosis and necrosis of leaves and needles, especially at the leaf margins; F^- intake can affect photosynthesis, respiration, cell membranes, DNA and metabolic processes. Maize and citrus crops are damaged by HF at air concentrations of 4 $\mu g\ m^{-3}$ and pine trees suffer synergistic effects of F gases and SO_2 (Mido and Satake, 2003). Plants have different tolerances to F pollution but damage starts to occur at tissue concentrations of around 20 mg kg^{-1} dry weight; levels significantly higher than this are observed in polluted areas (Kabata-Pendias, 2001). Animals are at risk from ingestion of fluoride-contaminated plants, as documented by an extensive literature on fluorosis in cattle. Effects of elevated F^- on freshwater organisms include genotoxicity in fish and reproductive decline in aquatic snails.

Ecotoxicity can also be caused by PFOS and PFOA; both chemicals can enter the environment via landfill leachate and industrial effluent and they have long atmospheric lifetimes. They are highly stable, resisting hydrolysis, photolysis and biodegradation. Some fish species bioconcentrate PFOS in their gills to concentrations of more than 1000 times that of the surrounding water (OECD, 2002). Toxicity in mammals appears to be related to liver dysfunction; perfluorinated compounds are also implicated in hormone disruption. Production is now restricted in some countries and PFOS is classified as a persistent organic pollutant under Annex B of the Stockholm Convention.

References

British Fluoridation Society. 2012. *One in a Million: The Facts About Water Fluoridation*, 3rd Edition. Available at: www.bfsweb.org (authoritative data sources, such as national health departments, are provided for individual countries).

Castro, K.S., de Araúgo Ferreira, A.C., Duarte, R.M., Sampaio, F.C. and Meireles, S.S. 2014. Acceptability, efficacy and safety of two treatment protocols for dental fluorosis: a randomized clinical trial. *Journal of Dentistry* 42(8), 938–944.

Dale, L. 2006. *Trace Elements in Coal*. Newsletter of the Australian Coal Association Research Program. Available at: www.acarp.com.au

Greta, E., Dai, S. and Li, X. 2013. Fluorine in Bulgarian coals. *International Journal of Coal Geology* 105, 16–23.

IPCC (Intergovernmental Panel on Climate Change). 2013. *Climate Change 2013: The Physical Science Basis*. Working Group II Contribution to the Fifth Assessment Report of the IPCC. IPCC, Geneva.

Jha, S.K., Nayak, A.K. and Sharma, Y.K. 2011. Site specific toxicological risk from fluoride exposure through ingestion of vegetables and cereal crops in Unnao district, Uttar Pradesh, India. *Ecotoxicology and Environmental Safety* 74, 940–946.

Kabata-Pendias, A. 2001. *Trace Elements in Soils and Plants*, 3rd Edition. CRC Press, Boca Raton, USA.

Li, J., Wang, Y., Xie, X. and Su, C. 2012. Hierarchical cluster analysis of arsenic and fluoride enrichments in groundwater from the Datong basin, Northern China. *Journal of Geochemical Exploration* 118, 77–89.

Mido, Y. and Satake, M. 2003. *Chemicals in the Environment*. Discovery Publishing, New Delhi.

Msonda, K.W.M., Masamba, W.R.L. and Fabiano, E. 2007. A study of fluoride groundwater occurrence in Nathenje, Lilongwe, Malawi. *Physics and Chemistry of the Earth* 32, 1178–1184.

OECD (Organisation for Economic Co-operation and Development). 2002. *Hazard Assessment of Perfluorooctane Sulfonate (PFOS) and its Salts*. Available at: http://www.oecd.org/chemicalsafety/risk-assessment/2382880.pdf

Rango, T., Kravchenko, J., Atlaw, B., McCornick, P.G., Jeuland, M., Merola, B. and Vengosh, A. 2012. Groundwater quality and its health impact: an assessment of dental fluorosis in rural inhabitants of the Main Ethiopian Rift. *Environment International* 43, 37–47.

Reddy, D.V., Nagabhushanam, P., Sukhija, B.S., Reddy, A.G.S. and Smedley, P.L. 2010. Fluoride dynamics in the granitic aquifer of the Wailapally watershed, Nalgonda District, India. *Chemical Geology* 269, 278–298.

Swanson, V.E., Medlin, J.H., Hatch, J.R., Colemen, S.L., Wood, G.H., Woodruff, S.D. and Hilderbrand, R.T. 1976. *Collection, Chemical Analysis and Evaluation of Coal Samples in 1975*. US Dept. of Interior, Geological Survey, Report No. 76-468.

WHO (World Health Organization). 2011. *Guidelines for Drinking Water Quality*, 4th Ed. World Health Organization, Geneva.

Wu, D., Zheng, B., Tang, X., Li, S., Wang, B. and Wang, M. 2004. Fluorine in Chinese coals. *Fluoride* 37, 125–132.

13 Lead

Lead in the natural environment

Lead (Pb) occurs mainly in the divalent form, Pb^{2+}, although tetravalent Pb^{4+} may be found in very oxic environments. Natural Pb compounds are inorganic and there is little known methylation of Pb in the environment. Biochemically, Pb is similar to calcium in terms of ionic radius and a stable 2+ oxidation state and it follows similar transport pathways in the biosphere.

The average crustal abundance of Pb is typically estimated to be around 15 mg kg^{-1}; acidic rocks and shales contain a little more (up to 25 mg kg^{-1}), but marine black shales, with plentiful organic matter and sulfides, contain up to 150 mg kg^{-1} (Thornton et al., 2001). The most common mineral is the sulfide galena (PbS; 87% Pb). The main secondary minerals are cerrusite ('white lead'; $PbCO_3$) and anglesite ($PbSO_4$). Other important Pb minerals include mimetite ($Pb_5(AsO_4)_3Cl$) and pyromorphite ($Pb_5(PO_4)_3Cl$). Natural occurrences of metallic Pb are rare. Lead minerals have low solubilities in water, dissolution increasing to some extent in acid conditions; Pb phosphates, especially pyromorphite, are particularly insoluble.

Published mean values for background Pb concentrations in topsoils are generally around 15 to 30 mg kg^{-1} with maximum values generally <100 mg kg^{-1} (Table 13.1). Such topsoil concentrations are typically higher than in the underlying rock, indicating deposition from the atmosphere. Pre-industrial concentrations are difficult to estimate because of the long history of Pb use by humans, its potential to travel long distances in the atmosphere and its strong binding in topsoils once deposited from the air; however, levels of <20 mg kg^{-1} seem likely. Lead has low solubility in soil and only small amounts exist in soil solution, mainly as the Pb^{2+} ion (particularly in acidic soils) and as dissolved organic and hydroxy complexes in alkaline soils. In anoxic soils (and sediments) PbS may precipitate. Cationic Pb^{2+} is strongly adsorbed by clays, Fe/Mn oxides and, more especially, by humus and there is little downward leaching, as indicated by the accumulation and persistence of Pb in the organic surface horizons. In addition to this strong binding, weathered particles of Pb minerals in soils have very low solubility; therefore it is less mobile and bioavailable in soils than most other metals. Mobility increases to some extent in acidic soils however, because of increased rates of dissolution and desorption.

Concentrations of Pb are generally low in natural waters because of its low solubility and strong binding in soils. It is stored mainly in sediments as particulate or adsorbed Pb, or possibly as Pb sulfide in anoxic sediments. Dissolved forms include free Pb^{2+} ions and complexes with dissolved organic carbon and inorganic species (e.g. chloride,

Table 13.1 Typical background concentrations of lead.

Environmental medium	*Typical background concentrations*[a]
Air	<0.1 $\mu g\ m^{-3}$
Soil	<100 $mg\ kg^{-1}$
Vegetation[b]	<4 $mg\ kg^{-1}$
Freshwater	<5 $\mu g\ L^{-1}$
Drinking water	<1 $\mu g\ L^{-1}$
Sea water	<5 $\mu g\ L^{-1}$
Sediment	<100 $mg\ kg^{-1}$

a These are *typical* values, encompassing the vast majority of reported concentrations, based on a large number of literature sources. Higher natural concentrations occur in some cases; for example in mineralised areas.
b Including crops. Not including metal-tolerant species.

hydroxide and carbonate in seawater). Concentrations are higher in surface seawaters, reflecting anthropogenic atmospheric inputs (see below) and scavenging in deeper waters by sinking particulate matter.

Nearly all of the naturally sourced Pb in the atmosphere originates from windblown soil and dust, with a minor contribution from natural biomass fires and even less from other natural sources such as volcanoes. Concentrations in areas that are very remote from terrestrial sources (e.g. mid-oceanic air) are generally <1 ng m^{-3}, whereas a typical 'background' level in rural areas may be two orders of magnitude higher.

Lead is not an important element in the natural biosphere because it has no known biological role and is not very bioavailable; however, where Pb is taken up by organisms, it has similar biochemical pathways to Ca, accumulating in bone tissue, for example.

Production and uses

Global reserves of Pb are estimated at 90 Mt and approximately 5 Mt of Pb are extracted and processed annually, half in China; of the other half, the largest producers are Australia, USA, Mexico and Peru (USGS, 2014). Lead is also recycled from scrap metal and Pb-acid batteries (Fig. 13.1). The dominant end-use is the Pb-acid battery, used mainly for vehicle ignition but also for back-up power supplies in strategic sectors such as telecommunications. Lead is used in the electrodes, which are alternating plates of metallic Pb and Pb dioxide (PbO_2) paste immersed in H_2SO_4. Despite reductions in Pb use in most other applications, because of concerns over its toxicity, vast numbers of Pb-acid batteries are still produced each year as there is no cost-effective alternative. Metallic Pb is also used as gutter, wall and roof flashing, cable sheathing in the petrochemical industry (and for undersea/underground high voltage cables) and in X-ray shielding. Alloys are used in Pb shot and in electronic solder, although non-Pb alloys are increasingly used. Lead is added to various other products including lead crystal (to improve the refractive index of the glass), ceramic glaze and rigid PVC (as a stabiliser). It is used in the screens of cathode-ray tube (CRT) TVs and computer monitors to reduce X-ray emission; however, CRTs have largely been replaced by LCD (liquid crystal display) and LED (light-emitting diode) technologies.

Figure 13.1 Lead being recycled from lead-acid batteries, the dominant end-use for the metal today.

Credit: National Institute for Occupational Safety and Health (Wikimedia Commons).

Because of the persistence of Pb in the environment, particularly in soils and sediments, historic uses of Pb are of interest in environmental pollution terms. Of particular importance in this context was the addition of performance-enhancing tetraethyl Pb ($(C_2H_5)_4Pb$) and tetramethyl Pb ($(CH_3)_4Pb$) to petrol between the 1920s and the last decades of the 20th century (Fig. 13.2). Volatile Pb halides were emitted from vehicles using leaded petrol because of the scavenging of tetraethyl Pb by halogenated compounds, which were added to prevent a damaging build up of Pb deposits in engines. Between the 1970s and 1990s, the amount of Pb in petrol was gradually reduced for health reasons and, during this period, unleaded petrol became a requirement in many countries to coincide with the introduction of catalytic converters, which are damaged by Pb. Today, tetraethyl Pb is banned for general vehicle use in nearly all countries (exceptions include Myanmar, North Korea and Yemen), but its legacy of contaminated soils and sediments remains worldwide. Similarly, the past use of Pb arsenate as an orchard insecticide has left some soils contaminated, albeit on a much more localised basis.

Other past uses of Pb, now reduced or banned, have also left a legacy of environmental pollution. Lead was used for many years in house paint and, despite a ban on this use in most developed countries, leaded paint still exists in many houses beneath wallpaper or newer paint and may become an exposure source during redecoration. Similarly, despite bans on the use of Pb in plumbing, Pb pipes are still found in older houses where they can contaminate drinking water. In the wider environment, ingestion of discarded Pb fishing weights and Pb shot remains an exposure risk to waterfowl despite some restrictions on their use.

Figure 13.2 Warning sign on an old petrol dispenser; tetraethyl lead was a major use of lead for many years before it was banned in most countries in the late 20th century.

Credit: Steve Snodgrass (Flickr).

Pollution and environmental impacts

Lead pollution

The historical extraction of Pb from the Earth's crust and its production, use and disposal, have significantly contaminated soils and water bodies. Once Pb has entered soils and sediments, a fraction of it tends to remain stored there, posing a hazard to the biosphere because of its toxicity, although this is mitigated to a large degree by its low bioavailability.

Lead is emitted to the atmosphere from Pb ore processing and smelting, areas of open mine waste, Pb-acid battery factories, fossil fuel combustion and waste incineration, especially where rigid PVC products are burned. It is generally emitted from these sources in fine particulate form. Most lead is deposited near to source but fine particles can be transported long distances; lake sediments, peat bogs and Arctic ice cores all show clear increases in atmospheric transport of Pb from ancient times, particularly in the Roman period. Such records also show distinct rises and falls in Pb concentrations coinciding with the introduction and banning, respectively, of leaded petrol. In the late 20th century, Pb concentrations in urban air were often $>0.5\ \mu g\ m^{-3}$, the WHO's recommended air quality guideline; today they have declined (Fig. 13.3) and the guideline level is exceeded relatively rarely away from the immediate surroundings of lead smelters, for example.

Despite reductions in Pb concentrations in the atmosphere, it is persistent in many topsoils because of its strong binding to clay and, in particular, soil organic matter; if

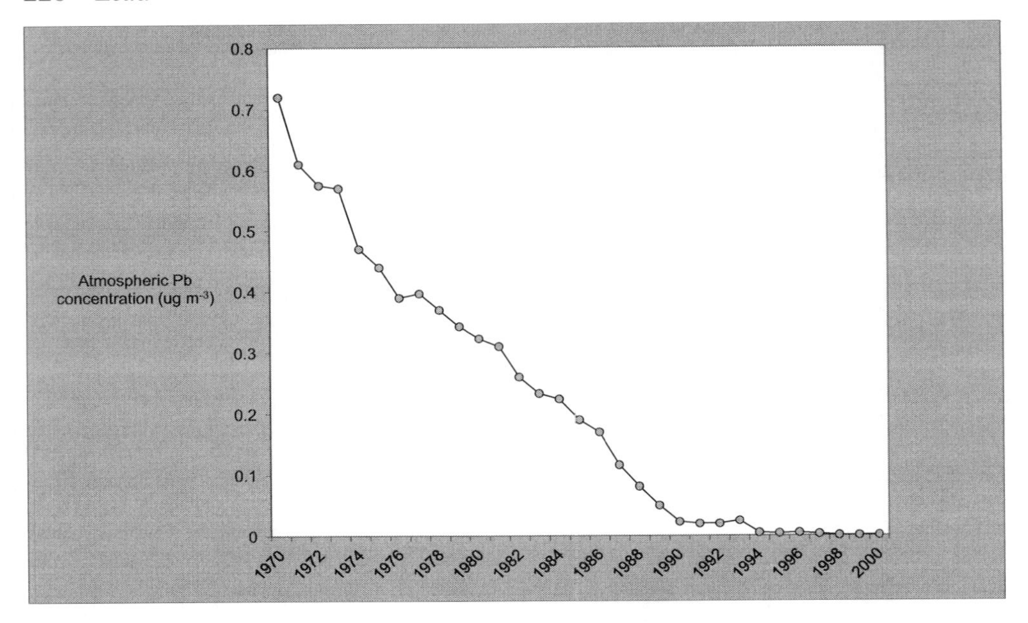

Figure 13.3 The effect of tetraethyl lead withdrawal from petrol on atmospheric lead concentrations. These national figures for Canada are broadly representative of the downward trend observed in most countries from the 1970s onwards.

Source: Adapted from Environment Canada, National Air Pollution Surveillance Program, www.ec.gc.ca/rnspa-naps.

these are present in very small amounts (e.g. in sandy soils), there is more likelihood of Pb leaching through the soil. In urban and industrial areas, Pb deposited from polluted air often accumulates in topsoils over time and can persist there, even in roadside soils where deposition may have ceased many years ago. Densely populated areas with a long history of vehicle use and nearby industry may have 'background' concentrations of >100 mg kg^{-1}, particularly in soils located <200 m from a busy road (Fig. 13.4). In areas adjacent to Pb-producing industries, particularly smelters, soil Pb concentrations can reach very high levels of >3% (>3×10^4 mg kg^{-1}) in exceptional cases (e.g. Rieuwerts and Farago, 1996). Sewage sludge inputs to soils are less problematic for Pb than Cd and Zn because of Pb's lower bioavailability and phytotoxicity. Localised soil contamination by Pb shot (e.g. near clay pigeon shooting grounds) can lead to Pb release from the pellets.

Sources of Pb pollution of freshwaters include: diffuse pollution and acid mine drainage {→S} in mining areas; industrial effluents; and leachates from unregulated or poorly engineered landfill sites. In urban areas, Pb enters drainage and sewage systems from road runoff (less important since the introduction of unleaded petrol), lead plumbing systems and drainage from leaded roofs and gutters. Virtually all Pb entering water bodies is in particulate form and its relative immobility in sediments offers some protection against excessive contamination of freshwater and seawater.

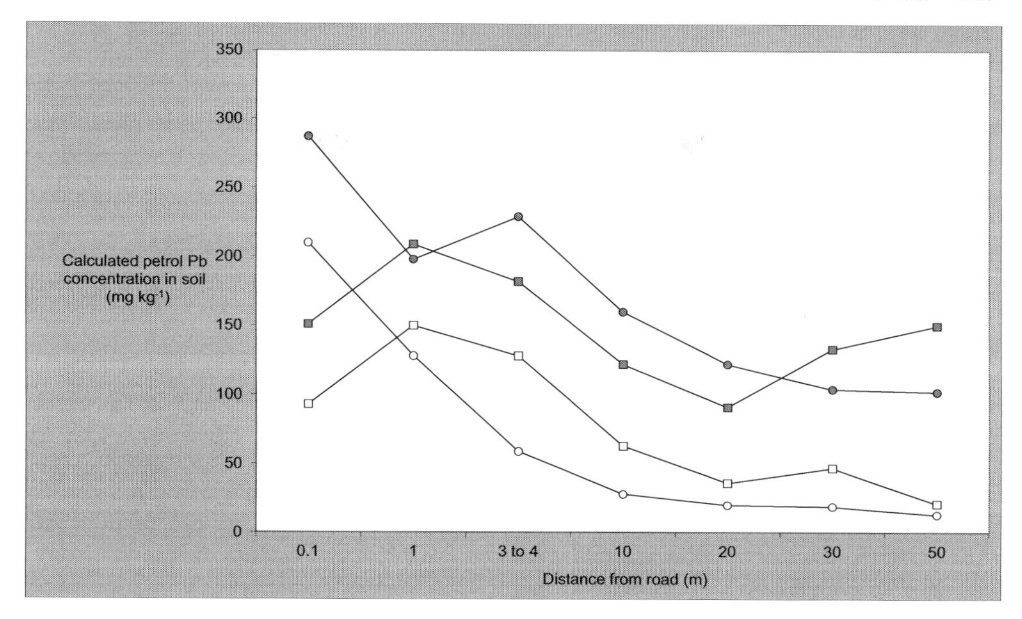

Figure 13.4 Total lead concentrations in farmland topsoils (0–3 cm depth) at increasing distances from a motorway (filled squares) and a minor road (filled circles) near Glasgow, UK. The open squares (motorway) and open circles (minor roads) show estimations of lead concentrations in soils attributed solely to emissions from vehicles, as calculated by the authors using lead isotope analysis. In all cases, the concentration shown is an average of two samples, one collected in 2001 (a year after the use of leaded petrol ceased in the UK) and the other in 2010. The authors noted that in most cases there was no systematic difference in total lead concentrations between the two years; this is not unexpected as lead tends to have limited mobility in most topsoils.

Source: Adapted from MacKinnon et al. (2011).

Human exposure and toxicity

Ingestion is the key exposure route for humans. Lead in the diet may result from the atmospheric deposition of lead onto the surfaces of vegetables in areas with industrial emissions, although such exposure is likely to be localised and controllable with adequate oversight. Also in polluted areas, fine particulate Pb is deposited in garden soils or house interiors via open windows or is 'tracked in' to houses on shoes or pets; this raises the potential for ingestion of contaminated dust and soil, especially by infants. Exceedances of the provisional WHO drinking water guideline for Pb (10 μg L^{-1}) are generally associated with Pb plumbing rather than environmental pollution; this is especially the case in areas with naturally soft and acidic water and where water has been standing in the pipes for some time. In serious cases, plumbing can be renewed or phosphate can be centrally added to the water to form highly insoluble Pb phosphate on the inner pipe surface; however, such actions are difficult and/or expensive. The presence of leaded paint in older homes becomes an ingestion hazard if the paint is exposed and peeling as infants are known to actively ingest such paint chips, the Pb giving them a sweet taste. Redecoration of such homes using paint sanders provides a more general inhalation hazard.

Up to 95% of excess Pb intake in adults is stored in the bones and teeth. This is mainly because Pb follows the biochemical transport pathways of Ca; indeed, more Pb is absorbed by the human body when Ca (and Fe) intake is low. Lead progressively accumulates in the bones during a human lifetime and is stable there, only becoming toxic if bone tissue dissolves, releasing Pb to the blood and soft tissues where it may cause harm; this may occur, for example, in people who have osteoporosis or other bone diseases, patients who are undergoing cortisone treatment or women who are pregnant or lactating. Foetal exposure to Pb is a concern because it passes readily through the placental and blood–brain barriers. Infants are vulnerable because they excrete less Pb than adults and retain approximately five times more in their blood and soft tissues; furthermore, they are particularly exposed to dust and soil Pb because they tend to play on the floor and engage in hand-to-mouth activity. In poorly nourished children, there is an increased risk of Pb absorption and toxicity (see above). Some children and adults exhibit pica (consumption of non-food items, particularly soil) and this can result in further exposure in polluted areas. Despite the insolubility (in water) of Pb in soil and dust particles, it becomes partly bioavailable when ingested because of increased dissolution in the acidic conditions of the stomach. While bone Pb represents long-term exposure, blood Pb indicates recent exposure (weeks rather than years) and is the key measure used by clinicians to assess the likelihood of subclinical toxicity in an exposed subject.

Occupational and accidental exposures to high levels of Pb have caused various clinical effects including kidney damage and effects on the nervous and reproductive systems; very high doses cause encephalopathy and death (Box 13.1). Other symptoms in occupational settings include anaemia, headache, insomnia, constipation, hypertension and low vitamin D production. Epidemiological studies are equivocal as regards cancer formation and Pb is listed as a *potential* carcinogen.

Chronic exposure of the general population to elevated levels of Pb in the environment is of concern mainly because of the possibility of initially subtle or subclinical effects. For example, in the human body, Pb binds to sulfhydryl groups in proteins and impairs enzymes involved in production of haemoglobin, potentially causing anaemia. An example of a subclinical effect in this context is the accumulation of δ-aminolaevulinic acid (ALA) in the blood. This occurs because ALA dehydratase (ALA-D), which forms the haem precursor porphobilinogen from ALA, is inhibited by Pb. Several other haem synthesis processes are similarly affected by elevated Pb intake, and other haem precursors accumulate in the same way as ALA.

Concern about the health effects of environmental Pb pollution (Table 13.2) is mainly focused on disruption of synapse formation and neurobehavioural impacts. Lead can impair the development of the central nervous system in the embryo and children, causing potentially irreversible learning and behavioural difficulties including attention deficit, hyperactivity, aggression and lowered intelligence quotient (IQ). Research in recent years has also shown associations between adult criminality and childhood Pb exposure. Estimates based on numerous studies indicate 2–6 point IQ reductions for each 100 $\mu g\ L^{-1}$ (10 $\mu g\ dL^{-1}$) increase in exposed children's blood Pb concentrations (WHO, 2010). Children's blood Pb levels are lower in polluted cities now than they were in the late 20th century, mainly because of measures taken to reduce exposure; however, perceptions of 'safe' levels of blood Pb in children have also altered over the years. Evidence in recent years indicates neurobehavioural damage occurs at $<5\ \mu g\ dL^{-1}$ and suggests there may be no safe threshold for Pb body burden.

Box 13.1 Unregulated lead exposure.

In 2007–08, 18 children in Dakar, Senegal died from the ingestion of Pb-contaminated soil (Haefliger et al., 2009). The children suffered encephalopathy after exposure to Pb-contaminated soil and dust caused by informal dismantling of lead-acid batteries. Cultural sensitivities precluded autopsies, but surviving children had very high blood Pb concentrations of between 40 and >600 μg dL^{-1} (n = 50). Symptoms among family members included convulsions, aggression and gastrointestinal disorders. Although the informal recycling activity was halted in 2008, researchers found extremely high Pb concentrations in the soils of the main working area of 2.1×10^5 mg kg^{-1} (21%) and of >30% in bags of sieved soil; Pb levels inside homes of up to 1.4×10^4 mg kg^{-1} (1.4%) were also observed. The authors of the study expressed concern that similar types of exposure may be occurring elsewhere in developing countries that lack resources for proper oversight and regulation.

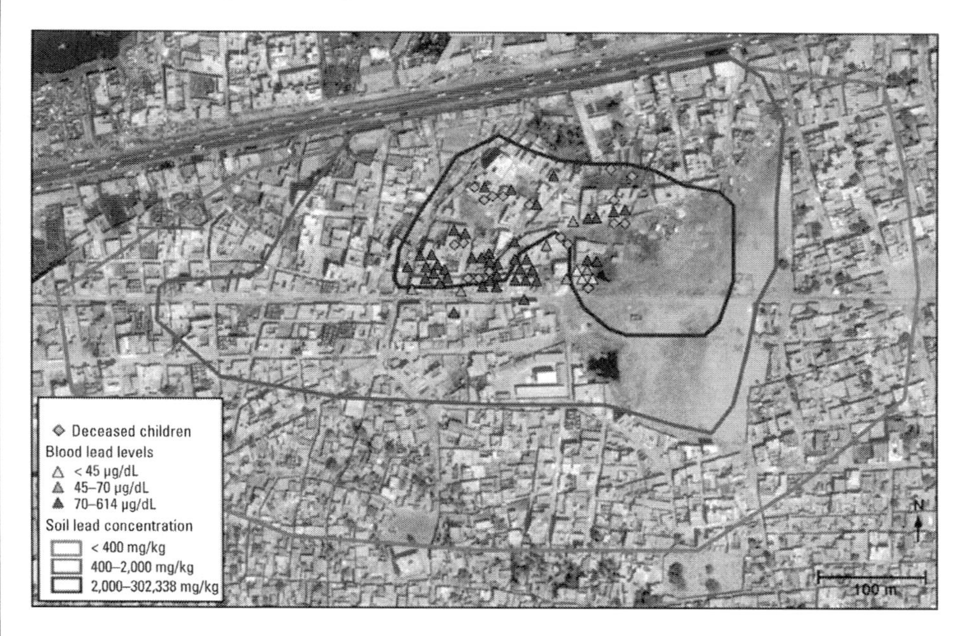

Figure 13.5 Satellite image of the Ngagne Diaw district of Dakar, Senegal showing distribution of fatalities and concentrations of lead in blood and environmental samples.

Credit: Haefliger et al. (2009), courtesy of Paul Haefliger.

Ecotoxicity

Lead bioaccumulates in aquatic organisms including phytoplankton, bryophytes, molluscs, crustaceans and fish, but does not biomagnify, probably because of its skeletal storage in vertebrates, thereby diminishing its passage up the food chain. In fish, excess intake of Pb can cause impairment of the gills. Some aquatic organisms

Table 13.2 Human health effects of lead.

Concentration of Pb in blood (μg dl^{-1})	*Effect*
<5	Neurodevelopmental effects
<5	Reduction of ALA-D levels
<10	Delayed sexual maturation
<10	Elevated blood pressure in adults
20 (approximately, in mother)	Reduction in birth weight
25–40	Anaemia in children
30	Slowing of nerve conduction velocity
40	Neurobehavioural effects in adults
40–50	Reduced fertility
50	Anaemia in adults
60	Colic in children
60	Kidney damage
80–100	Encephalopathy (see Box 13.1)

Source: Adapted from Thornton et al. (2001) and USDHHS (2007).

have an extraordinary capacity to accumulate Pb without apparent harm. For example, mussels containing a few grams of Pb per kg of tissue have been recorded without apparent ill-effects.

In Pb-contaminated soils, microbial enzymatic activity and organic matter breakdown can be impaired, together with reduced reproduction in nematodes. Plants take up less Pb from soils than more mobile and phytotoxic elements such as Cd and Ni. Nearly all Pb taken up remains in the roots and there is typically little translocation to the shoots and leaves. In this context, many plants are excluders of Pb, although they may not be able to prevent translocation at very high rates of exposure; indeed, transfer to the aerial parts of plants has been observed in very polluted soils probably because the capacity of root cells to bind Pb is surpassed in such circumstances. Where atmospheric pollution occurs, deposition of Pb onto leaf surfaces may be an important ecotoxicological exposure route; plant intake of Pb from leaf surfaces is thought to be very limited, but aerial deposition increases the likelihood of Pb transfer to herbivores. Controlled studies in laboratories and research plots have revealed effects on germination, growth, metabolism, photosynthesis and transpiration, for example when plants are exposed to soil Pb concentrations of 100–1000 mg kg^{-1}; however, such studies may not be reliable indicators of field conditions. On the other hand, where field studies in uncontrolled, environmentally polluted soils have shown correlations between soil Pb concentration and plant injury, a causal effect of Pb may not always be established because of the presence of other toxic elements.

In ecotoxicological terms, birds are perhaps the most well-documented recipients of Pb poisoning. For example, field studies undertaken over a number of years found Pb poisoning in environmentally exposed water fowl in a Pb mining and smelting area of Idaho, USA (Blus et al., 1999). Exposure was via ingestion of Pb-contaminated sediments. Symptoms in tundra swans (*Cygnus columbianus*) included decreases in ALA-D activity, haematocrit (volume of red blood cells in the blood) and haemoglobin concentration (Fig. 13.6). Additional symptoms of Pb poisoning in wild birds include neurobehavioural effects such as tremor, impaired coordination, inability to fly and moribundity. In addition to studies in polluted areas, research on avian Pb poisoning has also focused on the widespread exposure to Pb shot and fishing weights. Both can be taken up when birds are feeding in sediments and just one pellet may be fatal as the pellets partially dissolve in the acidic conditions of the stomach, releasing bioavailable Pb. An alternative route for the intake of Pb shot is the consumption of shotgun-hunted animals by prey birds and scavengers. Research into the causes of death of the endangered Californian Condor (*Gymnogyps californianus*) in southwest USA between 1992 and 2007 revealed that 23 deaths (of 65 cases where cause was established) were due to Pb exposure; Pb ammunition in prey was stated to be the main cause (Rideout et al., 2012).

Animals inhabiting mining and smelting areas are at particular risk of Pb exposure. Lead contamination of wild mammals and rodents inhabiting such areas has been associated with kidney disorders and reduced red blood cell counts and fatal Pb poisonings of livestock grazing in mining and smelting areas have been reported widely and over many years. In the UK, Pb poisoning accounts for the majority of livestock deaths reported to the country's Food Standards Agency. In 2008, 226 farm animals across the country showed Pb poisoning symptoms and 127 died, including six cows that had grazed near an old Pb mine; in polluted areas, cows' milk has also been declared unfit for human consumption because of Pb contamination (VLA/FSA, 2014).

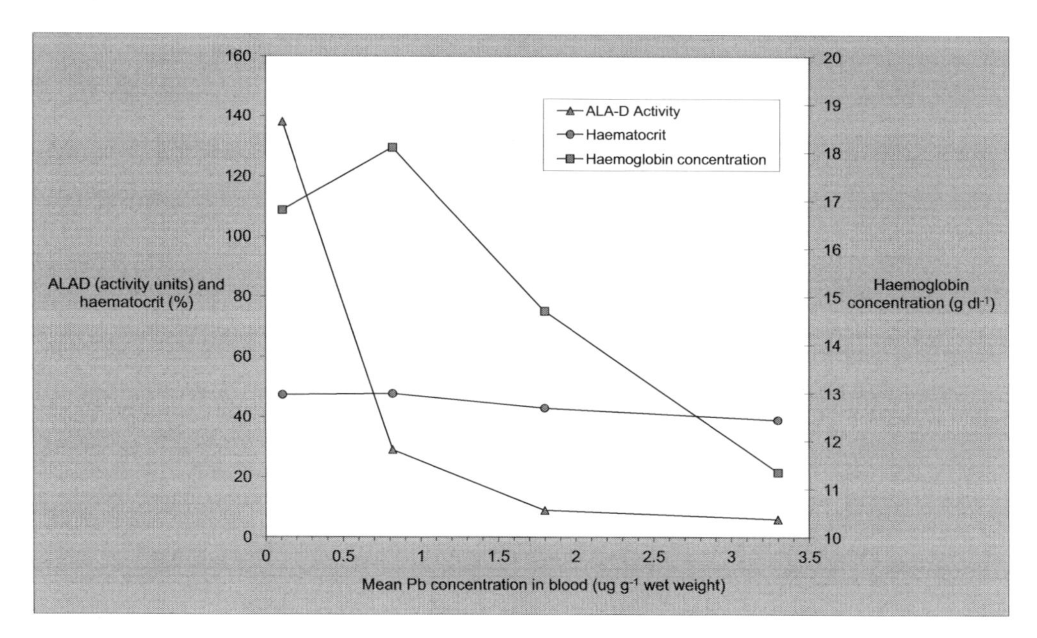

Figure 13.6 Effects of lead exposure in tundra swans (*Cygnus columbianus*). Based on 28 swans from a control area (with a mean blood lead concentration of 0.11 μg g^{-1}) and 39 from a lead-contaminated region in Idaho, USA (separate groups, with means of 0.82, 1.8 and 3.3 μg g^{-1}).

Source: Data from Blus et al. (1999).

References

Blus, L.J., Henny, C.J., Hoffman, D.J., Sileo, L. and Audet, D.J. 1999. Persistence of high lead concentrations and associated effects in tundra swans captured near a mining and smelting complex in Northern Idaho. *Ecotoxicology* 8, 125–132.

Haefliger, P., Mathieu-Nolf, M., Lociciro, S., Ndiaye, C., Coly, M., Diouf, A., Lam Faye, A., Sow, A., Tempowski, J., Pronczuk, J., Filipe, A.P. Jr, Bertollini, R. and Neira, M. 2009. Mass lead intoxication from informal used lead-acid battery recycling in Dakar, Senegal. *Environmental Health Perspectives* 117, 1535–1540.

MacKinnon, G., MacKenzie, A.B., Cook, G.T., Pulford, I.D., Duncan, H.J. and Scott, E.M. 2011. Spatial and temporal variations in Pb concentrations and isotopic composition in road dust, farmland soil and vegetation in proximity to roads since cessation of use of leaded petrol in the UK. *Science of the Total Environment* 409, 5010–5019.

Rideout, B.A., Stalis, I., Papendick, R., Pessier, A., Purschner, B., Finkelstein, M.E., Smith, D.R., Johnson, M., Mace, M., Stroud, R., Brandt, J., Burnett, J., Parish, C., Petterson, J., Witte, C., Stringfield, C., Orr, K., Zuba, J., Wallace, M. and Grantham, J. 2012. Patterns of mortality in free-ranging California Condors (*Gymnogyps californianus*). *Journal of Wildlife Diseases* 48, 95–112.

Rieuwerts, J.S. and Farago, M.E. 1996. Heavy metal pollution in the vicinity of a secondary lead smelter in the Czech Republic. *Applied Geochemistry* 11(1–2), 17–23.

Thornton, I., Rautiu, R. and Brush, S. 2001. *Lead: The Facts*. IC Consultants Ltd, London.

USDHHS (US Department of Health and Human Services: Public Health Service, Agency for Toxic Substances and Disease Registry, ATSDR). 2007. *Toxicological Profile for Lead*. ATSDR, Atlanta.

USGS (United States Geological Survey). 2014. *Mineral Commodity Summaries: Lead*. Available at: http://minerals.usgl.gov/minerals.

VLA/FSA (Veterinary Laboratories Agency and Food Standards Agency). 2014. Help stop on-farm lead poisoning (information leaflet). Available at: http://multimedia.food.gov.uk/multimedia/pdfs/publication/leadpoison0209.pdf

WHO (World Health Organization). 2010. *Childhood Lead Poisoning*. WHO, Geneva.

14 Mercury

Mercury in the natural environment

Introduction

Knowledge of the speciation of mercury (Hg) is important, particularly in gaining an understanding of its environmental impacts. The key inorganic species are metallic mercury, Hg°; the mercurous ion, Hg_2^{2+} and the mercuric ion, Hg^{2+}. Metallic Hg° and mercurous compounds are relatively rare in the natural environment, the more stable inorganic compounds being those containing the mercuric ion, e.g. the sulfide, HgS; hydroxide, $Hg(OH)_2$; nitrate, $Hg(NO_3)_2$; and chloride, $HgCl_2$. The main organic mercury species are alkylmercury compounds, particularly monomethylmercury (MMHg; CH_3Hg^+) and dimethylmercury (DMHg; CH_3HgCH_3). In these compounds the Hg atom is covalently bonded with the C atom of one methyl group (in MMHg) or two methyl groups (in DMHg). Because of mercury's volatility and its consequent cycling (and transport to remote areas) via the atmosphere, it is difficult to estimate pre-industrial concentrations, but Table 14.1 lists typical background concentrations.

Lithosphere

In common with other metals, the main natural store of Hg is the Earth's crust. However, even in the crust Hg is rare, with concentrations in most rocks of <50 µg kg^{-1}; levels in organic shales may be an order of magnitude higher, but the main deposits of Hg are in the ore mineral cinnabar, HgS, reflecting mercury's strongly

Table 14.1 Typical background concentrations of mercury.

Environmental medium	*Typical background concentrations*[a]
Air	<10 ng m^{-3}
Soil	<1 mg kg^{-1}
Vegetation[b]	<20 µg kg^{-1}
Freshwater	<0.1 µg L^{-1}
Sea water	<0.1 µg L^{-1}
Sediment	<0.1 mg kg^{-1}

a These are *typical* values, encompassing the vast majority of reported concentrations, based on a large number of literature sources. Higher natural concentrations occur in some cases; for example, in mineralised areas.

b Including crops. Not including metal-tolerant species.

chalcophilic (S-loving) nature. Cinnabar is a bright red mineral that darkens on exposure to light and is the source of vermilion, a vivid red dye used since antiquity. The ore mainly forms near the Earth's surface in volcanic regions because it crystallises at relatively low temperatures. It has been recorded in many parts of the world, including Europe, China, Mexico, USA and Russia. Other Hg minerals, such as livingstonite, are known but relatively rare. Mercury is also associated with ores of other metals, particularly lead, zinc and silver. Relatively small amounts of metallic Hg occur in places (including mines in Spain and California) as droplets in Hg ores (Fig. 14.1).

Mercury is found in soils as: MMHg (which, being cationic, is retained in soils); HgS and HgSe (mercury selenide) precipitates; organo-mercuric complexes; mercuric chloro ($pH < 7$) and hydroxo ($pH > 7$) complexes; and Hg° (in strongly reducing conditions). Background concentrations of Hg in soils vary quite widely, depending on degree of remoteness from volcanic regions or from anthropogenic sources, but are generally up to 1 mg kg^{-1}; i.e. higher than crustal values, reflecting the deposition of atmospheric Hg, both natural and anthropogenic. Soils in some mineralised areas have higher Hg concentrations than those of non-mineralised areas, reflecting their development on mineralised bedrock.

Atmosphere

The atmosphere is a more important reservoir of Hg than of other metals because of its volatility. Hg° vapour is the predominant atmospheric species (estimated to

Figure 14.1 Metallic mercury droplet on the surface of a cinnabar sample.

Credit: Parent Gery (Wikimedia Commons).

comprise 75–99% of total airborne Hg), but other inorganic (mainly Hg^{2+}) forms, such as mercuric chloride ($HgCl_2$) are present, as is DMHg and Hg associated with fine particulate matter. Background concentrations of Hg are approximately 3 ng m^{-3} in rural air and around 10 ng m^{-3} in urban air.

Hydrosphere

Around 25–50% of Hg in rivers is present as MMHg, while the corresponding percentage in the oceans, where DMHg is more prevalent, is <10%. Much of the remaining Hg in both cases is present as Hg^{2+}, not as the free ion but in Hg salts and in complexes with hydroxides, chlorides, sulfides and dissolved organic carbon (DOC). The chloro-complexes dominate in seawater and, except at low pH, Hg–DOC complexes are more prevalent in freshwaters. Dissolved Hg° is also present in the hydrosphere. In particulate form, Hg is mainly bound to organic colloids and inorganic phases such as Fe oxides.

Biosphere

Mercury has no known biological role but the biosphere plays a central role in the Hg cycle. Methylation is particularly important and one of the main forms of Hg found in the biosphere is the bioaccumulative MMHg.

Mercury cycling

The fate and pathways of Hg in the environment can be described in terms of several key processes: (i) degassing of Hg° from the Earth's surface and its long-range transport in the atmosphere; (ii) oxidation of atmospheric Hg° to Hg^{2+}; (iii) deposition of Hg^{2+} species from the atmosphere to soils and waters, including in remote areas; (iv) biotic and abiotic conversion of some Hg^{2+} in soils and sediments to volatile Hg° and DMHg, followed by their diffusion to the atmosphere (and subsequent re-deposition); (v) the formation in soils and sediments of MMHg (via methylation of Hg^{2+}) followed potentially by its bioaccumulation and biomagnification in the biosphere; (vi) retention of further Hg^{2+} in soils and sediments; (vii) the conversion of Hg° and Hg^{2+} species in sediments to HgS followed by, in some cases, permanent burial to the lithosphere. The following sub-sections explore these and other processes in more detail and Fig. 14.2 provides a summary.

Atmosphere

Many attempts have been made to estimate the amount of Hg entering the atmosphere from sources that do not include direct anthropogenic emissions. One of the more recent estimates (Pirrone et al., 2010) suggests outputs from the Earth's surface of 5.2 × 10^3 t a^{-1}, with 52% of this originating in the oceans, 33% from terrestrial degassing, 13% from biomass burning and 2% from volcanoes; note that degassing will include re-volatilisation of anthropogenic Hg. The ocean-derived Hg gases are likely to be in the forms of: (i) DMHg, which is created in sediments by bacterial methylation and is particularly prevalent in areas of upwelling; (ii) marine $HgCl_2$, emitted via seaspray; and (iii) Hg° vapour, formed in sediments by bacterial reduction of Hg^{2+} and in surface

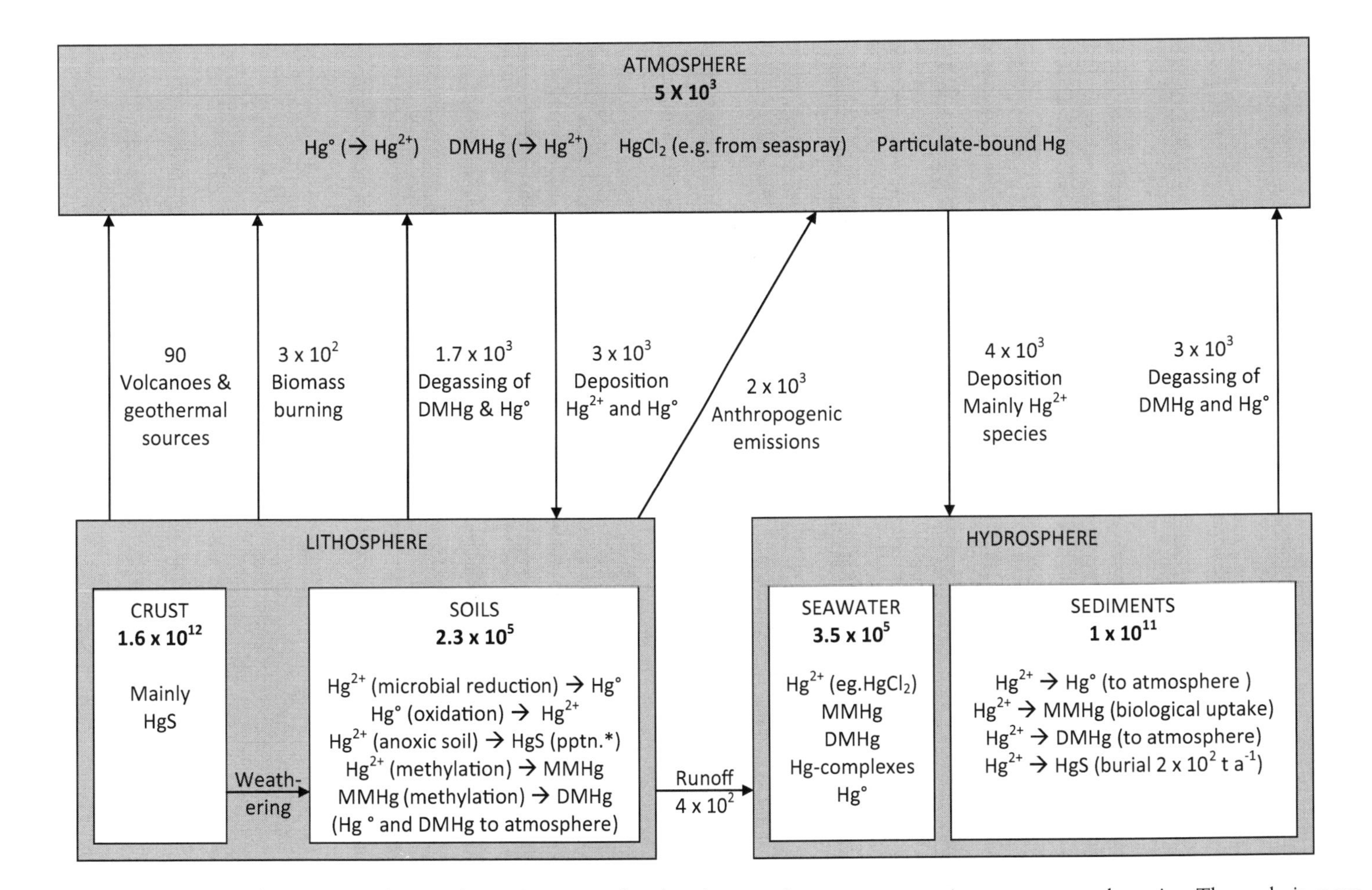

Figure 14.2 A summary of mercury cycling in the environment showing the most important reservoirs, processes and species. The cycle is more complex than shown here and the text should be consulted for more details. Because of mercury's volatility, it is difficult to estimate pre-industrial reservoirs and fluxes and, therefore, this cycle includes anthropogenic activity. Units for estimated reservoirs (in boxes) and fluxes (adjacent to arrows) are t Hg and t Hg a^{-1}, respectively. *pptn = precipitation.

Source: Adapted from various sources, primarily Pirrone et al. (2010) and Amos et al. (2013).

seawaters by photoreduction of methylmercury. Terrestrial degassing is of DMHg and Hg°, the former produced by bacterial methylation in soils and sediments. There are two main terrestrial sources of atmospheric Hg°: near-surface rocks containing metallic Hg and soils and freshwater/estuarine sediments where Hg^{2+} has been reduced to Hg°.

Estimates of the atmospheric lifetime of Hg° vapour, and its oxidation products, generally fall in the 6–36 months range, allowing sufficient time for global transport. Hg° is slightly soluble so may be both wet and dry deposited; however, much of the atmospheric Hg° is oxidised to Hg^{2+} compounds that are more water soluble than Hg° and thus more readily deposited, with shorter lifetimes of hours to months, depending on the compound. Oxidised Hg may also become attached to fine particles in the atmosphere, which have varied residence times based partly on their size and density; retention on particles is less likely to be an important fate for the Hg° form, because of its high vapour pressure. The other main Hg gas, DMHg, may be photolysed in the atmosphere to Hg° and two methyl radicals ($CH_3^{\cdot}$) and/or to MMHg and one methyl radical; MMHg is water soluble and readily wet deposited, as evidenced by studies of rainfall. As expected from the relatively short atmospheric lifetime of Hg, similar amounts are deposited from the atmosphere each year as are emitted into it. A greater proportion of the naturally derived atmospheric mercury is deposited to the sea than to land, reflecting the global nature of Hg transport in the atmosphere, as well as the sea's larger surface area.

Soils

In mineralised areas, pedogenesis on Hg-containing rocks introduces Hg-bearing mineral particles into soils. Additionally, Hg° degassing from the Earth's crust moves upwards through soils; a proportion is oxidised within the pore space to Hg^{2+} and retained in the soil matrix by adsorption. Most Hg in topsoils is derived from atmospheric deposition, mostly in the already-oxidised Hg^{2+} form, which will again have the potential for retention in soils; for this reason the Hg concentrations of soils are typically elevated above those in the parent rock. Aerially deposited Hg^{2+} can also be microbially reduced to gaseous Hg° and returned to the atmosphere by degassing. Alternatively, it can be methylated by other microorganisms, particularly in wetland soils (and aquatic sediments), to MMHg, which may be taken up by higher organisms, adsorbed on particle surfaces or methylated further to DMHg, which, being volatile, can diffuse into the atmosphere. Methylation to DMHg occurs microbially or by reaction of MMHg with H_2S in a two-step reaction:

$$2(CH_3)Hg^+ + H_2S \rightarrow (CH_3)Hg\text{–}S\text{–}Hg(CH_3) + 2H^+ \qquad [14.1]$$

$$(CH_3)Hg\text{–}S\text{–}Hg\ (CH_3) \rightarrow (CH_3)_2Hg + HgS \qquad [14.2]$$

The second product of equation 14.2 is HgS, a solid with low solubility that may represent a permanent sink for Hg when this reaction occurs in sediments that subsequently undergo lithification. Similarly, inorganic Hg^{2+} can react with H_2S to also form HgS.

Hydrosphere

The natural sources of Hg in the hydrosphere are: (i) direct atmospheric deposition, mainly of Hg^{2+} forms but also of MMHg, which is produced in the atmosphere by the photodegradation of DMHg; (ii) degassing of Hg° and DMHg from, and through, bottom sediments; (iii) biomethylation in sediments, producing MMHg; (iv) inputs from soils via terrestrial runoff of Hg-bearing soil particles, which is likely to be of greater importance than soil leaching, Hg being retained quite strongly in soils by organic matter (acid soils) and clays and Fe oxides (non-acidic soils). Most Hg entering rivers via runoff tends to accumulate in estuarine sediments with perhaps 10% directly reaching the open ocean; more Hg is subsequently transferred into seawater via remobilisation from sediments. The residence time of Hg in ocean waters is several centuries.

In near-surface waters, Hg^{2+} and methylmercury species may be photodegraded to Hg°, which diffuses into the atmosphere. Inorganic Hg may also be methylated to MMHg and DMHg. Methylation occurs biotically, particularly in bottom sediments, and also (more rarely) by the abiotic reaction of Hg^{2+} with methyl groups in dissolved organic matter. Methylated Hg is removed from the water column by biological uptake (MMHg) and degassing (DMHg), particularly following the upwelling of deep waters. MMHg is also removed from the water column by adsorption onto sediment particles and, near the surface, by photolysis. Sediment deposition is a removal mechanism for particulate Hg in the water column, including species that have entered the particulate phase by adsorption/reaction, either in the water column itself or previously in the atmosphere or in soils.

Biosphere

It will be apparent from the previous sub-sections that microbial processes are of fundamental importance when considering the fate of Hg in the environment. The processes of key significance are: (i) reduction of Hg^{2+} to Hg°, performed by many species of bacteria including *Pseudomonas* and *Bacillus* spp.; (ii) biomethylation of Hg^{2+} to MMHg and DMHg; and (iii) the demethylation of MMHg (CH_3Hg) to CH_4 and Hg°. These microbial transformations of Hg are best explained as detoxification mechanisms rather than as methods of obtaining energy or metabolites. In this context, Hg is effectively removed from the local microbial environment to the atmosphere by the production of: (i) volatile Hg°, either by the reduction of Hg^{2+} or the demethylation of MMHg and (ii) DMHg (also volatile), by the methylation of MMHg. The production of MMHg, if subsequently adsorbed onto particle surfaces, also removes Hg from the aqueous, microbial environment. Methylation is of further significance because MMHg is readily absorbed in the tissues of higher organisms; it is fat-soluble and has an affinity for free thiol groups (-SH) in proteins. MMHg has a long biological half-life and is bioaccumulative, unlike inorganic Hg, which is more readily excreted.

Biomethylation of Hg^{2+} to MMHg appears to be undertaken mainly by sulfate-reducing bacteria in partly anoxic, acidic sediments and soils; however, methanogenic microorganisms are also capable of Hg methylation. Methylation appears to be inhibited in completely anoxic environments because total sulfate reduction in such conditions favours the production of insoluble HgS, which cannot be methylated. In biomethylation, the co-enzyme methylcobalamine, a form of vitamin B12, is thought to be involved in the transfer of a methyl group from the amino acid, serine, to the Hg^{2+} ion that is undergoing methylation. The methylation of MMHg to DMHg appears to occur at higher pH levels (neutral and alkaline conditions). The relatively anoxic environments suitable for Hg methylation are found primarily in aquatic sediments and poorly aerated soils, although methylation may also occur in the gastrointestinal tracts of animals.

Production and uses

There is archaeological evidence that Hg was extracted in ancient times in India, China and the Middle East. In Anatolia (modern-day Turkey), cinnabar was being used as a pigment in Neolithic burials in 6000 BC and a small sample of metallic Hg was found in an Egyptian tomb dated to 1600 BC. Large amounts of Hg have been extracted from the mines in Almaden, Spain since Roman times (Fig. 14.3), but, despite still having the world's largest Hg reserves, the mines closed in 2000; the closure was partly because of environmental concerns, and the mines may now be used as a repository for the European Union's Hg waste. Much Hg is now recycled (see below) and the last major Hg mine still in operation, in Khaidarkan, Kazakhstan, is due for closure; however, small-scale Hg mines remain in China, producing Hg for domestic use.

Figure 14.3 Part of the Entredicho mercury mine, Almaden, Spain. Pure mercury was produced by heating cinnabar in these large ovens. One-third of the mercury extracted throughout history is derived from the mines of this one area.

Credit: amata_es (Flickr).

To produce metallic Hg, cinnabar is smelted: the sulfur is driven off as SO_2 and an intermediate Hg compound, mercuric oxide is formed (equation 14.3). This is heated further to produce metallic Hg (equation 14.4):

$$2HgS + 3O_2 \rightarrow 2HgO + 2SO_2 \quad [14.3]$$

$$2HgO \rightarrow 2Hg + O_2 \quad [14.4]$$

Historically, mercury has been used extensively and its use continues today (Table 14.2), albeit with more restrictions than in the past. Because of environmental and health concerns, it has already been largely phased out in some processes and products. For example, it was used to control slime mould in paper manufacture for many years, but waste effluent discharged into rivers often contained Hg and it is no longer used. Mercury compounds have been used in healing and medicine for centuries; for example, calomel (Hg_2Cl_2) was taken as a laxative, used against syphilis and was present in teething powder given to babies. Medicinal applications are largely discontinued because of health concerns, although Hg may still be found in some traditional Chinese medicines.

The predominant use of Hg today is in gold extraction; it is used by millions of 'artisanal' workers and their families worldwide, concentrated particularly in South America, sub-Saharan Africa and eastern Asia. Metallic Hg° is used to make an

Table 14.2 Current uses of mercury.

Use	*Description*	*Reduction measures*
Gold extraction	Amalgam to extract gold from sediments	Largely unregulated; used globally in 'artisanal mining'
Dentistry	Dental amalgam	Still used in most countries; alternatives are not problem-free
Measuring devices	Manometers (e.g. blood pressure); thermometers	Alternatives available but Hg types still widely used for their accuracy
Lighting	Fluorescent bulbs & tubes; street lights (USA & Asia)	Use not restricted; regulations on recycling to recover Hg
Chlor-alkali industry	Electrolysis of brine to produce Cl and NaOH	Target to replace Hg cathodes by membrane technology by 2020
Fungicide	Fungal control in vaccines ('Thiomersal')	Part-restrictions in some countries (i.e. routine childhood vaccinations)
Plastics manufacture	Catalyses formation of solid plastics (polyurethane, etc.)	Proposal in EU to reduce amounts of phenylmercury used
Electronics	Mercury switches in vehicles, thermostats, etc.	Being replaced by alternative switches in some countries
Medicine	Antiseptic liquids; Chinese medicine	Restricted in many countries
Cosmetics	Skin lightening cream/soap; preservative in eye make-up	Restricted in some countries but commonly used in many others
Batteries	'Button cells' in electronic items such as cameras	Restrictions on amounts of Hg permitted in batteries

amalgam with gold particles in river sediments; the heavy amalgam is gravity-separated from the sediments and the Hg is then driven off into air by heating the amalgam, leaving behind the accumulated gold. Dental amalgam, which is typically 50% Hg, 30% silver and 20% tin, copper and/or zinc, is still used in most countries. The main alternatives are resins, ceramics and gold alloy, but they all have disadvantages including toxicological hazards, clinical limitations and higher cost. A major use of Hg is in measuring devices, such as manometers, sphygmomanometers (blood pressure meters) and thermometers, in professional and medical practice. Alternatives for devices such as sphygmomanometers are available, but Hg instruments are valued for their accuracy and their use remains widespread; replacements for Hg thermometers (digital instruments and alternative liquids such as gallium alloys and alcohol dye) are more widely used. Also in the medical area is the continued use of Thiomersal (sodium ethylmercurythiosalicylate) as an antifungal agent in small amounts in vaccines (100 µg Hg per injection). While there are restrictions on the use of Hg in cosmetics, it is still used in skin-lightening creams and soaps and in eye make-up (as a preservative) in Africa and Asia.

These continuing uses mean that there is still a demand for Hg and, with growing restrictions on its extraction and processing, recycling of Hg has grown in recent years. For example, compact fluorescent lamps ('low-energy bulbs') and fluorescent tubes contain several milligrams of Hg and this is collected at many recycling centres. Specialist companies also recycle Hg from excess dental amalgam and old products such as electronic switches and mercury batteries. In other sectors, Hg is being replaced by alternatives. For example, in the chlor-alkali industry, Hg has long been used as the cathode in the electrolysis of brine (NaCl in water) to produce Cl and NaOH, both highly valued products in industry; however, Hg catalysts are now being replaced by more cost-effective membrane technology in all new plants and there is an international target to convert all plants to the new technology by 2020.

Pollution and environmental impacts

Human activities have contaminated the atmosphere, pedosphere and hydrosphere with Hg and continue to do so. The element, in its various chemical forms, is constantly cycled through the whole environment, including the biosphere. Exposure of humans and other organisms to Hg may occur in any part of the environment; however, pollution of the hydrosphere by Hg is particularly problematic because of the production of MMHg in sediments, followed by the bioaccumulation of this compound in the food chain.

Atmospheric pollution

Concentrations of Hg in lake sediments and peat cores provide evidence of the increased deposition of Hg from the atmosphere since the Industrial Revolution and monitoring in the Arctic indicates that Hg continues to be transported widely in the global atmosphere (Box 14.1). Anthropogenic emissions to the atmosphere have fallen to some extent in recent decades (to ~2000 $t\ a^{-1}$), but decreases in Europe, North America and Russia have been countered by rises in Africa, Asia and South America (Fig. 14.4); nearly one-third of global emissions are now from China. Globally, another 5000 t or more of Hg enters the atmosphere each year from the Earth's surface through volatilisation of natural Hg and re-volatilisation of anthropogenic Hg that has previously been deposited from the air; of the total input to the atmosphere (direct emissions and volatilisation combined), 90% is thought to be derived from anthropogenic activity (UNEP, 2013). This indicates that, despite the presence of natural sources, the vast majority of Hg emitted to the atmosphere each year is ultimately derived from human activity and will continue to be so for some considerable time, regardless of reductions in direct emissions.

Box 14.1 Mercury in the Arctic.

A clear indication of the global nature of the Hg cycle, due mainly to its volatile nature and worldwide atmospheric transport, is the contamination of fish and mammals in the Arctic, far remote from the major anthropogenic sources of Hg. The contaminated organisms contain sufficient MMHg to be a risk to the health of their indigenous consumers. Scientists (starting with a discovery by Schroeder et al., 1998) have hypothesised that atmospheric Hg° imported from lower latitudes reacts with elevated levels of Br and Cl that are released from ice crystals in polar stratospheric clouds in the northern springtime; this may result in the formation of oxidised Hg^{2+} species that are more readily dry-deposited to the Arctic surface than Hg°, much of which would otherwise remain aloft and/or drift by in the wind. However, this has not been verified definitively and alternative hypotheses have also been proposed.

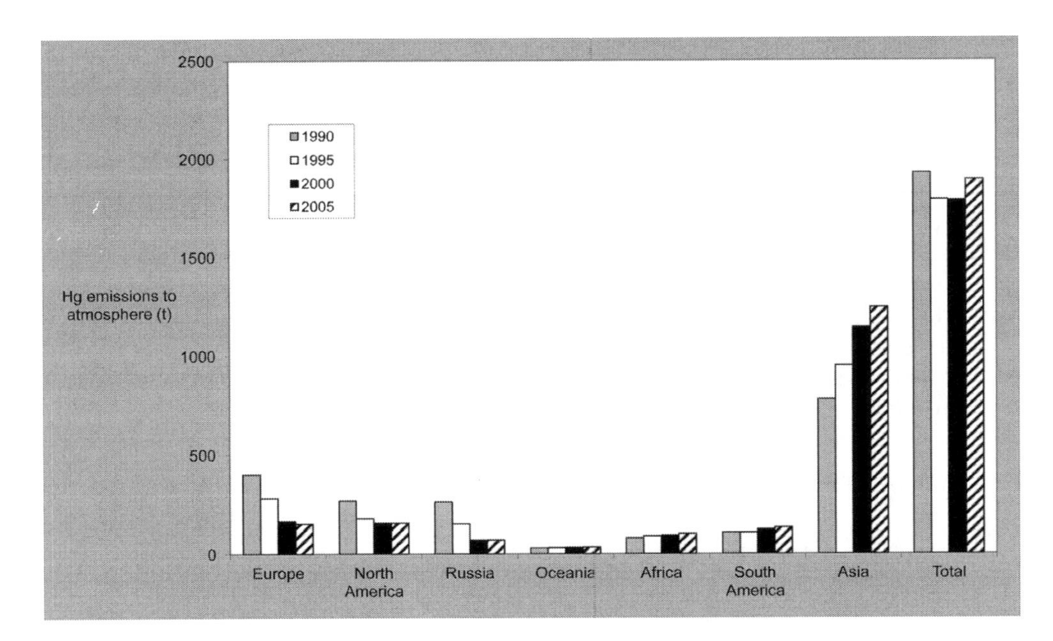

Figure 14.4 Global trends in mercury emissions.

Source: Adapted from UNEP (2013).

Fossil fuel combustion remains a significant source of Hg emissions (Table 14.3), despite a gradual decrease over recent decades that has resulted from better pollution control technologies and tighter regulations. Artisanal gold extraction, which is largely unregulated, is perhaps the major human source of atmospheric Hg today. Some of the smaller sources on a global scale may be of more importance in certain countries. For example, crematoria account for 16% of Hg emissions in the UK; a single crematorium emits several kilograms of Hg per year from the volatilisation of dental amalgam, although this may decrease following the recent introduction of filters to UK crematoria.

Table 14.3 Anthropogenic sources of mercury emissions to the atmosphere.

Source	*Emissions (t a^{-1}) (Pirrone et al., 2010)*	*Emissions (t a^{-1}) (UNEP, 2013)*
Fossil fuel combustion	800	484
Small-scale gold extraction[a]	400	727
Metals manufacturing	300	193
Cement production[b]	240	173
Waste incineration	190	NS
Chlor-alkali process	160	28.5
Large-scale gold mining	NS	97.5
Consumer product waste	NS	96
Contaminated sites	NS	83
Mercury mining & production	50	12
Iron and steel production	43	46
Coal bed fires[c]	32	NS
Vinyl chloride production	24	NS
Oil refining	NS	16
Cremation	NS	4
Other	65	NS
Total[d]	2304	1960

a The larger total for gold extraction in the more recent report may be a reflection (at least partly) of improved data-gathering.
b Attributed in large part to the use of Hg-containing wastes in kiln firing.
c Refers to uncontrolled fires of coal seams associated with human activity.
d The lower total in the more recent source does not necessarily indicate a reduction in emissions because it does not include sources for which emissions are not quantified, according to UNEP (including waste incineration for example).
NS = not stated.

Soil contamination

Soils are mainly contaminated by atmospheric deposition of anthropogenically derived Hg. Soils retain much of the deposited Hg, by complexation, adsorption and precipitation and, as a result, Hg concentrations are generally higher in topsoils than subsoils and higher in urban and industrial areas than in the countryside. For example, soils in urban areas of the UK have a mean Hg concentration of 350 μg kg^{-1} compared with 130 μg kg^{-1} in rural soils (Morgan et al., 2009). Direct inputs to agricultural soils of phosphate rock fertiliser and sewage sludge are further sources of Hg, which is present in small concentrations in both. Contaminated soils can also be a direct source of Hg pollution in waters, mainly via runoff of solid particles, with additional inputs from leaching of dissolved complexes and MMHg. Topsoil, as a major reservoir of anthropogenic Hg, is a source of constant Hg reintroduction to the atmosphere and therefore of its cycling in the wider environment.

Water pollution

UNEP (2013) has estimated the amounts of Hg released to water globally as a result of human activity. Today, most of the discharges into streams and rivers arise from artisanal gold mining, while the once important chlor-alkali industry is now a relatively small and decreasing source. In addition to these discharges, and others shown in Fig. 14.5, each year a much larger amount of Hg (several thousand tonnes) is deposited to waters, mainly the oceans, from the atmosphere; a small proportion of this atmospheric Hg is from natural degassing but most has resulted from human activity. Anthropogenic pollution has approximately doubled the Hg concentration in surface seawater over the last century; concentrations have continued to increase in the north Pacific in recent years, reflecting industrialisation in Asia, while decreases have been observed in the north Atlantic and the Mediterranean Sea, reflecting lower levels of discharge in North America and Europe, concurrent with de-industrialisation (UNEP, 2013). Human inputs of Hg to the oceans raise concern mainly because of the methylation of inorganic Hg to MMHg, which is bioaccumulative and toxic to humans and other organisms.

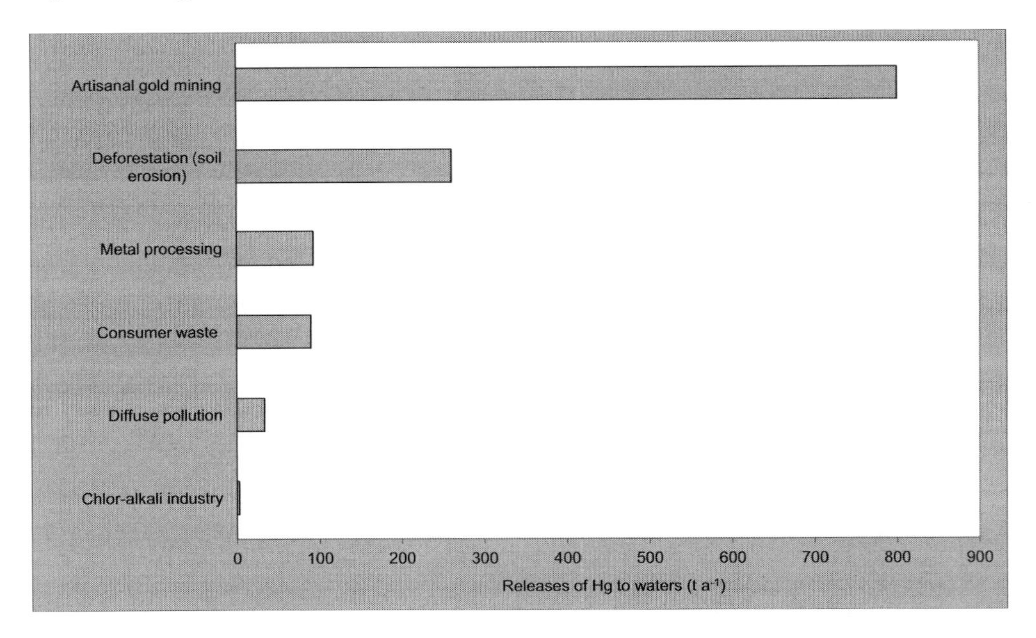

Figure 14.5 Estimated mercury releases to freshwaters (t a^{-1}). Diffuse pollution is from contaminated mine and industrial sites (the value shown is the estimated maximum). Consumer waste processing includes landfill sites and metals recycling facilities.

Source: Data from UNEP (2013).

Mercury bioaccumulation

The bioconcentration, bioaccumulation and biomagnification of MMHg occurs predominantly in the marine environment and wetlands, where MMHg is produced in bottom sediments and wetland soils, respectively. The problems arise because MMHg is readily absorbed into the blood and tissues of animals but is metabolised and eliminated slowly. For example, most MMHg ingested by fish (but only a small proportion of inorganic Hg) is absorbed and >95% of Hg in fish is typically methylmercury (USEPA, 1997). Therefore, MMHg accumulates in the food chain; the largest concentrations tend to occur in older individuals (because of bioaccumulation) and those organisms at the highest trophic levels (because of biomagnification), including large carnivorous fish such as tuna and marine mammals such as seals. MMHg is particularly strongly bioconcentrated by phytoplankton from marine waters, by factors of up to 10^5, and after subsequent biomagnification in the food chain (via zooplankton, prey fish and predator fish, for example), top predators can have tissue concentrations that are several million times higher than in the surrounding water. The biomagnification continues further through the consumption of predator fish by piscivorous birds and mammals.

Mercury ecotoxicity

Mercury has toxicological effects on various types of organisms, mainly because of its bioaccumulation and biomagnification (as MMHg) in the food chain. In fish, exposure to elevated Hg in the wild appears to mainly affect liver function and reproduction, including reductions in egg hatching and embryo fitness. Severe Hg pollution (e.g. Minamata Bay, Japan, see below) causes increased mortality and neurological defects (e.g. reduced locomotor activity) in fish. Exposure to elevated environmental levels of Hg reduces survival and/or reproductive success in some bird species, including common tern (*Sterna hirundo*), common loon (*Gavia immer*) and wood stork (*Mycteria americana*); this may be related to behavioural impacts of Hg, resulting in some birds spending less time fishing and protecting their eggs. Egg weight and size is also affected in some species; this was observed in Cartagena Bay, Colombia, which has Hg-contaminated sediments associated with a former chlor-alkali plant (Table 14.4). Reproductive success also appears to be affected in some terrestrial songbirds that feed on invertebrates in Hg-contaminated areas. In some amphibians, environmental exposure to MMHg affects development of larvae. Exposed mammals exhibit similar neurological symptoms to those observed in humans (see below); exposure to MMHg via the food chain has been suspected as the cause of death of carnivorous and piscivorous mammals such as otters, foxes and wild cats, where post-mortems showed elevated body burdens of Hg. Studies of polar bears indicate possible neurochemical effects linked to Hg exposure.

Mercury does not appear to be particularly toxic to terrestrial plants, probably because the main forms in soil are generally poorly bioavailable. Plants exposed to Hg tend to concentrate it in the roots (after soil Hg exposure) or the shoots (Hg° vapour exposure) and this may protect the other parts of the plant. However, exposure to very high concentrations of Hg affects photosynthesis, transpiration, cell permeability and genetic processes in some plants.

Table 14.4 Measurements of snowy egret (*Egretta thula*) eggs from two areas of northern Colombia; a coastal marsh with no industrial inputs and the bay of Cartagena city, which has mercury-contaminated sediments. *Egretta thula* is a piscivore and occupies a high trophic level.

Measured parameter	*Totumo Marsh (unpolluted)*; n = 20	*Cartagena Bay (polluted)*; n = 20
Hg concentration in eggshell (ng g^{-1})	7.3	19.9*
Hg concentration in egg contents (ng g^{-1})	27.1	29.7
Egg weight (g)	24.4	21.7*
Egg length (mm)	44.3	42.6*
Shell thickness (µm)	196	189*

* Statistically significant differences between the two areas ($p < 0.05$).

Source: Data from Olivero-Verbel et al. (2013).

Human exposure and toxicity

The main inhalation exposure source for many in the general population is volatilisation of Hg° from dental amalgam in filled teeth, which may lead to daily absorption levels of up to 17 µg Hg (Nielsen and Grandjean, 2000). Inhalation of Hg° vapour from the atmosphere, including in industrial areas, is not a particular hazard, but risks arise from the release of Hg° vapour in a confined area, in industrial settings or resulting from thermometer breakage, for example. Saturated air above a pool of metallic Hg contains 14 mg Hg° m^{-3} (at 20°C) because of its high vapour pressure, compared with the typical safety maximum of 0.1 mg m^{-3}. There is also a risk from the volatilisation of Hg from Hg-containing latex paint in houses where this now-banned substance was formerly used. The term 'mad as a hatter' derives from the neurological problems that befell Victorian felt hat makers who inhaled the fumes given off by mercuric nitrate solution, which was used to process animal skins. Similar problems were documented in goldsmiths, who used mercury–gold amalgam for gilding.

The main exposure source for most humans is the consumption of fish containing accumulated amounts of MMHg, up to 90% of which is absorbed by the human body. Freshwater fish typically have higher MMHg concentrations than marine fish, but consumption of either can be problematic. In some populations, marine mammals are an additional dietary source. Water ingestion is a potential source in the general population; drinking water standards for Hg are enforced in many countries (based on a WHO guideline value of 6 µg L^{-1}) to guard against this potential ingestion route.

Historically, the most severe cases of Hg exposure via the diet were in Minamata Bay, Japan in the 1950s and 1960s (Box 14.2); the symptoms of the Hg exposure in this case have become known as Minamata Disease. In Iraq, during a famine in 1970–71, >450 fatalities occurred when people made bread from seed (sent as aid) that had been intended only for planting and had been treated with methylmercury fungicide; similar problems occurred in other countries receiving similarly treated seed.

Box 14.2 A warning from history: Minamata Bay.

In 1932, Japan's largest chemical factory, owned by the Chisso Corporation, started producing acetaldehyde, an important raw material in the chemical industry. The process used mercuric sulfate ($HgSO_4$) as a catalyst and MMHg, formed in the production process, was emitted into Minamata Bay in the effluent stream (i.e. MMHg was not formed in the bay via bacterial action on discharged inorganic Hg, as sometimes stated). The discharges occurred for 36 years before a different production method was introduced in 1968. By the 1950s local people had shown symptoms of central nervous system (CNS) damage, in some cases leaving permanent disability and blindness; subsequent research suggests that the first symptoms may have occurred as early as 1942. A committee was established to investigate and heard that cats in the area had been showing CNS symptoms since 1950 and that seaweed and fish in the bay had been dying. In 1959, MMHg in locally caught seafood was identified as the cause of the illnesses and the company was told to install pollution control equipment. However, during the following decade many more cases became apparent and babies of exposed women were born with congenital diseases; subsequent research has shown elevated Hg concentrations in the umbilical cords of babies born during this period (Fig. 14.6). The company was later accused of not installing (following the 1959 committee hearings) sufficiently effective equipment to strip Hg from the effluent.

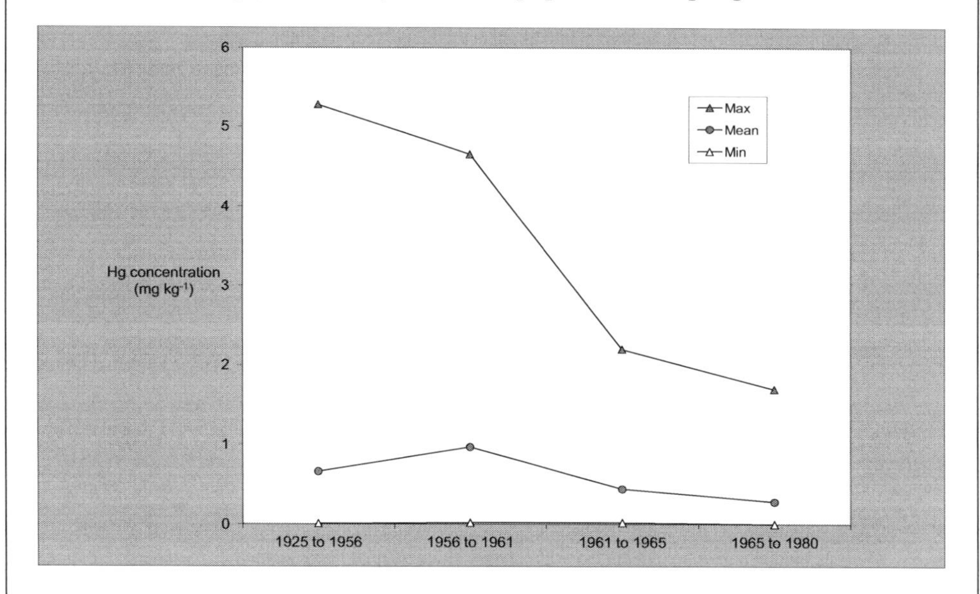

Figure 14.6 Mercury concentrations in umbilical cords of 267 residents of mercury-polluted areas in Minamata Bay, Japan, who were born between 1925 and 1980 (cords are retained by families in Japanese culture). No controls were included, but comparisons with other studies indicate that the mean concentrations shown here are several orders of magnitude higher than in unpolluted areas.

Source: Data from Yorifuji et al. (2009).

In the early 1970s the Japanese government acknowledged the extent of the problem and 1.5 mn m^3 of sediment were dredged from a 58 hectare area of the bay in a 14-year remediation project funded by the company and the government. An estimated 150 t of Hg had accumulated in the bay (most had formed HgS). Fishing resumed in the bay in 1997 and levels of Hg in locally caught fish now meet the national 0.4 mg kg^{-1} standard. More than 2000 people were officially certified as suffering health problems caused by Hg exposure, but the number of people with milder symptoms, including hearing loss, is thought to be much higher. More than 900 people died and 17 babies were born with congenital diseases with symptoms similar to cerebral palsy; a further 49 cases of congenital disease with milder symptoms were subsequently diagnosed. Compensation payments to surviving victims and fishermen began in 1995; these were funded by the company with financial support from the government. A comprehensive report by Yorifuji et al. (2013), which forms the basis of this case study, is recommended as further reading.

In addition to these historic incidents, there is ongoing concern about human health impacts of current Hg pollution. Among those considered to be most at risk are those with a diet rich in fish and the offspring of women exposed in this way during pregnancy. The WHO guideline for tolerable intake of MMHg via the diet is 1.6 µg per kg body weight per week, equal to a daily intake of 16 µg Hg for a 70 kg adult. This is unlikely to be breached in most consumers, but in polluted areas there is more cause for concern; for example, US authorities advise their citizens not to eat fish from some locations (Box 14.3). Precautions are also required for populations whose diet includes marine mammals because of Hg biomagnification; for example, large cohort epidemiological studies in the Faroe Islands indicate CNS effects in children of women who have high dietary Hg intake during pregnancy.

Box 14.3 Present-day mercury pollution: Great Lakes Region, USA.

The Great Lakes of the north-eastern USA, on the border with Canada, comprise the world's largest freshwater lake system and hold approximately 20% of the Earth's surface freshwater. They are a major source of recreation, with more that 10 million people sport fishing in the region each year, bringing an estimated \$20bn and 190,000 jobs to the local economy. Commercial fishing is still an important industry too, although much reduced from its heyday for a number of reasons. The lakes are polluted by Hg and the source has long been identified as the 144 coal-fired power stations located among the eight states that surround the lakes (Table 14.5). These power stations are thought to account for roughly one-quarter of Hg emissions in the USA. Approximately 20% of their Hg emissions to the atmosphere are deposited in the Great Lakes' catchment area, which also extends northwards into Canada. The deposited Hg accumulates in the lakes and their food chains, particularly in the more forested parts of the catchment because trees capture Hg from the air. In the same regions there tend also to be more wetlands – areas where MMHg is more likely to form compared with well-aerated, dry soils and where the lower pH of the lake water encourages its formation.

Table 14.5 Mercury emissions, projected reductions and prevalence of angling in the Great Lakes states, USA.

State	*Hg emissions from coal-fired power plants in 2010 (kg a^{-1})*	*Projected emissions reductions due to MATS[a] (%)*	*Anglers as percentage of state population*
Illinois	674	86	7
Indiana	987	56	12
Michigan	873	84	14
Minnesota	396	83	27
New York	109	65	5
Ohio	1301	61	12
Pennsylvania	1235	56	8
Wisconsin	576	85	23

a MATS = Mercury and Air Toxics Standards (see text). Projected for coal-fired power plants by 2016.

Source: Adapted from Stamper et al. (2012).

The main wildlife species affected are fish-eating birds, like the common loon (*Gavia immer*, with effects on reproduction and survival) and the great snowy egret (*Egretta thula*, liver damage), and mammals such as otter and mink (reduced reproduction rate). Six commonly eaten fish species have been studied and, across most of the Great Lakes region, found to have Hg levels higher than 0.3 mg kg^{-1}, the US Environmental Protection Agency (USEPA) safety guideline or 'reference dose'. Human health is a major concern; research by the Department of Health in one of the affected states (Minnesota) found elevated Hg in 8% of babies born between 2008 and 2010. Angling is a popular pastime in these states (Table 14.5) and, while 'fish consumption advisories' are declared by state authorities across the region, many locals are unaware of them. The advisories typically advise limiting the amount of affected species eaten to one per week or month in, respectively, the general or sensitive populations (e.g. women of child-bearing age); for some species no consumption is advised. The USEPA has introduced the Mercury and Air Toxics Standards, which require stringent air pollution controls in power stations, and has projected that Hg emissions from coal plants will be reduced by 72% on average across the eight states surrounding the Great Lakes. It says the costs of implementation will be far outweighed by the financial benefits of improved environment and health, saving several thousand premature deaths caused by air pollution each year. See Stamper et al. (2012) for a full account of this case study.

All forms of Hg are toxic. Inhaled Hg° enters the bloodstream and readily passes through cell membranes and through the blood–brain and placental barriers; Hg^{2+} passes these barriers to a lesser extent. Organomercury compounds are particularly toxic; MMHg is readily absorbed into the bloodstream and is metabolised slowly in the body (by demethylation). In the body, Hg attaches to the sulfhydryl groups of amino acids and can therefore disrupt enzymes; it inhibits protein synthesis and affects the permeability of cell membranes. Inorganic and organic Hg accumulates in the kidney and the brain. Like Hg°, MMHg crosses the placenta into the foetus and it is possible that this is actually a detoxification measure to protect the mother's body; in the mass Hg poisoning at Minamata Bay, Japan, mothers of children born with congenital defects had only minor symptoms, possibly because the MMHg had been sequestered in the foetus.

Acute Hg° vapour poisoning causes airway and lung disruption, while acute poisoning via ingestion disrupts the gastrointestinal tract, causing vomiting, diarrhoea and severe stomach pain. Chronic poisoning by Hg° vapour disrupts the CNS, manifested most commonly as erethism (irritability, insomnia, memory loss, depression), muscle tremor/spasm, gingivitis and salivation. Chronic Hg poisoning can also cause mild renal disruption and, in the past, a common symptom was pink-disease (characterised by red and flaky skin on the hands and feet), usually in infants who had been exposed to calomel in teething powder. Poisoning by ingestion of MMHg causes paraesthesia (tingling sensation) of the fingers and the mouth, followed by ataxia (lack of muscle coordination resulting from damage to the CNS) and dysphasia (loss of verbal communication skills). Foetal exposure may lead to congenital defects including cerebral palsy syndrome; severe cases can lead to total blindness or deafness. Mercuric chloride and methylmercury are listed as possible human carcinogens by the USEPA (2000).

References

Amos, H.M., Jacob, D.J., Streets, D.G. and Sunderland, E.M. 2013. Legacy impacts of all-time anthropogenic emissions on the global mercury cycle. *Global Biogeochemical Cycles* 27, 1–12.

Harada, M. 2007. Intrauterine methylmercury poisoning – congenital Minamata disease. *Korean Journal of Environmental Health* 33(3), 175–179.

Morgan, H., de Burca, R., Martin, I. and Jeffries, J. (UK Environment Agency). 2009. *Soil Guideline Values for Mercury in Soil: Science Report SC050021*. Environment Agency, Bristol.

Nielsen, J.B. and Grandjean, P. 2000. Mercury. In: Lippmann, M. (Ed.). *Environmental Toxicants*. John Wiley & Sons, New York.

Olivero-Verbel, J., Agudelo-Frias, D. and Caballero-Gallardo, K. 2013. Morphometric parameters and total mercury in eggs of snowy egret (*Egretta thula*) from Cartagena Bay and Totumo Marsh, north of Colombia. *Marine Pollution Bulletin* 69(1–2), 105–109.

Pirrone, N., Cinnirella, S., Feng, X., Finkelman, R.B., Friedli, H.R., Leaner, J., Mason, R., Mukherjee, A.B., Stracher, G.B., Streets, D.G. and Telmer, K. 2010. Global mercury emissions to the atmosphere from anthropogenic and natural sources. *Atmospheric Chemistry and Physics* 10, 5951–5964.

Schroeder, W.H., Anlauf, K.G., Barrie, L.A., Lu, J.Y., Steffen, A., Schneeberger, D.R. and Berg, T. 1998. Arctic springtime depletion of mercury. *Nature* 394, 331–332.

Stamper, V., Copeland, C., Williams, M. and Spencer, T. (contributing editor). 2012. *Poisoning the Great Lakes*. National Resources Defense Council, New York. Available at: http://www.nrdc.org/air/files/poisoning-the-great-lakes.pdf

UNEP (United Nations Environment Programme). 2013. *Global Mercury Assessment 2013: Sources, Emissions, Releases and Environmental Transport*. UNEP Chemicals Branch, Geneva.

USEPA (United States Environmental Protection Agency). 1997. *Mercury Study: Report to Congress. Volume III: Fate and Transport of Mercury in the Environment*. Available at: http://www.epa.gov/ttn/oarpg/t3/reports/volume3.pdf

USEPA (United States Environmental Protection Agency). 2000. *Air Toxics Web Site: Mercury Compounds*. Available at: http://www.epa.gov/ttn/atw/hlthef/mercury.html

Yorifuji, T., Kashima, S., Tsuda, T. and Harada, M. 2009. What has methylmercury in umbilical cords told us? – Minimata disease. *Science of the Total Environment* 408(2), 272–276.

Yorifuji, T., Tsuda, T. and Harada, M. 2013. Minimata disease: a challenge for democracy and justice. In: European Environment Agency (EEA). *Late Lessons from Early Warnings: Science, Precaution, Innovation*. EEA, Copenhagen.

15 Nickel

Nickel in the natural environment

Nickel (Ni) is a major component of the Earth's core but much less is present in the crust. Concentrations in sedimentary and acidic rocks are generally <100 mg kg^{-1} but up to 2000 mg kg^{-1} are recorded in basic and ultrabasic rocks such as serpentinite. The main Ni mineral is the sulfide pentlandite ($(Ni,Fe)_9S_8$), which is usually found in deposits of basic and ultrabasic rocks. Other major ores are millerite (NiS), the arsenides nickeline (NiAs) and chloanthite ($NiAs_{2-3}$) and the heavily weathered laterites or 'garnierites' ($(Mg,Ni)_6[SiO_4]_{10}(OH)_8$), some of which contain 10% Ni. The metal is also common in ferromagnesian minerals such as the iron oxide limonite, replacing Fe and/or Mg in such minerals. Nickel has an affinity for organic matter and is present in oil and coal. Meteorites are typically rich in Ni, containing 1–2% on average and form the major natural source of metallic Ni at the Earth's surface. In the natural environment Ni mostly occurs in the divalent form, Ni^{2+}.

Mean Ni concentrations reported in soils are generally <100 mg kg^{-1} (Table 15.1), but higher levels are observed in mineralised areas; soils that have developed on serpentinite contain up to several thousand mg kg^{-1}. Nickel in soils is associated with grains of Ni-bearing minerals and organic matter. It is not as strongly adsorbed in soils, especially acidic soils, as many other toxic metals, including Cr, Pb and Zn. Species present in the soil solution are mainly free ionic Ni^{2+} and complexes with dissolved organic matter and inorganic ligands such as sulfate and phosphate (acidic soils) and carbonate and hydroxyl (alkaline soils). Nickel is readily taken up by plants, especially in acidic soils, and there are many Ni hyperaccumulator species; there are records of plants growing on serpentinite soils containing nearly 2% Ni. Despite the apparent bioavailability of soil Ni, the element is not particularly mobile and, like many other divalent metals, is largely retained in topsoils.

Background concentrations in freshwaters and seawaters are generally low (Table 15.1); higher levels in drinking waters may be due to leaching from pipes rather than natural sources. Dissolved forms are mainly Ni^{2+}, $NiCl^{+}$ and complexes with dissolved organic matter. Natural sources of Ni in the atmosphere include windblown soils, volcanoes, forest fires, sea salt and meteoric dust.

Nickel is essential in plants for normal growth. The ionic radius of Ni is similar to Fe and some other micronutrients and it can replace them in enzymes. In plants, cyanobacteria and anaerobic bacteria, it is present in various enzymes including urease (which converts urea to ammonia), hydrogenases and dehydrogenases. Essentiality in animals and humans is not proven.

Table 15.1 Typical background concentrations of nickel.

Environmental medium	*Typical background concentrations*[a]
Air	<20 ng m^{-3}
Soil	<100 mg kg^{-1}
Vegetation[b]	<10 mg kg^{-1}
Freshwater	<10 μg L^{-1}
Sea water	<1 μg L^{-1}
Sediment	<60 mg kg^{-1}

a These are *typical* values, encompassing the vast majority of reported concentrations, based on a large number of literature sources. Higher natural concentrations occur in some cases; for example, in mineralised areas.
b Including crops. Not including metal-tolerant species.

Production and uses

The main ores mined commercially are deep-mined pentlandite and the lateritic garnierite, which is extracted by open-cast methods. Most extraction occurs in the Philippines and Indonesia, followed by Russia, Australia and Canada; global reserves are approximately 74 Mt, but additional resources include scrap metal and sea-floor deposits (USGS, 2014). The lateritic and sulfidic ores are processed using different methods but, broadly speaking, both involve the production, typically in an electric furnace, of a liquid matte comprising roughly 50% Ni and 50% slag containing other metals. The matte is refined using roasting and/or reduction methods to produce metallic Ni. Nickel may be purified by the Mond process: first Ni oxide (NiO) is reduced by reaction with H_2 to create impure Ni, which is then reacted with CO to produce Ni carbonyl ($Ni(CO)_4$); the latter is then decomposed by heating, leaving pure Ni. The intermediate gas, nickel carbonyl, is extremely hazardous.

The main uses of nickel are stainless steel (approximately 8% Ni) and the Ni-superalloys, especially Ni aluminide (Ni_3Al), which resist corrosion and are stable at high temperatures (Figure 15.1). Nickel also forms useful alloys with metals such as Fe, Cr and Cu and it is used extensively in electroplating. It is employed as a catalyst in hydrogenation, in processes such as vegetable oil production. An important end-use is in fuel cells and nickel-metal hydride (Ni-MH) batteries, which have become more common than Ni-Cd batteries. Nickel is also used in coins, jewellery and spectacle frames.

Pollution and environmental impacts

Most emissions of Ni to the atmosphere are from the combustion of oil and petroleum products; coal combustion and metal smelting are also important sources. There are further emissions from waste incineration (mainly due to the inclusion of batteries in domestic waste) and steel works. Atmospheric Ni concentrations of 100–200 ng m^{-3} are typical for industrial areas; i.e. 1–2 orders of magnitude higher than background levels. Airborne Ni is virtually all in the form of particulate matter and most emitted Ni is deposited close to source, as evidenced by high levels of Ni in soils in industrial areas and close to roads.

Figure 15.1 Jet engines utilise superalloys, which contain nickel as the main base alloying element.

Credit: Farhan Amoor (Flickr).

Nickel mining and smelting have long been the primary sources of Ni contamination in two areas considered to be among the most polluted on Earth: Norilsk in northern Siberia, Russia and Sudbury in Canada (Fig. 15.2), known as the 'Nickel Capital of the World'. Nickel has accumulated to high levels in the soils of both areas (Table 15.2 shows concentrations at Sudbury). This contamination, together with the presence of other pollutants (particularly SO_2), affects plants (see below) and microbial processes (e.g. reduced organic matter mineralisation). Away from industrial areas, agricultural soils can receive inputs of Ni from P fertilisers, which usually contain moderate amounts of the element, although concentrations of up to 1000 mg kg^{-1} are possible. Sewage sludge application can also increase soil Ni levels above background concentrations; bioavailable Ni (particularly in acidic soils) is readily taken up by plants so care has to be exercised with fertiliser applications because of the phytotoxicity of Ni.

The main environmental impact of soil contamination by Ni is phytotoxicity. In polluted soils plant uptake may be significant and lead to Ni concentrations in plants and crops that are an order of magnitude higher than background levels. Outcomes of Ni phytotoxicity include chlorosis and impairments to photosynthesis, transpiration, N-fixation and growth. Nickel interacts closely with Fe and appears to inhibit Fe translocation from root to shoot. Severe pollution, as observed in Ni smelting areas like Norilsk and Sudbury, can devastate vegetation over a radius of several kilometres (Box 15.1), but some species, such as *Alyssum* spp. and *Agrostis* spp., have some tolerance and are able to sequester and hyperaccumulate Ni.

Figure 15.2 Copper Cliff, one of three communities centred around the smelting complex of Sudbury, Canada.

Credit: P199 (Wikimedia Commons).

Table 15.2 Nickel concentrations in the topsoils (0–5 cm) of three communities in the vicinity of nickel smelters at Sudbury, Canada.

Sample location	n	*Mean concentrations*[a] *(mg kg^{-1})*	*Maximum concentration*[b] *(mg kg^{-1})*
Residential areas	752	336–1017	3700
Schools and daycare centres	9	97–1452	2500
Parks	35	300–959	3649
Background[c]	254	36	163

a Range of mean concentrations from the three communities; school surveys only undertaken in two of the communities.
b Maximum concentration of the combined sampling areas.
c Based on parent materials, which were analysed to estimate pre-industrial Ni concentrations in local soils. Compare also with table 15.1.

Source: Data from SARA Group (2004).

Box 15.1 Ecotoxicity at Sudbury.

Studies undertaken by SARA Group (2009) indicate that animal and plant communities surrounding the Ni-smelting complex at Sudbury, Canada are impacted by elevated soil concentrations of Ni and other pollutants (Table 15.3). Small mammals are at particular risk from Ni; for example, at least 10% of meadow voles (*Microtus pennsylvanicus*) at some sites are exposed to 1.3–2.7 times the 'toxicity reference value' for Ni, the concentration below which significant adverse effects are not expected. Impacts on plants are widespread and areas closest to the smelters are known as 'the barrens', being devoid of self-sustaining plant populations; in fact, reduced habitat quality here, and in the wider study area, is thought to be a more significant risk to small mammals (and nesting birds) than direct toxicity. A costly regreening programme over many years has involved grass seeding and the planting of thousands of trees across the impacted landscape. The affected soils have also been limed to increase soil pH and reduce the availability of Ni and other metals. In the first three decades of the programme, 3300 hectares of affected land had been revegetated, but 30,000 hectares of polluted soils still await treatment (City of Greater Sudbury, 2014).

Table 15.3 Ecological risk assessment at Sudbury, Canada. For each assessment criterion, the 19 sites that were surveyed in the smelting area were distributed into three impact levels.

Assessment criteria	*Impact in comparison with reference site*[a]		
	None or low	*Moderate*	*Severe*
Plant community survey including biodiversity	1	7	11
Toxicity to plants and earthworms[b]	1	3	9
Soil characterisation for healthy plant growth	2	13	4
Litter decomposition (microbial mineralisation)[c]	2	2	13
Final ranking[d]	0	8	10

a Reference sites (n = 3) were outside the area impacted by metal emissions.
b One site classed as low–moderate; five as moderate–severe.
c Two sites unclassified.
d One site unclassified.

Source: Data from SARA Group (2009).

The main source of Ni pollution in most surface waters is domestic wastewater; Ni is present in personal care products such as deodorants and toothpaste, but the main sources in wastewater are probably detergents. While such effluents are very widespread, their Ni concentrations are generally quite low and do not normally significantly raise background levels in rivers. More problematic, albeit more localised, are emissions from industry; for example, in recent years, rivers affected by emissions at Norilsk have had Ni concentrations that are several hundred times higher than accepted target levels and Ni levels in fish are much higher than in background areas (Fig. 15.3). Other sources of Ni in waters include leachate from landfill sites and from industrially contaminated soils, particularly in areas with acid soils and waters. The WHO drinking water guideline for Ni (70 $\mu g\ L^{-1}$) is set to protect against leakage from plumbing as well as the possibility of elevated levels from natural or industrial sources.

Nickel is a human carcinogen. In occupational exposure, insoluble Ni compounds can be deposited in the airways and remain there indefinitely, which may be a factor in the cases of lung and nasal cancer observed in Ni-industry workers. Gastric and renal cancers have also been reported. Nickel can substitute for Zn and Mg in DNA polymerase, causing mistakes in DNA replication. Another occupational hazard is inhalation of the very toxic Ni tetracarbonyl ($Ni(CO)_4$), which can cause fatal pulmonary oedema and other severe effects. In the general human population, health problems are mainly related to dermatological allergies (including 'nickel itch' and eczema) from jewellery and coins.

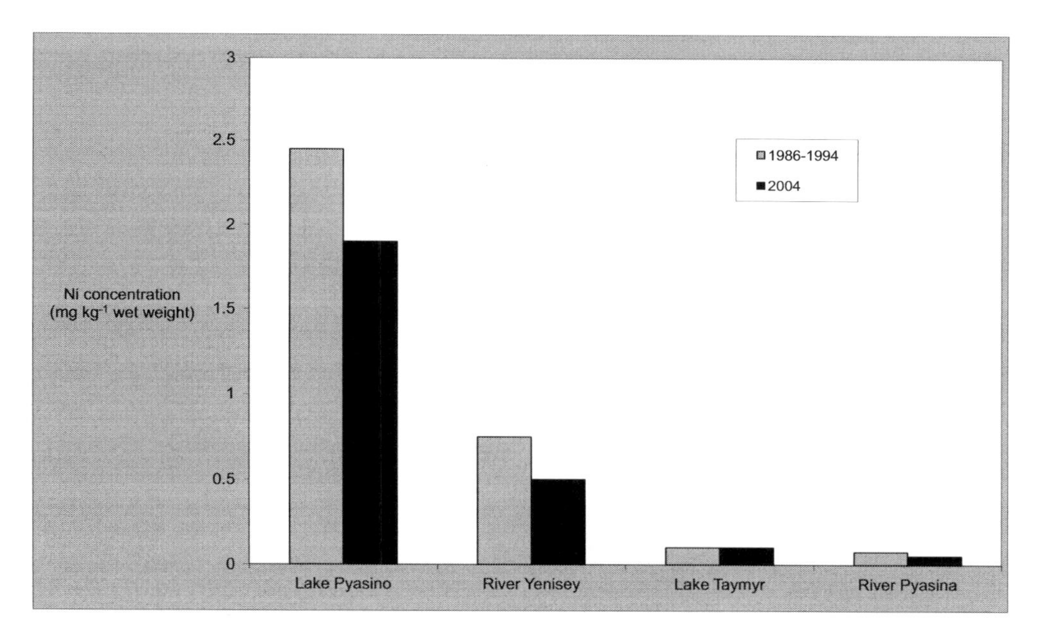

Figure 15.3 Nickel concentrations in fish at different distances from the Norilsk smelting complex, Russia, in two different sampling periods. Concentrations were determined in the livers of burbot fish (*Lota lota* L.), which are considered a delicacy in Russia. The approximate distances from Norilsk for the first three locations (from left to right) are 17, 80 and 700 km, respectively; neither the sampling location on the River Pyasina nor its distance from Norilsk are given but the site is described as being the most distant, together with Lake Taymyr (700 km). Lower concentrations in the more recent survey indicate a slightly improved situation following the adoption of pollution control measures before 2004, but Norilsk is still considered one of the most polluted locations on Earth.

Source: Adapted from Zhulidov et al. (2011).

References

City of Greater Sudbury. 2014. Timeline website. Available at: http://www.greatersudbury.ca/living/environmental-initiatives/regreening-program/timeline/

SARA Group. 2004. *Summary Report: Sudbury Soils Data*. SARA Group, Guelph.

SARA Group. 2009. *Sudbury Soils Study. Summary of Volume III: Ecological Risk Assessment*. SARA Group, Guelph.

USGS (United States Geological Survey). 2014. *Mineral Commodity Summaries: Nickel*. Available at: http://minerals.usgl.gov/minerals

Zhulidov, A.V., Robarts, R.D., Pavlov, D.F., Kamari, J., Gurtovaya, T.Y., Merilainen, J.J. and Pospelov, I.N. 2011. Long-term changes of heavy metal and sulphur concentrations in ecosystems of the Taymyr Peninsula (Russian Federation) north of the Norilsk Industrial Complex. *Environmental Monitoring and Assessment* 181, 539–553.

16 Selenium

Selenium in the natural environment

Selenium (Se) is a metalloid (semi-metal), having properties of both metals and non-metals. It is closely associated, chemically and biogeochemically, with sulfur. The main oxidation states in the natural environment are Se^{2-} (selenide), Se^0 (elemental Se), Se^{4+} (selenite) and Se^{6+} (selenate). Selenate, and to a lesser extent selenite, are the most common forms (Table 16.1). Concentrations in most rocks are <0.1 mg kg^{-1}, but much higher levels of up to 200 mg kg^{-1} occur as impurities in metal sulfide minerals, Se substituting for S (Kabata-Pendias, 2001). Higher concentrations also occur in some coals (e.g. as impurities in pyrite), carboniferous shales, volcanic rocks and phosphate rocks. It is found much less commonly as pure Se minerals such as stilleite (zinc selenide, ZnSe). Elemental Se is rare.

Soil concentrations are very variable, depending on underlying geology, but are typically <1 mg kg^{-1} (Table 16.2). Metal selenides in soils and sediments (together with any elemental Se) are not very soluble. However, weathering of mineral fragments and microbially mediated oxidation (by *Thiobacillus ferroxidans* and others) produce the more mobile selenite and selenate; these forms can subsequently be lost from soils to leaching and plant uptake. Selenite is strongly adsorbed to soil components, except in alkaline conditions where it is somewhat more mobile. Selenate, which predominates in oxic conditions, is much less strongly adsorbed and is more prone to leaching and plant uptake, both of which increase at higher pH levels. In anoxic environments, such as waterlogged soils and sediments, selenates are reduced to the less mobile selenite and elemental Se; microbial reduction to stable metal selenides also occurs. In such conditions Se is also methylated by microorganisms into volatile forms, particularly dimethylselenide, which can escape to the atmosphere. Conversely, in oxic conditions reduced species may be oxidised, ultimately to selenate.

In fresh and marine waters, Se is mainly found in sediments (Table 16.2) and is stored there as metal selenides or by retention in organic matter and sesquioxides; the main dissolved forms are selenate and, to some extent, selenite. In the atmosphere Se is present in both the gaseous and particulate forms. A natural source is the methylation, by plants and microorganisms, of inorganic Se to gaseous dimethylselenide and other volatiles; other sources of atmospheric Se are the burning of Se-rich biomass, volcanic emissions, wind-blown dust and sea spray.

Selenium is an important element in the biosphere; it is essential in humans and animals, although essentiality in most plants is not proven. It is biochemically similar to S; for example, the selenol group, R-Se-H, is analogous to the thiol group, R-S-H

Table 16.1 Summary of key chemical species of selenium.

Minerals and compounds	*Chemical formula (and typical environment)*
	Inorganic forms – Se^{2-}
Selenide	Se^{2-}
Selenium monohydride	HSe^- (anoxic, acidic conditions)
Hydrogen selenide	H_2Se (anoxic, very acidic)
	Inorganic Forms – Se^{4+}
Selenite	SeO_3^{2-} (moderate redox, alkaline conditions)
Hydrogen selenite	$HSeO^{3-}$ (moderate redox, acidic conditions)
Selenous acid	H_2SeO_3 (oxic, very acidic)
	Inorganic forms – Se^{6+}
Selenate	SeO_4^{2-} (oxic, neutral and alkaline conditions)
Hydrogen selenate	$HSeO^{4-}$ (very oxic, very acidic)
	Minerals
Metal selenides	E.g. CuSe, ZnSe, mainly as impurities in sulfide ores
Present in coal and shales	Organic and inorganic forms
Metallic selenium	Se^0 (rare)
	Organic forms
Dimethylselenide	$(CH_3)_2Se$ (volatile, from methylation of inorganic Se)
Selenomethionine	$C_5H_{11}NO_2Se$ (synthesised by plants/microorganisms)
Selenocysteine	$C_3H_7NO_2Se$ (synthesised by plants/microorganisms)

Table 16.2 Typical background concentrations of selenium.

Environmental medium	*Typical background concentrations*[a]
Air	~1 ng m^{-3}
Soil	<1 mg kg^{-1}
Vegetation[b]	<1 mg kg^{-1}
Freshwater	<0.5 µg L^{-1}
Sea water	<0.5 µg L^{-1}
Sediment	<2 mg kg^{-1}

a These are *typical* values, encompassing the vast majority of reported concentrations, based on a large number of literature sources. Higher natural concentrations occur in some cases; for example, in mineralised areas.

b Including crops. Not including metal-tolerant species.

(where R represents a carbon-containing group of atoms), and is present in amino acids such as selenomethionine and selenocysteine (Fig. 16.1). It is also present in a number of enzymes, including glutathione peroxidase, which has important antioxidant properties, and deiodinases, which regulate thyroid hormones. The protection of cells

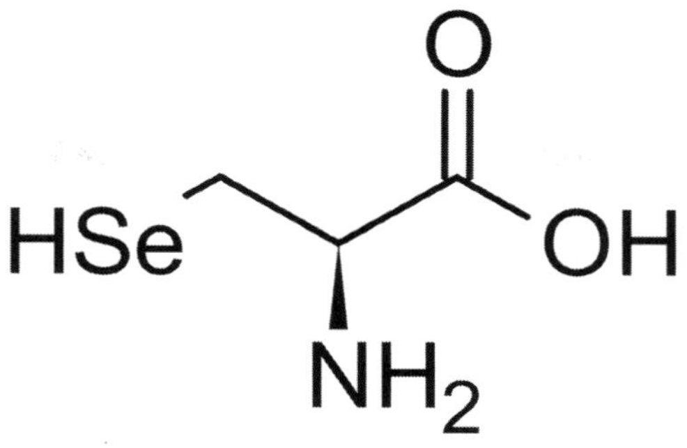

Figure 16.1 Molecular structure of selenocysteine, one of the 23 protein-building amino acids. Credit: Fuse809 (Wikimedia Commons).

from oxidative reactions conferred by glutathione peroxidase may protect organisms against carcinogenicity. While Se can be beneficial in these ways, excessive rates of intake and S replacement can be harmful (see below).

Plants take up selenate, and some selenite, from soils; uptake increases with soil pH and temperature. Root exudates are able to oxidise selenite in soil to the more mobile selenate, which is readily absorbed by plant roots. Selenate (SeO_4^{2-}) is taken up and translocated within plants in the same way as sulfate (SO_4^{2-}), its direct analogue. Selenite is also taken up; within the plant it is transformed to selenate and organic forms of Se, particularly dimethylselenide, which is volatilised from both leaves and roots. Selenium volatilisation releases an off-putting smell, which may be a defence against grazers. Plants such as *Astralagus* spp. (mainly milk-vetches and locoweeds) are able to accumulate Se and have been considered for phytoremediation of Se-rich or Se-polluted soils; some Se hyperaccumulators are known to accumulate concentrations of >1% dry weight.

Production and uses

Production of Se (approximately 2 Kt a^{-1}) is mainly concentrated in Europe, Japan and Russia; global reserves of Se total 0.1 Mt (USGS, 2014). Most Se is recovered from the anode slime wastes of Cu refining. The selenide impurities in the wastes are oxidised to Se dioxide and selenous acid, which is then reduced to produce Se metal. While some types of coal contain substantially more Se than Cu ores, extraction from the incidental wastes of Cu refining is more economically viable than extraction from coal.

Selenium is mainly used in glass-making, either as a pink-red colorant, to decrease solar glare or, in smaller amounts, as a clarifying agent. Other important uses include: copper indium gallium diselenide in solar cells; additives to animal feed in areas of Se-deficient soils; human mineral supplements and as an anti-fungal agent (Se disulfide, SeS_2) in some anti-dandruff shampoos. It is also used in a number of miscellaneous industrial processes, in catalysts and alloys for example.

Pollution and environmental impacts

Anthropogenic emissions of Se to the atmosphere derive mainly from coal burning and the smelting of Cu and Ni, although further releases occur from end-use manufacturing, particularly glass-making, and from waste incineration. Soils are contaminated by atmospheric deposition in such polluted areas, but the highest Se concentrations in soils probably result from the dumping to land of fly-ash, from coal combustion and waste incineration. Selenium is present in sewage sludge, but concentrations in treated soils, and the plants growing in them, are generally lower than some other metals and metalloids. Pollution of water bodies by Se has two main causes. Unusually, the first of these does not involve the dispersion of pollutants into the environment via emissions and wastes but the deliberate drainage of irrigated soils, naturally rich in Se, to evaporation ponds. This has created Se-polluted water bodies that create a hazard to local wildlife (see Box 16.1). The second main source is the disposal of coal fly-ash and bottom ash. In some cases ash is placed in piles and landfill sites, with a risk of rainwater ingress and leaching if uncontrolled; in other cases it is disposed as a slurry into ash ponds (see Box 16.2). Other sources of Se discharge into waters include industrial and sewage effluents and agricultural runoff.

Box 16.1 Selenium contamination at Kesterson Reservoir, California, USA.

In the 1980s, biological monitoring was instigated at 12 evaporation ponds (known as Kesterson Reservoir, a National Wildlife Refuge) in the San Joaquin Valley of California (Fig. 16.2). These ponds were intended to hold drainage water from irrigated fields in an intensively farmed area, but by the 1980s had become polluted by selenate draining from the naturally Se-rich farm soils. The Se concentrations in the water were >1 mg L^{-1} in some samples. Selenium was bioconcentrated by factors of >10^3 in invertebrate species, particularly dragonfly nymphs and midge larvae, and by similar factors in aquatic plants. Fish were exposed via their diet and birds and ducks consuming the fish and aquatic plants acquired very elevated Se concentrations in their tissues. Mean concentrations in Se-tolerant mosquitofish (*Gambusia affinis*) were typically >100 mg kg^{-1} (and >400 mg kg^{-1} as maxima) compared with 1–2 mg kg^{-1} in mosquitofish from a nearby control site. Selenium exposure caused high losses of grebe and coot eggs and hatchlings because of developmental abnormalities of many body parts; symptoms in adult birds included liver and beak damage and weight loss. Racoons and voles contained 10–500 times the Se concentrations recorded in animals at control sites; however, the health effects of Se exposure were not well-defined compared with those in aquatic species. A comprehensive account of Se ecotoxicity by Ohlendorf (2002), which forms the basis of this case study, is highly recommended as further reading.

Figure 16.2 Part of Kesterson Reservoir, USA. The algae accumulation in this image is caused by macronutrient elements, but Kesterson has also suffered from severe selenium contamination.

Credit: Gary Zahm, US Fish and Wildlife Service (Wikimedia Commons).

Box 16.2 Selenium pollution from ash ponds.

Coal-fired power stations need to dispose of large volumes of ash that are generated when coal is burned. This includes bottom ash and fly-ash; i.e. the particles scrubbed from the flue emissions that are generated during coal combustion. Frequently, the slurry containing the particles is directed to ash ponds that are built to store the ash. Such slurries contain a number of different toxic elements, but Se is of particular concern because of the high concentrations in the ash and its potential toxic effects. For example, between 1974 and 1986, the overflow from an ash pond in North Carolina, USA polluted the adjacent Belews Lake. Fish were exposed via the diet and suffered from reduced haemoglobin and swollen gill lamellae, reducing blood flow and respiratory capacity; there was also damage to the internal organs (including inflamed heart tissue) and the eyes. The Se reached the eggs of the fish and caused developmental abnormalities, resulting in serious declines in reproductive rate and whole populations were destroyed; of 20 species of fish originally present in the lake, all but one (the Se-tolerant mosquitofish, see Box 16.1) disappeared within a few years. The discharges stopped in 1986 but sediments remained contaminated, and continue to act as a long-term Se reservoir; Se toxicity, including teratogenicity (Fig. 16.3), was still evident 10 years later and it is estimated that the ecosystem may not fully recover for several decades. Concerns over ash ponds persist today at other locations in the USA and elsewhere. In 2008, 4×10^6 m^3 of coal ash slurry was released into the environment after the dam of a settling pond ruptured in Tennessee, USA. Subsequently the US Environmental Protection Agency pressed for greater regulation of coal ash disposal because of concerns over the environmental effects of Se and other toxic elements contained in coal ash; however, in 2014 another spill released up to 8×10^4 t coal ash into a river in North Carolina {→As}. See Lemly (2002) for a full account of the Belews Lake case.

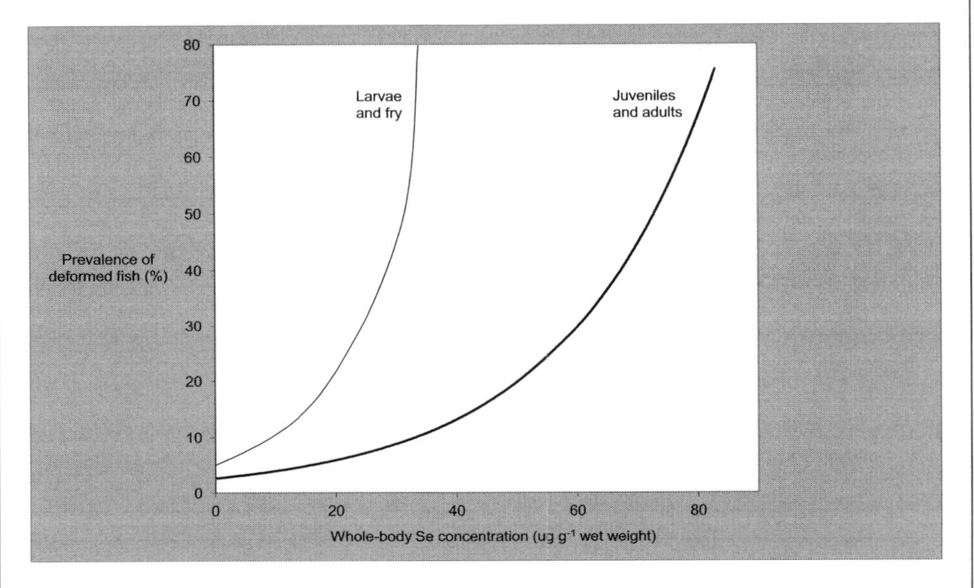

Figure 16.3 Relationship between tissue selenium concentration and prevalence of teratogenic deformities of the spine, head, mouth and fins in 20 fish species studied over two decades in Belews Lake, USA (1975–96).

Source: Adapted from Lemly (2002).

In the context of human health, it is important to note that Se *deficiency* can cause problems in many parts of the world because of low levels of the element in soils. For example, Keshan disease (swelling of the heart, heart failure and pulmonary oedema) and Kashin-Beck disease (bone disorders) were prevalent in people across a broad area of Se-deficient soils in China; these problems are now treated by mineral supplements. Furthermore, because Se is antagonistic to toxic metals such as Hg and Pb, Se deficiency may enhance the toxic effects of such metals. Despite widespread Se deficiency in many areas, in the context of environmental pollution there is also concern about toxic effects to both humans and animals; first, because of the unusually small difference between essential and toxic concentrations (e.g. in the diet, Fig. 16.4) and second, because of the similarity of Se to S and the failure of many organisms to discriminate between the two.

In humans, the symptoms of Se toxicity, generally referred to as selenosis, have been noted mainly, but not exclusively, in industry. In the general population, dietary exposure appears to be more important than inhalation. Typical symptoms of selenosis are hair and nail loss and damage to the neurological and gastrointestinal systems. Exposure to very high Se levels can damage the liver. Selenosis was recorded in hundreds of people in Huebi Province, China in the 1960s. The symptoms included hair and nail loss and redness, swelling, numbness and tingling in the limbs. The selenosis was attributed to the existence in the area of a 'stony coal' (carboniferous shale) with very high Se levels; one sample contained over 9% Se (Yang et al., 1983). Soils and crops may have been contaminated by Se leaching from the coal beds as traditional use of lime fertiliser increased the soil pH and thus the mobility of Se; however, it is likely that seleniferous coal smoke may also have contaminated food during cooking. Another pathway may have been provided by the application of coal ash to the soils.

Ecotoxicological problems occur mainly in aquatic environments contaminated by Se, typically relating to irrigation waters or ash settling ponds (Boxes 16.1 and 16.2). Selenium concentrations in various amphibian, reptile and mammal species in and around fly-ash landfills and ash settling ponds are elevated compared with control sites. Some zooplankton species bioconcentrate Se by factors of 10^3 and, in some cases, 10^5 (USDHHS, 2003). Selenium bioaccumulates in aquatic plants and

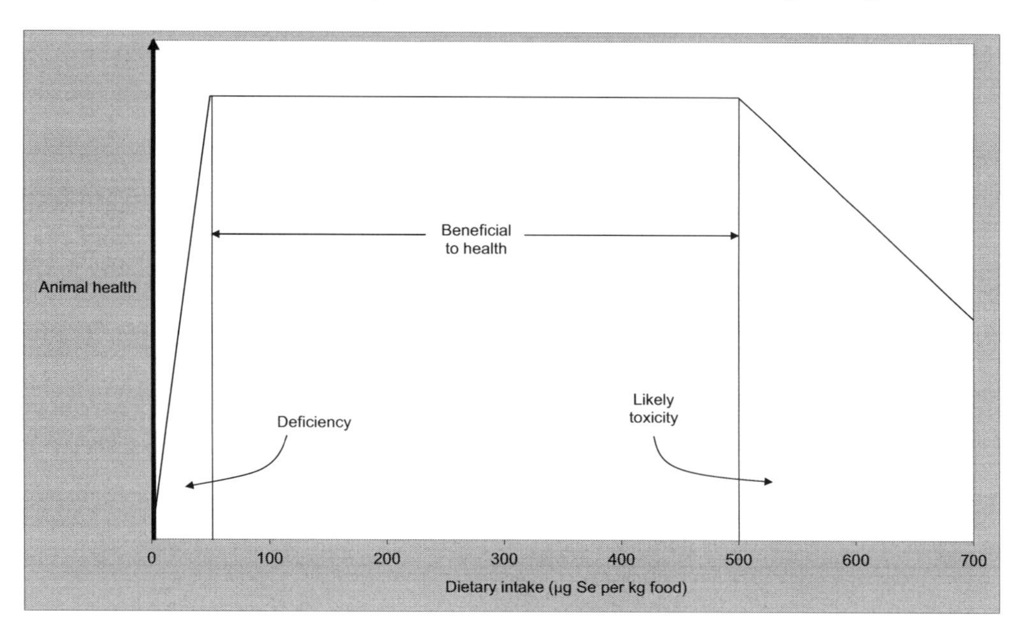

Figure 16.4 The difference between selenium deficiency and toxicity is small relative to many other elements and dietary requirements fall between 50 and 500 µg selenium per kg for most animals (Ohlendorf, 2002). The diagram is illustrative, based on these figures; the curve shape, particularly the slope on the right, will differ between species.

invertebrates and biomagnifies in food chains causing toxic effects at higher trophic levels in fish and piscivorous birds. Exposed fish suffer swelling of the gills and teratogenesis linked to protein and enzyme disruption; birds also suffer birth defects and losses of eggs and hatchlings (Table 16.3 and Boxes 16.1 and 16.2). Bioaccumulation also occurs in terrestrial animals, including mammals and invertebrates (Box 16.1).

Elevated Se concentrations are recorded in plants in polluted areas; for example, 200 mg kg^{-1} dry weight (DW) in clover growing on coal ash and up to 500 mg kg^{-1} DW in tree leaves near to a Cu refinery (Kabata-Pendias, 2001). These concentrations are significantly higher than background levels in plants of <1 mg kg^{-1}. Symptoms of selenosis in plants are discoloration and spotting of leaves and roots; high Se uptake may decrease levels of key nutrients in plant tissues. Selenosis in cattle has been recorded in areas contaminated by Se, both naturally and anthropogenically. A neurological condition in cattle, colloquially called 'blind staggers', was traditionally thought to be caused by grazing of Se-accumulating plants such as locoweed (*Astralagus* spp.), which gets its name from the Spanish 'loco' (insane). However, this has been questioned in recent years and it is thought that consumption of excess sulfates and alkaloids in these plants are the main causal factors. On the other hand, 'alkali disease' in cattle, characterised by hair and hoof loss and bone disorders, is attributed to chronic Se exposure.

Table 16.3 Effects of dietary selenium (as selenomethionine) on egg hatchability in mallards (*Anas platyrhynchos*). Adapted from Ohlendorf (2002), who compiled the data from six separate studies.

Dietary Se (mg kg^{-1})[a]	*Se concentration in egg (mg kg^{-1} DW)*	*Egg hatchability (%)*[b]	*Hatchability as a proportion of control*
Control	0.16	65.7	1
10	15.2	30.9	0.47
Control	0.59	59.6	1
1	2.74	70.7	1.19
2	5.28	60.0	1.01
4	11.2	53.4	0.896
8	36.3	36.9	0.619
16	59.4	**2.2**	0.037
Control	1.35	41.3	1
10	30.4	**7.6**	0.184
10	29.4	**6.4**	0.155
Control	1.16	44.2	1
10	25.1	24.0	0.543
Control	1.4	88.2	1
10	37	**20.0**	0.227
Control	0.89	62.0	1
3.5	11.6	61.0	0.984
7.0	23.4	**41.0**	0.661

a Control diets typically contained 0.4 mg kg^{-1} Se.
b Figures in bold: hatchability significantly lower than control.
DW: dry weight.

References

Kabata-Pendias, A. 2001. *Trace Elements in Soils and Plants*, 3rd Ed. CRC Press, Boca Raton.

Lemly, A.D. 2002. Symptoms and implications of selenium toxicity in fish: the Belews Lake case example. *Aquatic Toxicology* 57, 39–49.

Ohlendorf, H.M. 2002. Ecotoxicology of selenium. In: Hoffman, D.J., Rattner, B.A., Allen Burton, G. Jr. and Cairns, J. Jr. (Eds.). *Handbook of Ecotoxicology*. CRC Press, Boca Raton.

USDHHS (US Department of Health and Human Services: Public Health Service, Agency for Toxic Substances and Disease Registry, ATSDR). 2003. *Toxicological Profile for Selenium*. ATSDR, Atlanta.

USGS (United States Geological Survey). 2014. *Mineral Commodity Summaries: Selenium*. Available at: http://minerals.usgl.gov/minerals

Yang, G., Wang, S., Zhou, R. and Shuzhuang, S. 1983. Endemic selenium intoxication of humans in China. *The American Journal of Clinical Nutrition* 37, 872–881.

17 Tin

Tin in the natural environment

Tin (Sn) is present in the natural environment mainly in inorganic forms that have relatively low toxicity, although some bacterial and algal species can methylate it. The main mineral is cassiterite (SnO).

Production and uses

Unlike the other metals and metalloids included in this book, the extraction and processing of Sn ores and the manufacture of metallic and inorganic Sn products does not normally create problems of environmental pollution and the use of metallic Sn in food containers attests to its low toxicity. However, organic Sn ('organotin') compounds that are manufactured as biocides are also toxic to some non-target organisms and have caused environmental pollution problems. Organotins are molecules containing a central Sn atom, covalently bonded to between one and four organic groups. The main categories of organotins used as biocides include trialkyltin and triphenyltin compounds, which occur in the form of monovalent cations with three covalent bonds to alkyl and phenyl groups, respectively (Fig. 17.1). Compounds in both categories are classed as persistent organic compounds. The most commonly used trialkyltin compound is tributyltin. Other organotin compounds, mainly less toxic monoalkyl and dialkyl forms, are used as stabilisers in PVC products and this is the main end-use of organotins, others being biocides (see below), industrial catalysts and glass coatings.

Environmental concern is focused on the use of organotin biocides in the marine environment in antifouling treatments, although they have also been used in wood pulp, paper, leather and textile production, as crop pesticides and as slimicides in water treatment. The most commonly used organotin biocide is tributyltin oxide, commonly abbreviated (including here) to TBT, although other tributyltin compounds are also manufactured, including halide and methacrylate forms. TBT is used as a wood preservative, particularly in the tropics, but was mainly developed as an antifouling agent for ships' hulls (and lobster pots and harbour walls, for example) to prevent build up of crustaceans and algae. These can accumulate to considerable masses, particularly on large ships, significantly increasing drag and fuel consumption; for example, TBT treatment of large cruise liners is estimated to decrease fuel costs by >10% (Cima et al., 2003).

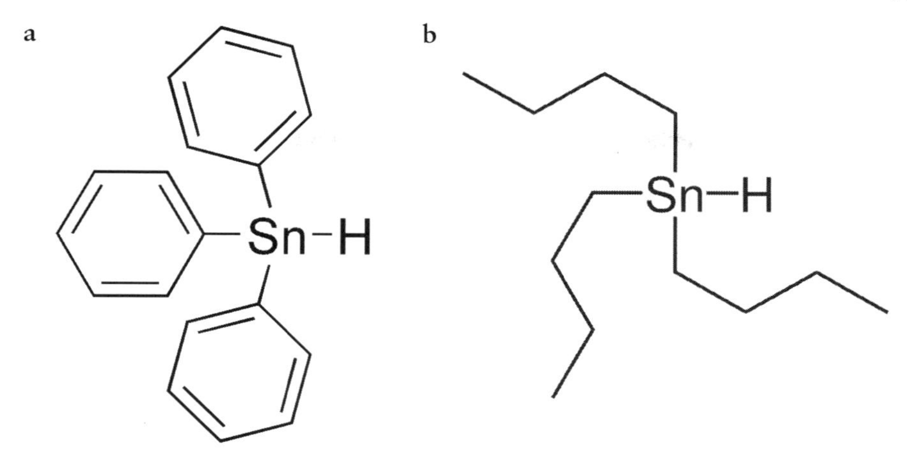

Figure 17.1 The molecular structure of (a) triphenyltin; (b) tributyltin (TBT, an example of a trialkyltin compound). In both examples the cation is bonded with a hydrogen atom, to form the hydride. The three branched arms of TBT each represent a butyl group with four central carbon atoms (shown by the joints and end of each arm).

Credit: Edgar181 (Wikimedia Commons).

Pollution and environmental impacts

TBT treatments are designed to slowly release surface molecules into the water over a period of years so that newer and more effective biocide molecules are continually present at the surface to inhibit fouling. This active leaching process releases daily a few micrograms of TBT from every square centimetre of the treated surface, amounting to several hundred grams per week for a large vessel. This pollutes waters and particularly sediments, where TBT is persistent because of its low aqueous solubility and, as a cationic molecule, strong adsorption; its half-life in anoxic sediment is several years before it is ultimately biodegraded and/or photodegraded. TBT is quite lipophilic and therefore bioaccumulative. Bioconcentration factors of up to 6×10^4 have been recorded in field studies of marine invertebrates (Table 17.1) and there is evidence for biomagnification, large amounts being recorded in the tissues of sharks, dolphins and otters in affected areas; the latter may suffer impairment to the immune system as a result of exposure.

TBT has significant toxic impacts on marine invertebrates at very low environmental concentrations; 20 ng TBT per litre of water can be fatal and at lower levels it causes reduced growth, atrophy in muscles and shell deformation. Some species are particularly sensitive, especially benthic feeders taking in sediment particles; at very low concentrations of <2 ng L^{-1} the dog whelk (*Nucella lapillus*) develops imposex, the development of male sexual organs (a penis and sperm duct) in the female, which blocks the genital opening. This results from TBT-induced hormone disruption and reduces reproduction rate, causing local declines in populations. Some other gastropod species and bivalves are similarly affected by TBT contamination (Fig. 17.2). Another known effect is reduced growth (because of shell thickening) in the young of the Pacific Oyster (*Crassostrea gigas*) living in waters with TBT concentrations of approximately 10 ng L^{-1} (USEPA, 2003). This has badly affected the oyster industry in polluted areas, notably the Bassin d'Arcachon in France some years ago. Death and sterility have been

Table 17.1 Bioconcentration factors of tributyltin in marine invertebrates.

Study	*Common name*	*Latin name*	*Tributyltin conc. in water* ($\mu g\ L^{-1}$)	*Exposure time (d)*	*Bioconcentration factors*
1	Atlantic dogwinkle	*Nucella lapillus*	0.07	529–634	17,000
2	Blue mussel (adult)	*Mytilus edulis*	<0.1	60	11,000
	Blue mussel (juvenile)	*Mytilus edulis*	<0.1	60	25,000
3	Blue mussel (juvenile)	*Mytilus edulis*	0.452	56	23,000
	Blue mussel (juvenile)	*Mytilus edulis*	0.204	56	10,400–27,000
	Blue mussel (juvenile)	*Mytilus edulis*	0.079	56	37,500
4	Blue mussel (juvenile)	*Mytilus edulis*	<0.105	84	5000–60,000

Source: Data from USEPA (2003); collated data from four previously published field studies.

noted in some species of marine fish exposed to TBT. In the terrestrial environment, the main environmental effects of TBT arise from its use as a wood preservative.

The primary modes of organotin toxicity appear to be: (i) inhibition of several important enzymes including those involved in the mitochondrial ATP synthesis that is vital for the normal functioning of cells and (ii) impairment of Ca regulation in cells, causing cell death and affecting skeletal or shell development.

Organotins are metabolised relatively easily in humans, but exposure to high levels can lead to neurological, gastrointestinal, renal and hepatic effects. Exposure is most likely to be via a seafood-rich diet, but direct contact with organotins may cause irritation of the skin, eyes and respiratory system. Occupational exposure to inorganic Sn may cause stannosis, a respiratory condition.

The use of TBT and other organotin compounds as anti-foulants is banned by signatories to the International Convention on the Control of Harmful Anti-fouling Systems on Ships. By 2012, the Convention had been ratified by 63 countries. This reduction in use will decrease TBT pollution but its usage has not been completely eliminated by the Convention; additionally, the more widespread use from the past will continue to have effects for some time because of the persistence of TBT in sediments. The main replacements for organotin in anti-fouling treatments are copper-based, an ironic turn of events because organotins were first developed to replace Cu, which is itself a toxic element at elevated concentrations {→Cu}; organic 'booster' biocides are also used on small vessels.

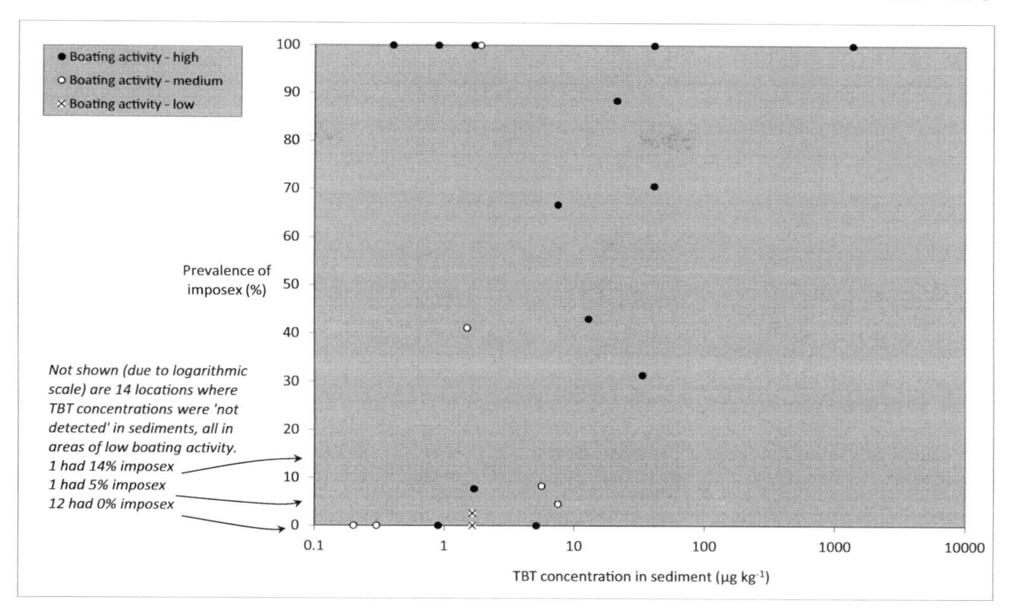

Figure 17.2 Association between tributyltin (TBT) concentration in marine sediments and incidence of imposex in gastropods along 4700 km of the Argentinian coastline. The figure shows that all sites with >10 µg TBT per kg of sediment are in locations of 'high boating activity' (high volume marine traffic and/or ports with ship painting) and have imposex incidences of >30%, in contrast to locations with lower TBT concentrations. The 100% imposex incidence at four sites with lower TBT concentrations in sediments (<10 µg kg^{-1}) is attributed by the authors to exposure to surface-adsorbed TBT in the very fine-grained sediments found at these sites.

Note: Only 23 of 41 original sampling locations are included in the figure because of missing data at some locations. At some of these 23 locations, more than one gastropod species was measured, giving 35 data points altogether. At six of the 23 sites a range of TBT concentrations was listed rather than a single value and in these cases the maximum concentration has been included in the graph.

Source: Data from Bigatti et al. (2009).

References

Bigatti, G., Primost, M.A., Cledon, M., Averbuj, A., Theobold, N., Gerwinski, W., Arntz, W., Morriconi, E. and Penchaszadeh, P.E. 2009. Biomonitoring of TBT contamination and imposex incidence along 4700 km of Argentinean shoreline (SW Atlantic: from 38S to 54S). *Marine Pollution Bulletin* 58, 695–701.

Cima, F., Craig, P.J. and Harrington, C. 2003. Organotin compounds in the environment. In: Craig, P.J. (Ed.). *Organometallic Compounds in the Environment*. Wiley, Chichester.

USEPA (United States Environmental Protection Agency). 2003. *Ambient Aquatic Life Water Quality Criteria for Tributyltin (TBT)*. USEPA, Washington DC.

18 Uranium

Introduction

Uranium (U) is chemically toxic, affecting the kidney, liver and lungs but, in pollution terms, the main interest in this element is its radioactivity and more particularly that of its decay products. In this regard it differs from the other elements considered in this book, which have only chemical pollution and toxicity as their focus. Uranium is the most widespread radioactive element occurring in concentrated forms in the Earth's crust and is used extensively by humans. To appreciate the importance of U as a naturally occurring element, and as a pollutant, it is necessary to first understand a little about the fundamentals of radioactivity (see Box 18.1).

Box 18.1 Fundamental concepts of radioactivity.

Radioactive elements Elements with an atomic number of ≥83 are unstable because of the repulsive electromagnetic forces that exist between the large numbers of protons they contain; such unstable atoms are called radionuclides or radioisotopes. In lighter elements the repulsive force is balanced by another fundamental force of nature, the strong nuclear force (SNF), which attracts protons (and neutrons) to each other; however, it is also a very short-range force that becomes weaker as the sizes of nuclei increase in the heavier elements. In radionuclides the repulsive force between protons overcomes the attractive SNF and this instability leads to disintegration into one or more daughter atoms, each containing a different number of neutrons (new isotopes) and/or protons (new elements). These decay products may themselves be radioactive so that the disintegrations continue until stable isotopes are produced; this forms a decay series containing several different radionuclides.

Radiation With each disintegration, one of the following forms of ionising radiation is emitted (so called because the radioactive particle or ray is able to ionise atoms – this ionisation can damage biological molecules and cells).

- Alpha particle, ${}^{4}_{2}\alpha$: a particle containing two protons and two neutrons (i.e. the same as a He nucleus and therefore often denoted as ${}^{4}_{2}He$). Alpha particles are the slowest and heaviest of the three main types of radiation encountered but have the highest charge and are the most strongly ionising.

They can travel only a few micrometres through air and tissues, but can affect internal tissues if alpha-emitters are inhaled or ingested, severely damaging cells and molecules that they encounter, including DNA. When an unstable radionuclide emits an alpha particle, it is left with two fewer protons and therefore becomes a different element; in the following example, U becomes thorium:

$^{238}_{92}U \rightarrow ^{234}_{90}Th + ^{4}_{2}He$ (or $^{4}_{2}\alpha$) + energy

- Beta particle, $^{0}_{-1}\beta$ or $^{0}_{-1}e$: an electron that is emitted from a neutron – this leaves behind an atom with one less neutron and one extra proton; therefore its mass does not change but, as with alpha emission, it becomes a different element. In this example, thorium becomes protactinium:

  e.g. $^{234}_{90}Th \rightarrow ^{234}_{91}Pa + ^{0}_{-1}\beta$

  Another type of beta particle is a positron β^+, which may be emitted from a proton to increase nuclear stability in some circumstances. Beta particles, with very little mass, less charge than alpha particles and with greater kinetic energy, can travel 1 m in air and a few centimetres in tissues; they can ionise atoms, although to a lesser extent than alpha particles.
- Gamma ray, $^{0}_{0}\gamma$: very short wavelength electromagnetic radiation, which is released as the excess energy retained by a nucleus after emission of an alpha or beta particle. Gamma rays have no mass or charge and are very energetic, travelling (at the speed of light) deep into, and through, most materials including biological tissues, so do not have to be inhaled or ingested to have internal impacts on the body. Their electrical fields can ionise atoms, although their energy per unit path length is lower than that of alpha and beta particles.
- Two other types of radiation associated with unstable nuclei are neutrons and X-rays. Neutron emission is used in nuclear reactors to deliberately produce unstable, fissile nuclei. X-rays are emitted as a consequence of another process occurring in unstable nuclei, called electron capture.

Activity This is the amount of radiation occurring per unit time, measured in bequerels (Bq); 1 Bq = 1 nuclear disintegration per second.

Half-life Radionuclides decay at different rates, the most unstable ones decaying the fastest. The rate of decay of a radionuclide is measured as its half-life, i.e. the time taken for half of the atoms of the radionuclide in a sample to decay. A radionuclide with a very short half-life (e.g. a fraction of a second) is highly active but does not last long enough to do much harm, whereas one with a very long half-life (e.g. millions of years) persists in the environment but has low activity and is less hazardous because the chance of a radioactive decay occurring per unit time is much less. Radionuclides with half-lives of days to tens of years (e.g. I-131 – 8 days, Sr-90 – 29 years, Cs-137 – 30 years) are sufficiently radioactive to damage tissues and persist long enough to cause widespread environmental damage. For example, after nearly a century the activity of Cs-137 released from the Chernobyl nuclear reactor (see below) will still be 12.5% of the original value (3 half-lives = 90 years). By contrast, U-238 has a half-life of 4.5×10^9 years; therefore, it is still relatively abundant in the Earth's crust but, in comparison with the radionuclides mentioned above, has a much lower level of activity.

Dose Exposure of humans and other organisms to radiation is expressed in terms of radiation dose. The amount of energy from radiation that is absorbed by biological tissue is measured in grays (Gy), where 1 Gy = 1 joule of energy deposited in 1 kg of tissue. This is referred to as the physical or absorbed dose. However, because alpha particles, for example, travel only short distances through biological tissue, their energy is deposited in a concentrated area; therefore 1 Gy of alpha radiation causes more damage to biological tissue than 1 Gy of beta or gamma radiation. To account for such differences in the biological damage caused by the various forms of radiation, an 'equivalent dose' is calculated by the application of weighting factors to the absorbed dose; the equivalent dose is expressed in sieverts (Sv), where 1 Sv of alpha radiation has the same biological effect as 1 Sv of beta or gamma radiation.

Uranium in the natural environment

More than 99% of U in the Earth's crust is present as the U-238 isotope, which has the longest half-life (~4.5 billion years) of the U isotopes. The remaining fraction is virtually all U-235, but there are also small amounts of U-230 and U-234. U-238 is an alpha-emitter and its decay products are isotopes of the elements thorium (Th), protactinium (Pa), radium (Ra), radon (Rn), polonium (Po) and lead (Pb). The Pb isotope formed in this decay series (Pb-214) itself heads a number of decay series, containing isotopes of astatine (At), bismuth (Bi), thallium (Tl), mercury (Hg) and other Pb isotopes, each culminating in a stable Pb isotope, Pb-206. Among these decay products, Rn is of particular interest because, being gaseous, it can enter homes; Rn isotopes emit alpha particles, forming isotopes of Po that can attach to fine particles in the air; inhalation of such particles poses an increased risk of radiation and lung cancer (Fig. 18.1).

There are two other decay series headed by isotopes present in the Earth's crust: the relatively rare U-235 and Th-232; both series end with a stable Pb isotope. Another radioactive isotope, potassium-40 (K-40) is not part of a decay series but exists in the natural environment as a rare isotope of this element and is a dietary source of radioactivity to organisms. Other radionuclides, such as C-14 and P-32, are formed by the action of cosmic rays that continually bombard the Earth. Therefore, radionuclides are present in the natural environment and are sources of radiation to lesser or greater extents. In an extreme example of this, 1.7 billion years ago a natural deposit of U ore in Gabon, west Africa appears to have become saturated with groundwater and formed a natural 'nuclear reactor', as evidenced by lower-than-normal levels of U-235 observed in the ore at Oklo mine in the 1970s.

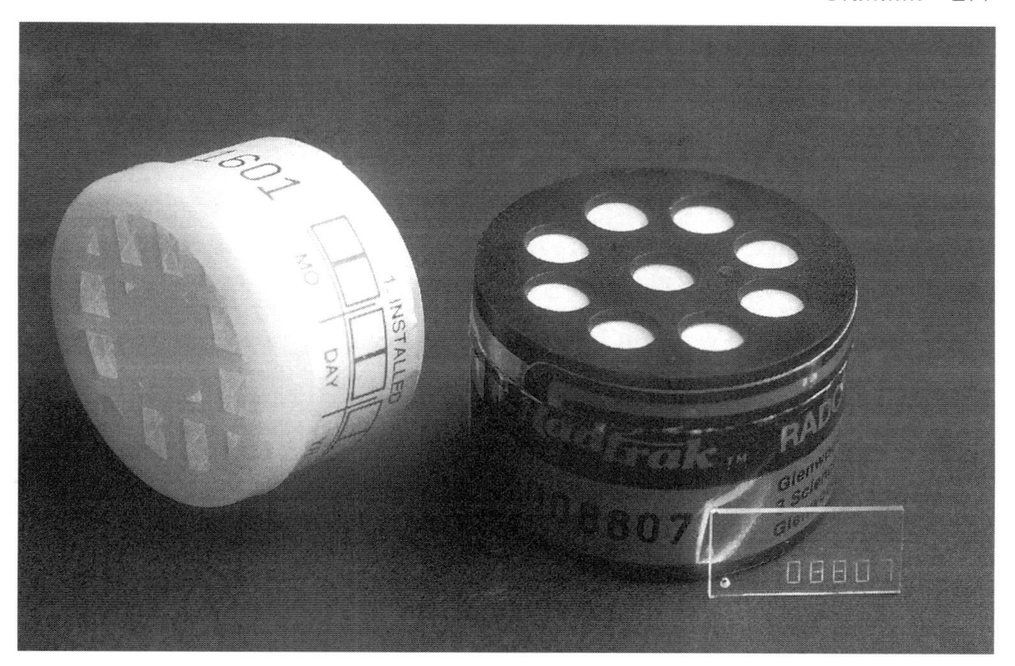

Figure 18.1 A radon test kit, used to test for naturally elevated levels of the gas in houses built on uranium-bearing rocks, particularly granite.

Credit: National Institutes of Health, USA (Wikimedia Commons).

Most major rock types contain 0.5–5 mg kg^{-1} U, so it is not a particularly rare element; some rocks, including granite and apatite, contain more than others and U is also present in coal. The main U ore is uraninite (commonly known as pitchblende), which contains uranium dioxide (UO_2) as the main repeating unit. Carnotite ($K(UO_2)(VO_4)H_2O$) is another important U mineral. Mean concentrations in soils are typically <5 mg kg^{-1}. Uranium oxides are quite insoluble but oxidise in soils and sediments to mobile uranyl (UO_2^{2+}) ions. Therefore, there is some mobility across the typical pH range of oxic soils, although less in anoxic soils. Some mobile U is adsorbed by soils, particularly by the organic fraction. Typical concentrations of U in freshwaters and seawater are <5 µg L^{-1}, although higher levels are detected in areas of uraniferous geology. Waterborne U may exist as weathered particulate or as dissolved forms, the latter especially in areas of mineral soils (i.e. with relatively low OM content). Uranium contributes to the natural low-level radioactivity of seawater, which is, however, mainly attributable to seawater's considerable K-40 content.

Uranium has no known biological role, although 300-fold bioconcentration has been observed in some bacterial species; similarly, a species of lichen at a former U mine in the UK was found to be storing the element. Plants and animals take up background U in small amounts and in animals it concentrates in the bones. Radiation from U-238 and its daughter radionuclides is able to damage tissues in organisms because it can ionise the atoms it contacts. This can directly and indirectly (by the formation of free radicals) damage molecules, although at low doses cell repair mechanisms can operate. If irreparable damage is done to DNA, mutations and birth defects can result. Proteins may also be damaged, potentially impairing enzymatic processes.

Production and uses

Uranium ores are extracted from opencast or underground mines and in some areas leaching agents such as sulfuric acid and sodium bicarbonate are used to leach U in situ from ore bodies. Large amounts of U are produced in some countries from phosphate extraction. The main centres of production include Australia, Canada and Kazakhstan.

The main use of U globally (in >30 countries) is in nuclear power generation (Fig. 18.2). Mined U is enriched until its U-235 content is approximately 3%. Nuclei of U-235, contained within fuel rods, are bombarded with neutrons, forming unstable nuclei that spontaneously decay to daughter radionuclides (e.g. Ba-142, Kr-36, Sb-133 and Nb-99 in equations 18.1 and 18.2), in the process releasing energy and additional neutrons ($^{1}_{0}n$) that produce further nuclear fission in a chain reaction.

$$^{235}_{92}U + ^{1}_{0}n \rightarrow ^{142}_{56}Ba + ^{91}_{36}Kr + 3^{1}_{0}n + \text{energy} \qquad [18.1]$$

$$^{235}_{92}U + ^{1}_{0}n \rightarrow ^{133}_{51}Sb + ^{99}_{41}Nb + 4^{1}_{0}n + \text{energy} \qquad [18.2]$$

Figure 18.2 Sizewell nuclear power stations, UK. Sizewell A (centre) is now being decommissioned and Sizewell B (right) is the UK's newest nuclear reactor, coming online in 1995. There are plans for a third plant to be commissioned at this site.

Credit: Ted and Jen (Flickr).

In a nuclear reactor the chain reaction is controlled by the use of neutron-absorbing control rods made of cadmium or boron. Cooling water surrounding the fuel rods also serves to slow down neutrons to increase their likelihood of splitting a U-235 nucleus. One gram of U-235 can yield the equivalent energy of 2.7 t of coal. Spent fuel reprocessing is undertaken to recycle unused U-235 and provide Pu for nuclear fuel and weapons production. Nuclear weapons incorporate U-235 or Pu-239, which is produced in nuclear reactors by neutron bombardment of U-238; this forms U-239 which undergoes transformation to neptunium and Pu via beta decay:

$$^{238}_{92}U + ^{1}_{0}n \rightarrow ^{239}_{92}U \rightarrow ^{239}_{93}Np + ^{0}_{-1}\beta \rightarrow ^{239}_{94}Pu + ^{0}_{-1}\beta \quad [18.3]$$

Depleted U, which remains after U-235 enrichment for nuclear fission, is extremely dense and is used for ships' ballast and in armour-piercing weapons.

Pollution and environmental impacts

The main source of pollution from human use of U is nuclear power generation. The process of nuclear fission produces environmentally hazardous radionuclides such as Sr-90, Cs-137, I-131 and Pu-239 (detailed later; Table 18.3) and a great deal of hazardous waste, including spent U fuel. There are five main pollution hazards associated with nuclear power generation:

(i) U mining, processing and refining, producing tailings with the potential for soil, water and sediment pollution by U-238 and its decay chain radionuclides;
(ii) fossil fuel combustion that occurs during the extraction, processing and enrichment of U ores and the construction and operation of nuclear power plants – however, no fossil fuel is used in the production of nuclear energy via nuclear fission in the reactor;
(iii) operational effluents and emissions containing small amounts of radioactive isotopes, mainly cooling water irradiated by neutrons from the fuel elements; these generate a small fraction of the total radiation dose experienced by an average person from all radiation sources;
(iv) long-lived radioactive wastes (see Box 18.2);
(v) accidental releases and large-scale nuclear accidents (Table 18.1).

Box 18.2 Radioactive waste.

Wastes generated during operation of a nuclear power plant fall into three categories:

- Low-level wastes (LLW). These relate mainly to materials (e.g. protective clothing) that have come into contact with radioactive materials and comprise the largest volumes. In the past, LLW was sometimes dumped at sea but it is now mainly sent for near-surface land disposal.
- Intermediate wastes (ILW). Typically these are reprocessing and decommissioning wastes; for example, the structures remaining after plant closedown, which remain radioactive and require safe disposal. Formerly, ILW was dumped in deep ocean waters but is now mainly stored on land indefinitely, awaiting decisions on ultimate disposal.

Table 18.1 Notable nuclear accidents.

Location and date	*Cause*	*Environmental impact*
Windscale, UK 1957	Human error leading to fire in core	• Soils across Northern Europe contaminated by I-131 and Cs-137 • Human exposure via milk sourced from contaminated fields • Estimate of >200 excess deaths in following decades
Chelyabinsk, USSR 1957	Cooling system failure leading to explosion	• Release of Cs-137 and Sr-90 across several thousand square kilometres of central Russia • Thousands of residents evacuated • Unknown number of deaths from exposure – estimates from <100 to several thousand
3-Mile Island, USA 1979	Mechanical and procedural failures causing a loss of coolant water and a partial meltdown of the core	• Release of xenon-135 (negligible inhalation risk) and a small amount of I-131 to atmosphere • Release of radioactive water to river • Considered to have caused no deaths or major health effects
Chernobyl, USSR 1986	Human error leading to reactor explosion	• Release of I-131, Cs-137, Pu-139, Sr-90 • Fallout contaminated soils over most of eastern Russia and Europe • Large areas of land removed from agricultural and forestry use • Mass evacuation of 135,000 people from a 30-mile radius • Those in radiation plume had short-term exposure to I-131 (thyroid cancer) and longer term exposure to Cs-137 (mutagenic effects in newborns) • 31 official deaths (emergency workers) but thousands estimated from radiation exposure
Fukushima, Japan 2011	Earthquake and subsequent 13 m tsunami flooded 5 reactors disabling power and pumps; cooling water protection of fuel rods failed; explosions at 3 reactors.	• Radionuclides emitted to atmosphere (gases deliberately vented to relieve pressure) and to seawater (coolant) • Immediate and indefinite 20-km evacuation zone • Release to air was mostly Xe (negligible inhalation risk), I-131 and Cs-137 • Seawater contaminated with Cs-134, Cs-137 and I-131 and fishing banned indefinitely; in 2013, radioactive water found to be leaking from storage tanks into sea • 4–7% increase in the lifetime risk of some specific cancers is expected for infants (the most at-risk group) and a 70% increased risk of thyroid cancer in infant girls (WHO, 2013)

- High-level wastes (HLW). The most hazardous wastes are mainly spent fuel rods and reprocessing wastes. Such materials remain highly radioactive for hundreds to thousands of years and deep underground burial of vitrified waste is the preferred option. However, final decisions on suitable locations for deep burial facilities are typically held back by lack of political acceptance. In the meantime, HLW is stored above ground in spent fuel water tanks or, after an initial period of cooling, in dry cask storage (Fig. 18.3). Table 18.2 shows approximate amounts of radioactive wastes generated each year.

Figure 18.3 Dry storage of spent nuclear fuel.
Credit: US Nuclear Regulatory Commission (Flickr).

Table 18.2 Radioactive waste generation.

Description	*Approximate amount*
LLW and ILW from 1000-MW nuclear power station	300 $m^3\ a^{-1}$
HLW from 1000-MW nuclear power station	30 $t\ a^{-1}$
Ash from 1000-MW coal-fired power station (comparison)	$3 \times 10^5\ t\ a^{-1}$
Global LLW and ILW	$2 \times 10^5\ m^3\ a^{-1}$
Global HLW	$1 \times 10^4\ m^3\ a^{-1}$

Source: Data from IAEA (2014).

Other sources of radionuclide pollution include the detonation of nuclear weapons, mainly at test sites, which gives rise to Sr-90, Cs-137 and Pu-239 in soils. Most nuclear nations carried out above-ground weapons testing between the 1940s and 1960s; subsequently testing was undertaken underground and this largely ceased in the 1990s, with only North Korea continuing underground weapons testing in more recent years. Other pollution hazards include the sinking of nuclear submarines and the dismantling of medical or scientific equipment containing radionuclides. On a wider scale, soils receive low-level U contamination from both P fertiliser application and deposition of ash from fossil fuel combustion.

The extent of biological damage caused by a radionuclide depends on several factors:

(i) its radioactive half-life and therefore its persistence and activity in the environment (see Box 18.1)
(ii) its chemical form and mobility in the environment; for example, it may be strongly adsorbed/accumulated by soils and sediments
(iii) its route of exposure to the organism and the intensity and duration of exposure
(iv) the type of radiation emitted when it decays (see Box 18.1)
(v) its rate of absorption and excretion in organisms, i.e. its *biological* half-life.

Radionuclides with intermediate half-lives, that are both persistent and sufficiently active, can cause serious biological damage. In this context, several radionuclides associated with pollution, mainly from nuclear accidents and weapons testing but also from operational releases, are of particular concern (Table 18.3); for example, Cs-137 and Sr-90 have half-lives of about 30 years, meaning their radioactivity will not diminish to <1% until two centuries have elapsed. Uranium itself is both chemically and radiologically toxic at sufficient doses, but the main concerns in nuclear accidents and pollution incidents tend to be focused on these more potentially damaging radionuclides.

Inhalation and ingestion, as exposure routes, are of more concern than external contact as they can lead to dangerous internal radiation in humans and animals. Radionuclides occurring in the gaseous and/or fine particulate phases (e.g. radon and its progeny) can readily enter the lungs and those that are taken up by plants and transported up the food chain are ingested via the diet. An example of an exposure risk via ingestion is the consumption of the seaweed *Porphyra umbilicalis* (an algae that accumulates Cs-137 and other radionuclides) by communities living near nuclear power stations, such as those on the west coast of Britain. Ingestion of radionuclides via water ingestion can also be important in polluted catchments. It should be noted however that most exposure to radioactivity is attributable to background radiation from natural sources and, in some cases, to medical exposure (Table 18.4).

Upon entry into the body, radionuclides are transported in the same way as their stable counterparts and can accumulate in specific organs where their radioactivity is then likely to cause damage; a good example is provided by I-131. This isotope is concentrated (for a short time, before it decays) in the thyroid gland in the same way as stable iodine, which is mainly required by the body to produce the essential metabolic hormone thyroxine. The short half-life of I-131 (8 days) means that the main risk of exposure is restricted to people living in the vicinity of the source; for example, thyroid-related diseases, including cancers, were elevated in people who were exposed as children (via dairy products) to radiation from the Chernobyl accident in 1986 (Fig. 18.4). Some radionuclides follow similar biochemical pathways to essential elements with which they share chemical properties; for example Sr-90 accumulates in bones and bone marrow, mimicking Ca, which is in the same chemical

Table 18.3 Environmentally damaging radionuclides associated with uranium use.

Radionuclide	*Half-life and radiation*	*Main exposure pathways*	*Environmental and health concerns*
Caesium-137	30 years β and γ emitter	Ingestion of contaminated crops, meat, milk and water; inhalation of contaminated dust also possible.	Persistent so widely transported in atmosphere upon release and deposited onto soils over a very wide area. Environmentally mobile and bioavailable. In body, follows similar biochemical pathways to K (present in all cells) so distributed throughout body; accumulates and biomagnifies. Highly carcinogenic.
Iodine-131	8 days β and γ emitter	Inhalation of gas and aerosol; ingestion of contaminated food and water.	Relatively short half-life so less long-distance transport; adsorbed by soils when deposited. Accumulates in thyroid gland and can cause thyroid cancer.
Plutonium-239	24,000 years α emitter	Inhalation of contaminated dust; ingestion less important	Very persistent and widely transported. Strongly adsorbed by soils; relatively limited aqueous mobility. Very long biological half-life; accumulates in bone in particular. Causes anaemia and is highly carcinogenic (especially leukaemia and lung cancer). Chemically toxic, especially affecting kidneys.
Strontium-90	29 years β emitter	Mainly ingestion of contaminated food and water	Persistent, so widely transported in atmosphere upon release and deposited onto soils over a very wide area. Some soil adsorption, but quite mobile in environment. Accumulates in bones and bone marrow; causes anaemia, bone cancers and leukaemia and, on inhalation, lung defects.

Table 18.4 Radiation doses in context.

Dose (mSv)	*Source*
0.3	Total dose received by each resident of Europe for 20 years after Chernobyl
2.4	Average annual background radiation globally
9	Total dose received by the 6 million residents in Chernobyl-contaminated areas (>37 kBq m^{-2}) of the former USSR
9	One computed-tomography (CT) scan
9	Annual exposure of airline crew flying regularly between New York and Tokyo
30	Average total dose of external radiation received by evacuees from Chernobyl plant and surrounding area
120	Average total dose received by liquidators at Chernobyl (1986–90)

Source: Adapted from Peplow (2011).

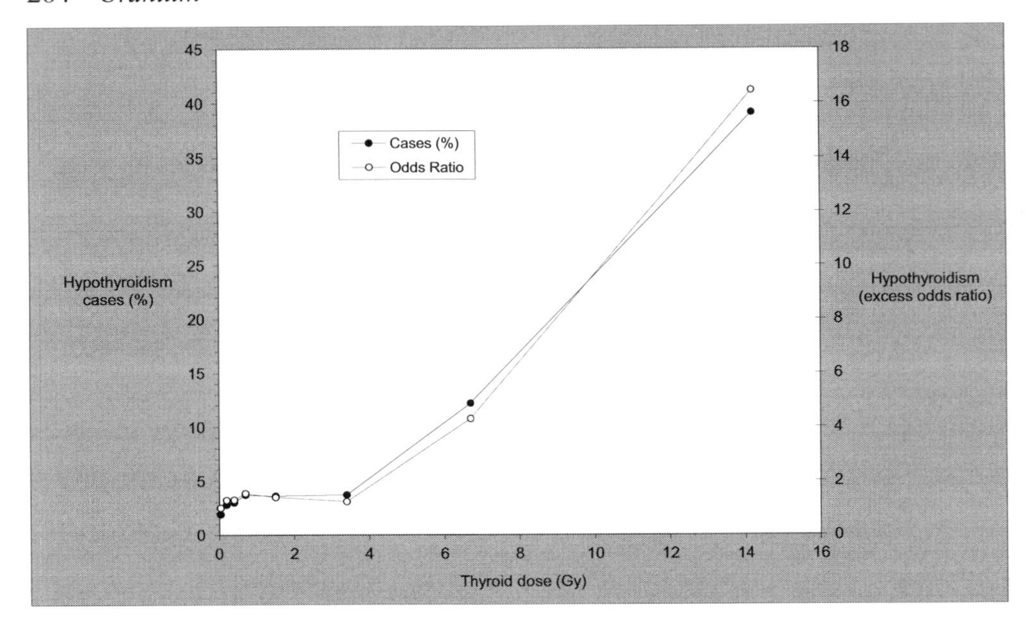

Figure 18.4 Association between iodine-131 dose and prevalence of hypothyroidism (an endocrine disorder) in a cohort of 10,827 Belarusian citizens exposed to radiation from the Chernobyl accident as children. Dose based on thyroid radioactivity, measured within 2 months of the accident; hypothyroidism prevalence measured over the following 7 years. Left axis: percentage of subjects receiving stated dose who were subsequently diagnosed with hypothyroidism. Right axis: excess odds ratio (no excess = 1).

Source: Data from Ostroumova et al. (2013).

group. The chemical similarity also means that Sr-90 uptake by plants can be reduced by ensuring a plentiful supply to the soil of Ca, which plants take up in preference. A further example is the chemical similarity of Cs and K.

The severity of biological damage caused by absorbed radionuclides depends to a large extent on the nature of the radiation they emit; alpha, beta or gamma (see Box 18.1). Radiation physically damages molecules and, in some cases, whole cells, so processes that depend on rapid cell division, such as the production of blood cells in the bone marrow, are especially vulnerable. As well as direct damage, radiation can also form reactive free radicals in the body, causing indirect damage to molecules and cells. Acute exposure to radionuclides associated with nuclear power and weapons production causes skin burns, radiation sickness and cancers. Chronic exposure causes cancers of the blood, bone, thyroid, lungs and other organs; it also causes birth defects from genetic damage in exposed parents. Most of the human health impacts caused by anthropogenic use of U is known from the aftermath of nuclear bombing in the 1940s and from major accidents at Chernobyl and elsewhere. There is less understanding of the effects caused by long-term exposure to lower environmental levels.

Most of the evidence for ecotoxicological impacts caused by human use of U is from Chernobyl and other nuclear accidents and from nuclear weapons test sites like Semipalatinsk in Kazakhstan; for example, at the latter location reproductive impairment in plants was noted in radiation-affected areas. Similarly, near Chernobyl pine trees were badly affected by the release of radioactivity and suffered high rates of dieback; further afield, plants and animals were also exposed to high levels of Cs-137 and other radionuclides. For example, in Scandinavia, fish were exposed via their prey (bottom feeding invertebrates) and exposure in deer was via ingestion of radionuclide-contaminated lichen. In upland areas of the UK, sheep grazing on Cs-137-contaminated grass absorbed radiation levels that left them unsafe for consumption; some areas remain affected more than a quarter of a century after the accident at Chernobyl and hundreds of British farmers still face restrictions on the sale of their livestock. In the immediate surroundings of Chernobyl some opportunist species appear to have benefited from the human evacuation, but ecological studies have concluded that the most contaminated areas have less abundant and biodiverse wildlife.

References

IAEA (International Atomic Energy Agency). 2014. *Managing Radioactive Waste*. Available at: www.iaea.org/publications/factsheets

Ostroumova, E., Rozhko, A., Hatch, M., Furukawa, K., Polyanskaya, O., McConnell, R.J., Nadyrov, E., Petrenko, S., Romanov, G., Yauseyenka, V., Drozdovitch, V., Minenko, V., Prokopovich, A., Savasteeva, I., Zablotska, L.B., Mabuchi, K. and Brenner, A.V. 2013. Measures of thyroid function among Belarusian children and adolescents exposed to Iodine-131 from the accident at the Chernobyl nuclear plant. *Environmental Health Perspectives* 121(7), 865–871.

Peplow, M. 2011. Chernobyl's legacy. *Nature* 471, 562–565.

WHO (World Health Organization). 2013. *Health Risk Assessment from the Nuclear Accident after the 2011 Great East Japan Earthquake and Tsunami based on Preliminary Dose Estimation*. WHO, Geneva.

19 Zinc

Zinc in the natural environment

Zinc (Zn) occurs in the natural environment almost entirely in the Zn^{2+} oxidation state; Zn^0 (native metal) and Zn^+ are rare. Typical concentrations in the major rock types are in the 10–120 mg kg^{-1} range. The most common ore mineral is sphalerite (ZnS); other important ores include smithsonite ($ZnCO_3$); hemimorphite ($Zn_4Si_2O_7(OH)_2.H_2O$); and zincite (ZnO), which is relatively rare. Smithsonite and zincite are secondary minerals, formed near the surface from weathering of primary Zn minerals.

Soil Zn concentrations are typically 20–120 mg kg^{-1}, but levels of up to (and in some cases, exceeding) 300 mg kg^{-1} are possible, depending on geology (Table 19.1). The free ion, Zn^{2+}, can be adsorbed, particularly by clay minerals and sesquioxides; however, a proportion of adsorbed Zn^{2+} is readily exchangeable, especially at lower pH, and zinc is one of the most mobile metals in soils. Zinc is also associated with soil organic matter and typically accumulates in surface horizons. It forms inorganic and organic complexes, some of which are soluble and mobile, others insoluble and bound to the solid phase. Zn also occurs in the solid phase of soils in mineral fragments. Zinc occurs in waters in suspended and settled particles and, in the dissolved phase, mainly as the free, hydrated Zn^{2+} ion, but also as complexes with organic and inorganic ligands. In the atmosphere, Zn occurs in particles derived from volcanoes, aeolian dust, forest fires and sea spray.

Table 19.1 Typical background concentrations of zinc.

Environmental medium	*Typical background concentrations*[a]
Air	<1 µg m^{-3}
Soil	<300 mg kg^{-1}
Vegetation[b]	<100 mg kg^{-1}
Freshwater	<20 µg L^{-1}
Sea water	<5 µg L^{-1}
Sediment	<200 mg kg^{-1}

a These are *typical* values, encompassing the vast majority of reported concentrations, based on a large number of literature sources. Higher natural concentrations occur in some cases; for example, in mineralised areas.

b Including crops. Not including metal-tolerant species.

Zinc is an essential micronutrient in living organisms; it is an active catalytic component of approximately 300 enzymes including dehydrogenases, phosphatases, and RNA polymerases and has many physiological roles including DNA replication, CO_2 regulation and metabolism of fats, carbohydrates and proteins. It also has a structural role in protein folding and it is involved in the regulation of gene expression. Deficiency of Zn in soils is widespread in some countries and shortage of Zn in many plants and animals slows growth and sexual development and causes skin lesions. In excess, it is less hazardous than some other metals, but becomes toxic at sufficiently high concentrations (see below); however, some organisms (plants in particular) are able to tolerate, and actively accumulate, very high levels of Zn. For example, tissue levels of >5% Zn have been recorded in Brassicaceae species such as *Thlaspi caerulescens* (Zhao et al., 2003), which sequesters Zn in the central vacuole of leaf cells, away from metabolic activity.

Production and uses

Global reserves of Zn are estimated at 250 Mt and approximately 13 Mt are extracted annually; most is mined in China, Australia and Peru, but use is also made of Zn recycled from wastes (USGS, 2014). Zinc is mainly used in the galvanising of iron and steel (Fig. 19.1), for protection against corrosion, and in the production of alloys, including brass. Zinc oxide is used as an additive in rubber, paints and plastics. Other end-uses include alkaline batteries, skincare products, mineral supplements and agricultural fungicides.

Figure 19.1 The main use of zinc is in galvanisation (zinc-plating) of iron and steel objects, particularly those exposed to the elements, to protect against rusting.

Credit: Rob Young (Wikimedia Commons).

Pollution and environmental impacts

Anthropogenic emissions to the atmosphere are mainly from coal burning, metal smelting, steel works and waste incineration; atmospheric concentrations in close proximity to industrial sources can be several orders of magnitude higher than background. Emissions to air are of concern mainly because of subsequent deposition of Zn onto soils and uptake by plants; for example, maximum Zn concentrations of >10,000 mg kg^{-1} (>1%) have been recorded in soils adjacent to metal smelters (Kabata-Pendias, 2001).

Agricultural soils are contaminated by the application of P fertilisers, some of which contain >1000 mg Zn kg^{-1}, and sewage sludge, which has higher concentrations of Zn than any of the other common toxic elements. Zinc levels of up to 5% have been recorded in sewage sludge and to protect against excessive Zn contamination of agricultural soils, the European Union, for example, allows sludge application only if this will not increase the soil's Zn concentration above 300 mg kg^{-1} (at pH 6–7); concentrations of 450 mg kg^{-1} and 200–250 mg kg^{-1} are allowed in alkaline and acid soils, respectively (CEC, 1986). This has resulted in lower amounts of sewage sludge

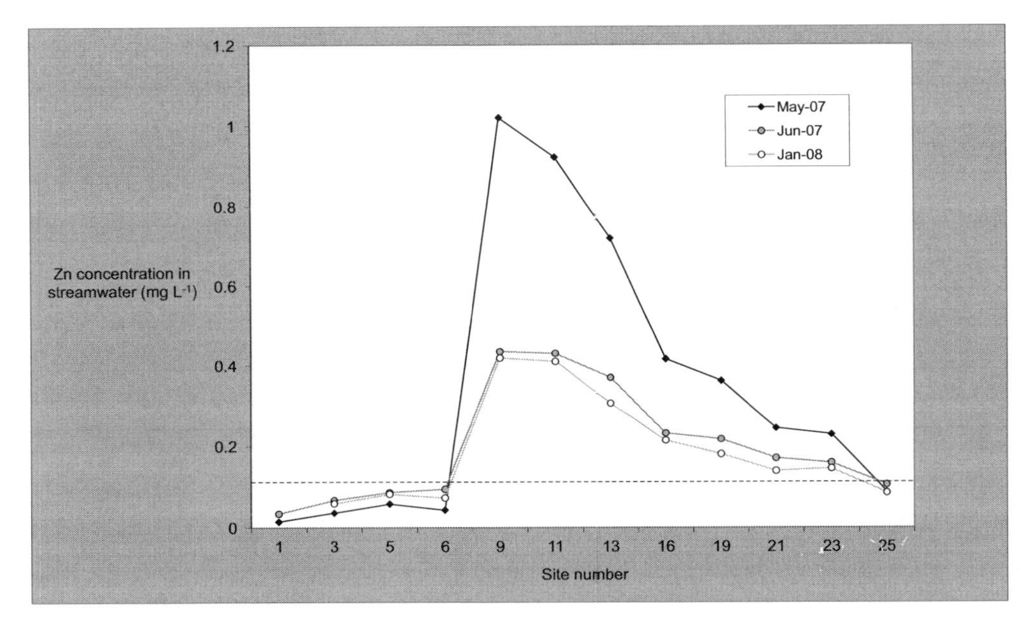

Figure 19.2 Zinc concentrations in streamwater samples from Rookhope Burn, UK, which is located in an area of former lead-zinc mining. The horizontal dashed line denotes the environmental quality standard (EQS) relating to protection of aquatic life (see Appendix, Table A1). Site 1 is near the headwaters of the catchment and site 25 is approximately 11 km downstream of site 1. Samples were also taken from mine outflows and waste leachates (not shown). Immediately prior to sampling in May 2007 an outburst of minewater occurred from a shaft between sites 6 and 9, possibly caused by an underground collapse. As a result, Zn concentrations downstream increased to ~1 mg L^{-1}. The shaft was capped and subsequent sampling (Jun 2007 and Jan 2008) showed a reduction in Zn levels; however, concentrations remained above the EQS and the authors conclude that Zn is the key contaminant of ecological concern in the study area.

Source: Data from Banks and Palumbo-Roe (2010).

being applied to soils, especially in areas with naturally elevated Zn levels, and this is problematic because sludge that cannot be used as fertiliser (amounting to >60% in parts of the EU) has to be landfilled or incinerated.

Key sources of Zn pollution in freshwaters are mine workings, industrial effluents (e.g. from smelting and refining) and leaching/runoff from contaminated soils. An additional source is urban runoff containing Zn from the abrasion of galvanised roofs and tyre rubber. Zinc concentrations of >1 mg L^{-1} have been recorded in streams in mining areas (e.g. Fig. 19.2).

Zinc is less toxic to humans than some of the elements it is often associated with, particularly Cd. For example, no drinking water or air quality guidelines are set for Zn by the WHO, although caution is expressed regarding Zn inputs to drinking water from galvanised water pipes. Reports of human toxicity relate to accidental or occupational exposure; for example, 'zinc fever', or 'metal fume fever', characterised by nausea, chest pains and ordinary fever symptoms, is reported after inhalation of fine ZnO particulate by workers. Excessive Zn intake in such circumstances decreases Cu absorption, causing anaemia and reducing immune function, and appears to cause neurotoxicity. Zinc is not generally considered to be carcinogenic, although the possibility has been raised of a link with prostate cancer.

In polluted environments, Zn is known to cause toxic effects in microorganisms (including N-fixing *Rhizobium* spp.), fish and aquatic invertebrates; an example of the latter is demonstrated by low survival rates in crustaceans in river water polluted by smelter wastes (Fig. 19.3). In Lake Ontario, species richness of *Ephemeroptera*

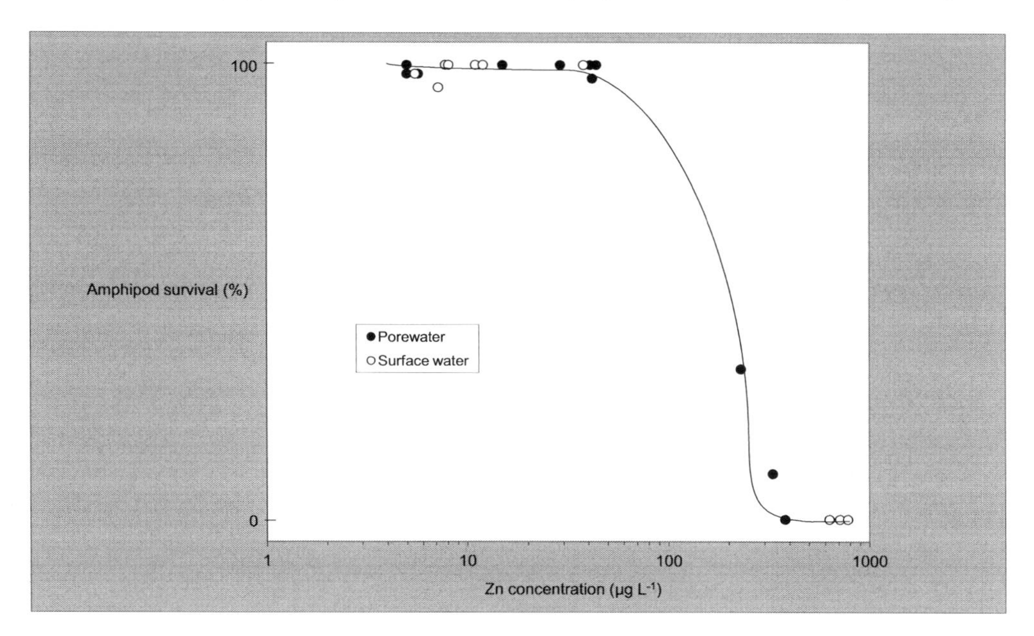

Figure 19.3 Tests on river water and sediment porewater polluted by smelter wastes at Palmerton, Pennsylvania, USA showed that where zinc concentrations were <50 μg L^{-1} the widespread crustacean *Hyalella azteca* was virtually unaffected (mostly 98–100% survival), but where zinc levels were >230 μg L^{-1}, a survival rate of <33% was recorded; concentrations of other metals were low, although some samples contained up to 3 μg Cd L^{-1}.

Source: Adapted from Besser et al. (2009).

(mayfly), *Plecoptera* (stonefly) and *Tricoptera* (caddisfly) was inversely related to Zn concentrations in sediments (CCME, 1999). In fish, Zn is bioconcentrated in the liver and the gills, causing tissue damage and reduced respiratory capacity. Fish in the Upper Enoree River, South Carolina, USA were still being affected by an accidental release of Zn from a galvanising plant when monitoring was undertaken two decades later in 2006; Zn concentrations of 7 mg L^{-1} (nearest to the original spill site) were significantly affecting fish mortality (Giardina et al., 2009). Records of mammalian toxicity are fewer, but there are historic records of livestock grazing near Zn smelters showing weakness and jaundice.

An important environmental impact of Zn is its phytotoxicity. In Zn-contaminated soils (concentrations in excess of ~400 mg kg^{-1}), plant uptake of other essential elements, particularly boron, iron and manganese, is inhibited. Toxic symptoms, in those plants managing to survive in grossly contaminated soils, include stunted plant growth and chlorosis of new leaves. Zn concentrations in plants in polluted areas can reach several hundred mg kg^{-1} and in some cases >1000 mg kg^{-1} (Kabata-Pendias, 2001). Most plants show toxic effects at tissue levels of 100–500 mg kg^{-1}, but others can tolerate much more (see above).

References

Banks, V.J. and Palumbo-Roe, B. 2010. Synoptic monitoring as an approach to discriminating between point and diffuse source contributions to zinc loads in mining impacted catchments. *Journal of Environmental Monitoring* 12, 1684–1698.

Besser, J.M., Brumbaugh, B. and Ingersoll, C. 2009. Ecotoxicology studies with sediment, pore water and surface water from the Palmerton zinc site. Available at: www.fws.gov/contaminants/restorationplans/palmerton/Besserposter.pdf

CCME (Canadian Council of Ministers of the Environment). 1999. Canadian sediment quality guidelines for the protection of aquatic life: zinc. In: *Canadian Environmental Quality Guidelines*. CCME, Winnipeg.

CEC (Council of the European Communities). 1986. Council Directive 86/278/EEC.

Giardina, A., Larson, S.F., Wisner, B., Wheeler, J. and Chao, M. 2009. Long-term and acute effects of zinc contamination of a stream on fish mortality and physiology. *Environmental Toxicology and Chemistry* 28(2), 287–295.

Kabata-Pendias, A. 2001. *Trace Elements in Soils and Plants*, 3rd Ed. CRC Press, Boca Raton, USA.

USGS (United States Geological Survey). 2014. *Mineral Commodity Summaries: Zinc.* Available at: http://minerals.usgl.gov/minerals

Zhao, F.J., Lombi, E. and McGrath, S.P. 2003. Assessing the potential for zinc and cadmium remediation with the hyperaccumulator, *Thlaspi caerulescens*. *Plant and Soil* 249, 37–43.

20 Additional environmental pollutants

Aluminium (Al)

Aluminium is the third most common element – and the most abundant metal – in the Earth's crust (concentration 8.2%). It is used in a wide range of important products, being highly valued in manufacturing for its properties of lightness, conductivity and resistance to corrosion. To meet the demand for Al products, from packaging and cooking utensils to transport and construction applications, it is extracted from the ground in large quantities. However, pollution impacts associated with Al production are associated with the considerable energy inputs required for its extraction and not by the dispersal of Al itself into the environment. In this context it contrasts with other lithogenic elements considered in this book, which have direct impacts because of their transfer from the lithosphere to the atmosphere, pedosphere and/or hydrosphere. In fact, Al toxicity is observed in fish, aquatic invertebrates and other animals; however, these toxic impacts occur when Al that is *naturally present* in soils and sediments is released to waters in bioavailable form because of catchment acidification. Thus, Al toxicity in acidified catchments is a symptom of S and N pollution (see Box 5.1). The energy-intensive process of Al extraction transfers C, S, N, Hg and other elements to the atmosphere, but does not cause Al pollution. Similarly, the production of Al metal, in the Hall-Heroult process, does not appear to create harmful Al pollution; the main concerns about the environment in areas surrounding Al smelters are related to other pollutants, particularly fluorides {→F}.

There have been concerns about Al toxicity via the direct input of Al into the environment, but these have been specific accidents or incidents and not the result of widespread anthropogenic activities. For example, in 1988, Al sulfate was accidentally added to the water supply in Camelford, UK and over the following decades concerns about ill-health in the local population have been investigated. Following the accident, a large number of fish were killed by Al that was flushed out of the water mains into local waterways. In 2010, a massive failure of a waste sludge dam at an alumina plant in Hungary deluged downstream areas, drowning a number of people and causing caustic burns (because of the waste's high pH) before reaching the River Danube a few days later; the sludge was approximately 10% alumina (Al_2O_3), but this compound is very insoluble and no toxic effects were reported. An example of potential ecological harm from non-accidental anthropogenic activity is the deliberate application of Al salts to eutrophic lakes to reduce the release of P from sediments; such treatments have increased Al concentrations in fish, although direct toxicity has not been reported.

Boron (B)

The metalloid boron occurs in nature mainly as borates (anionic molecules of B combined with oxygen) and is essential to plants and some animals. It is used mainly in the production of borosilicate glass (e.g. Pyrex™) and in various other products including fertiliser, detergents and flame retardants. Boron is emitted into freshwaters in sewage effluent (in detergents) and in effluents from industries in which it is used. It may also be a soil contaminant in areas of industrial atmospheric pollution, irrigation (using wastewaters) and fertiliser use (either as foliar sprays for trees or as sewage sludge applications). Boron is readily available to plants and while some tolerate high amounts, phytotoxicity problems are known to occur in soils with naturally elevated B; the latter may be exacerbated by anthropogenic inputs. Exposure via drinking water is also a concern because of the symptoms observed in accidentally and occupationally exposed humans; these include gastrointestinal and skin disorders. In some Mediterranean countries, including Israel and Italy, naturally B-rich water has to be mixed with cleaner water to meet drinking and irrigation water standards. Another Mediterranean nation, Turkey, has the world's largest B reserves and studies there show the potential for B pollution. In particular, geothermal plant wastewaters and drainage waters from B mines are discharged to rivers, from which water is taken for crop irrigation; this increases soil levels of B to above that required by plants. There are also concerns about the effect of these waters entering lakes (in the B mining region) that are internationally recognised for their wildlife value.

Cyanide (CN^-)

The main form of cyanide is hydrogen cyanide (HCN), which readily dissolves in water to form hydrocyanic acid (also called prussic acid) and its salts, including sodium and potassium cyanides, NaCN and KCN. There are also less toxic cyanates (e.g. sodium cyanate, NaOCN), thiocyanates (e.g. sodium thiocyanate, NaSCN) and organic cyanides (nitriles). One of the main applications of cyanide is in a process called cyanidation, where a Na, K or Ca cyanide solution is added to low-grade metal ores to extract gold and silver, utilising the ability of cyanide to dissolve and complex these elements. At some mine sites the resulting cyanide waste is converted to less toxic cyanate, which is then stored in a tailings pond. Cyanides are also used in metal cleaning, electroplating, steel and plastics manufacture and in the synthesis of organic compounds. HCN has also been used as a fumigant in pest control.

Cyanide is very toxic and its release into the environment, for example from ore processing, electroplating factories, gas works and coke plants, can have serious impacts. In addition to direct effects, cyanide left in processed wastes can dissolve toxic metals present in the wastes and release them to the environment. Another source of pollution is cyanide fishing, where NaCN is added to reef waters in Indonesia and some other tropical countries to stun fish for capture and sale as exotic pets or food. Cyanide can be inhaled, ingested and, in some cases, absorbed through the skin. Hydrogen cyanide is released by some plants and invertebrates as a defence mechanism. Its toxic nature is mainly due to its inhibition of respiratory enzymes: the CN^- anion binds strongly with Fe^{3+} in the enzyme ferricytochrome oxidase, preventing its reduction to cytochrome oxidase (containing Fe^{2+}), which is used in electron transport and aerobic respiration in tissues. Cyanide also inhibits enzymes involved in amino acid biotransformation and thyroid hormone production. Chronic exposure causes weakness, nausea and heart pains, while acute exposure can cause degeneration of the central nervous system, respiratory paralysis, unconsciousness and cardiac arrest.

Cyanidation was used in Baia Mare, Romania in the 1990s to extract gold and silver from mine tailings. The tailings dam failed in 2000, leaking toxic metals and 50–100 t of cyanide into the River Somes, affecting 2000 km of the River Danube catchment (Csagoly, 2000). Cyanide is very soluble and the plume of water contaminated with CN^- reached the Black Sea about a month after the spill; in contrast, most metals were attenuated by sediments near to the source. All water-borne plankton and most fish along the most polluted stretch of the river were killed, including >1000 t of fish in Hungary and large amounts of fish in Romania and former Yugoslavia. Plankton from unpolluted stretches of the catchment recolonised the affected area quite quickly, but this was not the case for fish populations, which took much longer to recover.

Manganese (Mn)

Manganese is abundant in the Earth's crust and in soils and is an essential element. The largest reserves are in South Africa and Mn ores such as pyrolusite (MnO_2) are extracted in large quantities, mainly for use in steel production. It is also used in cleaning products as potassium permanganate ($KMnO_4$). In some countries it is still used, in the form of methylcyclopentadienyl manganese tricarbonyl (MMT), as a replacement for tetraethyl Pb as an anti-knock additive in petrol. The main sources of Mn pollution are mining activities and metallurgy plants. Its concentrations in drinking waters sourced from anoxic groundwaters can be naturally elevated. Levels of Mn reported in some drinking waters are higher than the WHO standard of 0.4 mg L^{-1}, a guideline that is based on neurological impairment and other health outcomes observed in occupational studies involving exposure to much higher levels of Mn. In 2011, a study of Mn in drinking waters in Quebec, Canada reported a six-point difference in intelligence quotient (IQ) between children with the highest (median 216 µg L^{-1}) and lowest exposures (1 µg L^{-1}) (Bouchard et al., 2011). These findings are related to naturally elevated Mn, but contamination of groundwaters by land disposal of metal wastes has also been recorded. Human exposure to Mn is increased in urban air polluted by the petrol additive MMT. In the aquatic environment, Mn is bioconcentrated by plankton, algae, invertebrates and fish, and can cause ecotoxicity in excessive amounts; freshwater crustaceans and molluscs appear to be particularly sensitive. Gill and shell defects in crustaceans living in polluted estuaries have been attributed to Mn toxicity. Known impacts also include iron deficiency in some plankton species, although in others Mn is thought to lessen the toxicity of some metals. Field observations of salmonid fish have shown correlations between Mn concentrations and mortality related to gill precipitation of Mn. Recorded effects in terrestrial plants include impairments in growth and enzyme and hormone activities; typical symptoms of Mn phytotoxicity include chlorosis and necrosis.

Molybdenum (Mo)

Molybdenum is present at low levels in most rocks and soils (approximately 2 mg kg^{-1} or less) and is concentrated in its main ore minerals: molybdenite (MoS) and the molybdate mineral, powellite ($CaMoO_4$). In neutral and alkaline soils Mo is quite mobile and bioavailable, occurring mainly as the molybdate anion, MoO_4^{2-}, and low availability of Mo in acidic agricultural soils is treated by soil liming. Molybdenum is an essential element, a vital component of several enzymes including nitrogenase, the enzyme required by plants for N-fixation. Most Mo is mined in China; it is extracted both in the primary form and as by-products from the processing of ores of metals such as Cu. It is mainly used in alloy and steel production. Sources of air pollution include metal mines and processing plants, oil refineries and brick kilns. Deposition of Mo from these sources can contaminate soils, as can the application of sewage sludge in agricultural areas. The main impacts of Mo pollution are also related to air and soil pollution. At particular risk are livestock grazing Mo-polluted areas, including mine wastes. In such cases, excessive Mo intake depresses the absorption of Cu at micronutrient levels; the resulting condition, known as molybdenosis, is characterised by diarrhoea, weight loss, reduced reproductive capacity and, in extreme cases, death. The toxic effects can be ameliorated by providing Cu supplements (Steinke and Majak, 2010). In areas with toxic levels of Cu in soils and vegetation, excess Mo intake can reduce the toxicity of Cu to livestock by the same action of depressing Cu absorption.

Silver (Ag)

Most silver is mined in Mexico and Peru, typically as acanthite (AgS), which is the low-temperature form of the vein mineral argentite; it also occurs in association with deposits of other metal ores, particularly Pb. Its main uses are in electronics, coinage, decoration (jewellery and silverware) and photography, although the latter use continues to decline. It is also used as a bactericide. Silver nanoparticles (1–100 nm sized particles of elemental Ag) are present in a number of consumer products, including sunscreens, disinfectants and clothing (e.g. odour-free socks), and the particles consequently enter the sewage system, where they are converted to AgS nanoparticles, becoming concentrated in sewage sludge. The application of sewage sludge to soils is therefore a potential source of Ag pollution, others being silver plating (waste effluents to rivers), metal processing and fossil fuel combustion. Generally, Ag has low solubility and, as such, is not considered a major environmental toxicant; however, elevated concentrations of mobile forms, particularly the free Ag^+ ion, which binds to sulfhydryl groups, can impair bacterial processes in contaminated soils and sediments, including respiration and denitrification. Laboratory experiments show Ag solutions and nanoparticles to be toxic to soil and aquatic organisms, including plants, invertebrates and microorganisms; for example, Ag nanoparticles can cause lysis of bacterial cells and reduce earthworm growth and reproduction. Levels of Ag in some sewage sludges are high enough to impair microbial function in treated soil. Some plants bioconcentrate Ag and it can be phytotoxic; soil concentrations in excess of 2 mg kg^{-1} can cause toxicity to plants, mainly because of changes to the permeability of cell membranes. Environmental exposure is not known in humans, but excessive exposure via medication or mineral supplements can cause argyria, an irreversible blue-grey discoloration of the skin.

Thallium (Tl)

Thallium is not abundant in the Earth's crust and Tl minerals, such as crookesite ($TlCu_7Se_4$) and lorandite ($TlAsS_2$) are rare. Small amounts occur in the sulfide ores of

Zn and other metals, and it is mainly extracted from the wastes of ore smelters. It is also present in small amounts in coal. Its main uses are in medical imaging, scintillometers, optics and superconductors; it has also been used as a rodent and insect poison. Thallium pollution, from metal smelting, coal combustion and other sources, is not as widespread as that of most toxic metals, but is of concern because many Tl compounds are soluble (and therefore mobile and bioavailable) and it is also a very toxic element. Monovalent Tl has a similar ionic radius and charge to the essential element potassium and it can therefore replace K^+ in brain cells and other tissues. For this reason, guideline levels have been established in some countries for soils (e.g. 1 mg kg^{-1} in Canada) and drinking water (2 μg L^{-1} in the USA).

High levels of Tl have been recorded in soils near metal sulfide mines and cement factories and in freshwaters near coal-ash reservoirs and metal processing works. There is a well-documented case of Tl pollution that occurred in the 1980s in the vicinity of a cement plant in Lengerich, Germany that had used ores containing Tl. Elevated levels of Tl were observed in the urine and hair of local people, who appear to have been exposed via the ingestion of locally grown fruits and vegetables. Those with the highest Tl levels suffered insomnia, anxiety and weakness, but not hair loss, which is a common symptom of Tl toxicity. In another case in the same decade, soils in Chernivtsi, Ukraine were contaminated by Tl from an unknown source and >100 local children exhibited Tl toxicity symptoms, including hair loss and nervous disorders. Thallium is also toxic to microorganisms, plants and animals. Laboratory experiments on plants indicate that it inhibits germination and growth, especially in some leguminous species. Livestock and predators that have ingested Tl rodenticides have exhibited gastrointestinal and neurological disorders. Otherwise, the effects of Tl intake on flora and fauna inhabiting polluted areas is under-researched and, as with other low-level toxicants, difficult to ascertain because of confounding factors, not least the typical presence of other toxicants.

Vanadium (V)

Vanadium is essential in some organisms, particularly species of algae. Some species of sea squirt (*Ascidiacea*) bioconcentrate V to very high levels in haemovanadin, which is an analogue of haemoglobin; the function of haemovanadin is not clear, but it may be used as a toxic defence mechanism against predation. Vanadium ores such as vanadinite ($Pb_5(VO_4)_3Cl$) are quite rare, but V is associated with the ores of other metals, particularly titanium-rich magnetite (Fe_3O_4). It is one of the main metals (along with Fe and Ni) present in crude oil and it also occurs in elevated levels in some types of coal and in some uraniferous deposits. Vanadium is mostly recovered from steel slags and the fly-ash of heavy oil combustion and most V is used to produce ferrovanadium, a hard steel alloy. Pollution sources include heavy oil and coal combustion, industrial effluents, and the land disposal of power station ash and steel industry slag. Vanadium accumulates in some plant species and phytotoxicity has been observed, mainly in laboratory experiments, in the form of reduced growth and necrosis. Soil microbial activity may also be impaired in polluted areas. Occupational exposure to V dust can cause irritation of the eyes, skin and respiratory tract. In the 1990s, elevated levels of V were recorded in the hair of children living near to a V processing plant in the Czech Republic, and particularly in those whose parents worked at the plant; V concentrations in local soils were also elevated compared with controls (Lener et al., 1998). Alterations to the blood and immune systems (e.g. reduced red blood counts and higher incidences of infection) were observed in the exposed children, but exposure to other toxicants was not controlled for in the epidemiological studies undertaken.

A note on some remaining elements

The elements included in this text are limited to those that are added to the natural environment by human activities in sufficient quantities to have observable environmental and health impacts. This excludes elements that are known from poisonings, experimental studies and/or occupational epidemiology studies to be toxic but are not widespread contaminants in the environment. On the other hand, some elements that are emitted into the environment do not cause major environmental impacts because of their low reactivity, solubility and/or bioavailability. Notable examples of elements falling into either of these categories include **antimony, barium, beryllium, bismuth, cobalt** and **tellurium.** It should be noted, however, that some elements that traditionally have been considered as non-toxic or environmentally benign, possibly because of lack of detailed investigation, are sometimes reconsidered as 'emerging contaminants' as more research is undertaken and data collected.

Some of the elements mentioned above are sometimes included in lists of published safety guidelines for potential exposure; however, in these cases the reasons are not related to environmental pollution. Examples include **antimony**, which is sometimes included in drinking water guidelines, mainly because of the risk of leaching from plumbing and not because toxic amounts are thought likely to leach into groundwaters from polluted soils, for example. Concentrations of **barium** in drinking waters are regulated in some countries, based on studies that show a link to increased hypertension in those exposed to naturally elevated levels of the element. The case of **beryllium** (Be) is also illustrative in this context: Be is sometimes included in air quality standards, but Be toxicity (berylliosis), albeit very serious, is known from occupational, rather than environmental exposure. In this case, it is worth mentioning that a major accidental release of Be occurred in the former Soviet Union in 1990 after an explosion at a metallurgy plant and the resulting cloud of Be dust, which passed over the nearby city of >100,000 people, is not reported to have caused serious health effects in the downwind population.

Iron (Fe) is released into the environment from anthropogenic activities in greater quantities than any other metal. However, iron is ubiquitous in the environment, is required in large amounts by many organisms and is not toxic at environmental concentrations that are typically much higher than those of other metals; however, physical damage can be caused by Fe oxyhydroxide (ochre) precipitation associated with acid mine drainage {→S}. An even more ubiquitous element in the natural environment is **silicon** (Si). While asbestosis, a potentially fatal lung disease, is caused by silicate minerals, it is generally a consequence of accidental or occupational exposure, rather than environmental pollution. The toxicity of silicon tetrachloride, which is used in the production of photovoltaic cells and semiconductors, is caused by the compound's chlorine, illustrating that the mere presence of an element in a toxic compound does not mean that it is itself toxic.

Two other groups of elements of potential interest are the **rare earth elements** (REE) and the **platinum group elements** (PGE). Rare earth elements are now mined extensively, particularly to provide the raw materials for mobile phones and other electronic goods. Their extraction can create mildly radioactive waste materials because of the presence of U and thorium in their ores; however the REEs themselves are not known, at least at the present time, to cause serious environmental or toxicological problems to environmentally exposed humans or wildlife. There is some evidence from laboratory and occupational studies that **cerium** may damage cardiac tissues, but data on this and other REEs are relatively limited at present. **Platinum** and some other PGEs are used in industry and in products including vehicle catalytic converters and anti-cancer drugs; the latter uses have led to emissions to the atmosphere and surface waters, respectively. Occupational toxicity, in the form of increased rates of allergic reaction to some forms of PGEs, has been observed, but environmental toxicity has not been recorded.

Non-chemical pollution

The Elements of Environmental Pollution focuses mainly on chemical pollution, but forms of non-chemical pollution are worthy of mention; they are primarily related to biological pollutants such as pathogenic bacteria and various physical effects caused by fine particulate matter (PM), litter and excess heat.

Pathogens

In the poorest parts of the world, particularly those areas where sewage treatment is not undertaken, water contaminated by human and animal excreta is likely to be the biggest pollution hazard of all. Approximately 2.5 billion people have no access to proper sanitation facilities and ~750 million people have to use unsafe drinking water (WHO/UNICEF, 2014), putting them at high risk of contracting pathogenic illnesses. The main pathogens in polluted waters are bacteria, viruses and parasites (mainly protozoa and nematodes). Exposure to these pathogens can cause serious diseases like typhoid, dysentery and cholera (bacteria), hepatitis A and polio (viruses) and schistosomiasis (parasites). The World Health Organization estimates that globally 760,000 children under the age of five die of diarrhoea every year (WHO, 2013). The main challenge is to improve sanitation for all; practical solutions mainly involve the provision of sanitary facilities and the treatment of unsafe water by water purification, generally using hypochlorite salts and/or filtration systems.

Particulate matter and foams

Atmospheric particulate matter (PM) has been discussed in relation to poor air quality, global warming and 'global dimming' {→C →S}, but some other types of PM pollution are worthy of separate mention. Asbestos, if not professionally removed (from derelict sites, for example), can pose a major health risk via inhalation of the fibrous particles. Excess PM in waters is associated with activities such as dredging, china clay effluents and the soil erosion associated with vegetation clearance and intensive farming. In such conditions organisms may be adversely affected by turbidity, which impedes photosynthesis, and blanketing of stream beds, which can affect spawning and feeding, for example; direct physical harm, including suffocation, may also occur. The deposition of 'ochre' on stream beds can cause similar problems {→S}. Detergents in domestic and industrial effluents can harm aquatic life by generating surface foam in watercourses. This can reduce the amount of light for photosynthesisers and limit oxygen diffusion into the water, potentially affecting respiration.

Litter

The occurrence of persistent wastes such as plastics in both terrestrial and marine environments can be a cause of external injury, starvation and/or suffocation in marine animals that swallow the materials or become ensnared in them. In some parts of the world, rivers are used as conduits for physical waste materials as well as for sewage and chemical pollutants; a stark example is the Citarum River in Indonesia, which is virtually choked in parts with plastics and other litter. Out at sea, vast accumulations of litter are known to form in oceanic gyres. In addition to the physical environmental damage posed by such large-scale wastes, concerns are growing about the ingestion of micro-plastics (tiny particles of physically degraded plastic litter) by organisms at the bottom of the food chain and their transfer to higher trophic levels. Plastics may be sources of chemical pollutants such as bisphenol A {→C} to soils and waters containing them and to organisms that ingest the litter.

Thermal pollution

Cooling waters from industry, particularly power plants, can create unnaturally high water temperatures in watercourses close to the effluent source. One of the main impacts is a decrease in dissolved oxygen (DO), which can affect aerobic organisms {→C}. The metabolic rate of aquatic organisms is likely to increase at higher temperature, adding to the stress caused by a drop in DO concentrations. Temperature changes may also have effects at the cellular level, disrupting enzymatic processes, for example. Some species are likely to benefit from the warmer conditions, but this may be at the expense of overall biodiversity; furthermore, those benefiting are likely to include pathogens, increasing the risk of disease in higher organisms. It is worth noting that cooling of natural waters, particularly by releases of reservoir waters in warmer countries, is also likely to have deleterious effects on some aquatic organisms.

Noise, light and 'visual' pollution

The existence of noise and artificial light in areas where they are unwanted is often termed 'pollution'. 'Visual pollution', the accumulation of graffiti, road signs and advertising hoardings in urban areas, is also documented. These types of pollution have little in common with the chemical pollutants detailed in this book, but, in the case of noise pollution at least, human health impacts can occur.

References and further reading

Bouchard, M.F., Sauve, S., Barbeau, B., Legrand, M., Brodeur, M.-E., Bouffard, T., Limoges, E., Bellinger, D.C. and Mergler, D. 2011. Intellectual impairment in school-age children exposed to manganese from drinking water. *Environmental Health Perspectives* 119(1), 138–143.

Csagoly, P. 2000. The Cyanide Spill at Baia Mare, Romania: Before During and After. Summary of a report by the United Nations Environment Programme and the United Nations Office for the Coordination of Humanitarian Affairs. The Regional Environmental Center for Central and Eastern Europe, Szentendre.

Lener, J., Kucera, J., Kodl, M. and Skokanova, V. 1998. Health effects of environmental exposure to vanadium. In: Nriagu, J.O. (Ed.). *Vanadium in the Environment, Part 2: Health Effects*. John Wiley & Sons, New York.

Steinke, D.R. and Majak, W. 2010. Cattle can tolerate high-molybdenum forage grown on reclaimed mine tailings: a review. *International Journal of Mining, Reclamation and Environment* 24, 255–266.

WHO (World Health Organization). 2013. *Diarrhoeal Disease: Fact Sheet No. 330* (April 2013). Available at: http://www.who.int/mediacentre/factsheets/fs330/en/

WHO (World Health Organization)/UNICEF. 2014. *Progress on Drinking Water and Sanitation: 2014 Update*. WHO, Geneva.

Appendix 1. Scientific notation and units of measurement

Scientific notation

Environmental science involves descriptions of both miniscule and enormous quantities, from the amount of mercury in ambient air to the total mass of carbon in the Earth's crust for example. To enable clear and concise descriptions of such concentrations, scientific notation is used in preference to long strings of digits. Scientific notation takes the form:

$A \times 10^b$, where A is a decimal number (typically ≥ 1 and <10) and b, the exponent, is a whole number and is the amount by which 10 is raised. For example:

One million, 1,000,000, is written as 1×10^6, i.e. 1.0 multiplied by 1,000,000. The latter number is the product of $10 \times 10 \times 10 \times 10 \times 10 \times 10$, and is therefore written as 10^6.

2,000,000 is therefore 2×10^6 and 1,500,000 is 1.5×10^6.

Notice that in each case, to convert scientific notation to the original number, the decimal place is moved forward (to the right) six times because, in these examples, the exponent is 6.

Taking this a step further, 36,000,000 becomes 36×10^6 or, more correctly, 3.6×10^7.

In this example, using the latter form of 3.6×10^7 is preferred because, otherwise, scientific notation itself becomes less concise; i.e. ten thousand million is written in scientific notation as 1.0×10^{10}, not the longer form, $10{,}000 \times 10^6$. Within reason, exceptions can be made for purposes of clarity and, in this text, ranges of values or comparable values are sometimes expressed using a single order of magnitude (power of ten) as the exponent; for example, when describing a range of 1 million to 110 million megatonnes, the notation 1×10^6 to 110×10^6 Mt may be slightly easier to grasp at first glance than 1×10^6 to 1.1×10^8 Mt, at least for those new to scientific notation.

For numbers <1, exponents with negative values are used. For example:

0.00001 becomes 1×10^{-5} and 0.000037 becomes 3.7×10^{-5}

In these examples, to convert scientific notation to the original number, the decimal point is moved five places to the left.

In correct scientific notation, *significant figures* are also used so that digits that add to the original number's precision are incorporated. This means, for example, that:

0.004107 becomes 4.107×10^{-3}, not 4.1×10^{-3}

and 340,200 becomes 3.402×10^{5}, not 3.4×10^{5}.

Units of measurement

Mass

The base unit of mass is the gram.

- g = gram
- pg = picogram = 1×10^{-12} g
- ng = nanogram = 1×10^{-9} g
- µg = microgram = 1×10^{-6} g
- mg = milligram = 1×10^{-3} g
- kg = kilogram = 1×10^{3} g
- Mg = Megagram = 1×10^{6} g or 1 tonne
- Gg = Gigagram = 1×10^{9} g or 1 thousand tonnes
- Tg = Teragram = 1×10^{12} g or 1 million tonnes (1 megatonne, Mt*)

* In this text, the megatonne (Mt) is used in preference to the equivalent Tg and is used extensively in relation to the enormous, global-scale masses that are frequently described. The recurring Mt unit is more likely to be remembered by the average reader as referring to a million tonnes, a measure that may at least be grasped in a general way compared with the more abstract Tg, which is not in general use in everyday life.

In chemistry, masses of elements or compounds are sometimes expressed as moles, where 1 mole of a substance is an unchanging but very large number of atoms or molecules of that substance (the number of atoms or molecules in a mole is always 6.022×10^{23}, known as Avogadro's number). Individual atoms are too small to hold, of course, so using moles (very large numbers of atoms) allows us to make everyday measurements of chemicals in the laboratory. The mass of 1 mole of a substance is simply its relative atomic or molecular mass (read from the periodic table) expressed in grams. For example, 1 mole of carbon (atomic mass of 12) weighs 12 g and 1 mole of lead (atomic mass of 207) weighs 207 g. Another way of looking at this is that 6.022×10^{23} atoms of C have a mass of 12 g while the same (Avogadro's) number of Pb atoms (each of which is much heavier than a C atom) has a mass of 207 g. For molecules, the individual atomic masses of the component elements are added together; e.g. 1 mole of water, H_2O, has a mass of 18 g, derived from the atomic masses of the 2 H atoms (each with an atomic mass of 1) and the O atom (atomic mass of 16). Multiples or fractions of moles just require simple arithmetic; e.g. 3.85 moles of water have a mass of 69.3 g ($18 \times 3.85 = 69.3$) and a millimole (mmol, a thousandth of a mole) of water has a mass of 0.018 g (18g/1000).

Volume

The most commonly used units are:

- L = litre
- nL = nanolitre = 1×10^{-9} L
- μL = microlitre = 1×10^{-6} L
- mL = millilitre = 1×10^{-3} L
- m^3 = cubic metre

Length

The most commonly used units are:

- m = metre
- mm = millimetre, 1×10^{-3} m
- cm = centimetre, 1×10^{-2} m
- km = kilometre, 1×10^{3} m

Concentration

Concentration is the mass of a substance per unit mass or volume of the medium that contains it. For example:

- The concentration of copper in soil may be expressed in units of mg kg^{-1}; i.e. mg of copper per kilogram of soil.
- The concentration of xylene in drinking water may be expressed in units of mg L^{-1}; i.e. mg of xylene per litre of drinking water.
- The concentration of ozone in the atmosphere may be expressed in units of mg m^{-3}; i.e. mg of ozone per cubic metre of air.

Concentrations in this text are generally expressed in these terms. Concentrations of soil and water contaminants are often described in the literature in 'parts per' terms. Parts per million (ppm) is equivalent to mg kg^{-1} or mg L^{-1} and parts per billion (ppb) is equivalent to μg kg^{-1} or μg L^{-1}. Atmospheric concentrations are often expressed by scientists as 'parts per million (or billion or trillion) volume' (ppmv, ppbv and pptv, respectively). Concentrations in waters are sometimes described in units of millimolar (mM) or micromolar (μM); these are simply the number of millimoles (thousandths of a mole) or micromoles (millionths of a mole) of a pollutant contained in a litre of solution (see above, under 'Mass', for a brief explanation of the 'mole').

Time

In this text, the unit for time, used in descriptions of annual fluxes for example, is the year or *annum*, a. In descriptions of geological timescales and long-term residence times the unit 'Ma', or million years, is used.

Appendix 2. Environmental quality standards

Environmental quality standards (EQS) are concentrations of pollutants that should not be exceeded, in the interests of environmental protection and human health. They are based on scientific evidence and guidance from bodies such as the World Health Organization and are implemented by authorities, typically on a national basis. Tables A1 to A3 list EQS for inorganic and organic pollutants in air, soils, sediments, waters and other environmental media.

Table A1 Environmental quality standards (EQS) or guideline values for metals and metalloids.

Medium	*Source*[1]	*Unit*[2]	*As*	*Cd*	*Cr*	*Cu*	*Pb*	*Hg*[3]	*Ni*	*Se*	*Zn*
Air	WHO[a]	µg m^{-3}	NSL[4]	0.005[5]	NSL[4]	–	0.5	1	NSL[4]	–	–
Soil	EA[b]	mg kg^{-1}	32/43/640	10/2/230	–	–	–	1/26/26[6]	130/230/1800	350/120/13000	–
Soil	EPA[c]	mg kg^{-1}	0.4	78	390	–	400	23	1600	390	23,000
Crops	FAO[d]	mg kg^{-1}	–	0.05–0.4	–	–	0.1–0.3	–	–	–	–
Freshwater	EU[e7]	µg L^{-1}	*50*[8]	0.08–0.25	*5–50*[8]	*1–28*[8]	7.2	0.05	20	–	*8–125*[8]
Freshwater	EPA[f]	µg L^{-1}	150/340	0.25/2	11/16	–	2.5/65	0.77/1.4	52/470	5/–	120/120
Drinking water	WHO[g]	µg L^{-1}	10	3	50	2000	10	6	70	40	–
Saltwater	EU[e]	µg L^{-1}	*25*[8]	0.2	*15*[8]	*5*[8]	7.2	0.05	20	–	*40*[8]
Saltwater	EPA[f]	µg L^{-1}	36/69	8.8/40	50/1100	3.1/4.8	8.1/210	0.94/1.8	8.2/74	71/290	81/90
Sediment[9]	CCME[h]	mg kg^{-1}	5.9/17	0.6/3.5	37.3/90	36/197	35/91	0.17/0.5	–	–	123/315
Sediment[10]	CCME[h]	mg kg^{-1}	7.2/41.6	0.7/4.2	52/160	19/108	30/112	0.13/0.7	–	–	124/271

1 Sources:

a) World Health Organization: Regional Office for Europe. 2000. *Air Quality Guidelines for Europe*, 2nd Edition.

b) Environment Agency, UK. Soil Guideline Values (SGVs). Available at www.environment-agency.co.uk. For each element three SGVs are given for the following land uses: residential, allotment (domestic food growing) and commercial, respectively. The EA states: 'The SGVs should only be used in conjunction with the information contained [in the accompanying technical notes] and with an understanding of the [separately published] exposure and toxicological assumptions'.

c) Environmental Protection Agency, USA. 1996. *Soil Screening Guidance: Technical Background Document*, 2nd Edition. USEPA, Washington DC. Values shown are soil screening levels, based on protection of human health via the soil/dust ingestion pathway.

d) Food and Agriculture Organization of the United Nations. *CODEX General Standard for Contaminants and Toxins in Food and Feed* (CODEX STAN 193/1995). Applies to fruits, vegetable and cereals.

e) European Union, Directive 2008/105/EC (relating to the Water Framework Directive, WFD). Figures shown are environmental quality standards for the protection of human health and the aquatic environment, based on chronic exposure (annual averages).

f) Environmental Protection Agency, USA. National Recommended Water Quality Criteria. Available at www.water.epa.gov. The 2 figures shown are the highest concentrations not causing an 'unacceptable effect' to aquatic organisms, based on chronic and acute exposures, respectively. Cr figures are for Cr(VI).

g) World Health Organization. 2011. *Guidelines for Drinking Water Quality*. Provisional guidelines. Zinc 'not of health concern at levels found in drinking water'.

h) Canadian Council of Ministers of the Environment. Interim Sediment Quality Guidelines, for the protection of aquatic life.

2 Dry weights for soils and sediments; wet weight for crops.

3 Inorganic mercury, except where stated otherwise in these footnotes.

4 No safe level (as carcinogens). Concentrations giving an estimated excess lifetime cancer risk of 1:10,000 for As, Cr(VI) and Ni are, respectively, 66, 2.5 and 250 ng m^{-3}.

5 For effects other than cancer (for which there is no reliable unit risk for Cd).

6 For elemental mercury. Corresponding figures for inorganic Hg^{2+} and methylmercury are, respectively: 170/80/3600 and 11/8/410.

7 Based on the most sensitive freshwater organisms. Ranges relate to water hardness, from minimum value (relating to hardness of 0–50 $CaCO_3$) to maximum value (relating to hardness of >250 mg L^{-1} $CaCO_3$).

8 Elements with EQSs in italics not currently included in Water Framework Directive but listed as 'Annex VIII' substances for which appropriate EQSs are being sought. Basis for discussions are current EQSs as shown in italics. EQS for Cr applies to Cr(VI) only (the other concentrations in this column apply to total Cr).

9 Freshwater sediments. The first and second figures are, respectively, threshold effect levels and probable effect levels. Figures rounded in some cases.

10 Estuarine and marine sediments. The first and second figures are, respectively, threshold effect levels and probable effect levels. Figures rounded in some cases.

Table A2 Environmental quality standards (EQS) or guideline values for organic compounds described in the text[1].

	Air[a]	*Soil*[b]	*Soil*[c]	*Freshwater*[d]	*Drinking water*[e]	*Saltwater*[d]	*Sediment*[f]
Aldrin	–	–	0.04/3/0.02	Σ0.01[2]	0.03[3]	Σ0.005[2]	–
Benzene	NSL[4]	0.3/0.07/95	22/0.8/0.002	10	10	8	–
Bromodichloromethane	–	–	10/3000/0.03	–	60	–	–
Bromoform	–	–	81/53/0.04	–	100	–	–
Carbon tetrachloride	–	–	5/0.3/0.003	12	4	12	–
Chlordane	–	–	0.5/20/0.5	–	0.2	–	4.5/9
Chloroethene	NSL[4]	–	0.3/0.03/0.0007	–	0.3	–	–
Chloroform	–	–	100/0.3/0.03	2.5	300	2.5	–
Chlorpyrifos	–	–	–	0.03	30	0.03	–
DDT	–	–	2/–/2	0.025	1	0.025	1.2/4.8
Dichloromethane	3000	–	85/13/0.001	20	20	20	–
Dieldrin	–	–	0.04/1/0.0002	Σ0.01[2]	0.03[3]	Σ0.005[2]	2.8/6.7
Dimethoate	–	–	–	–	6	–	–
Dioxins and Furans[5]	NP[6]	0.008/0.008/0.24	–	–	–	–	[0.85/21.5][7]
Endosulfan	–	–	470/–/0.9	0.005	NE[8]	0.0005	–
Endrin	–	–	23/–/0.05	Σ0.01[2]	0.6	Σ0.005[2]	2.7/62
Heptachlor	–	–	0.1/4/1	–	NE[8]	–	0.6/2.7
Hexachlorobenzene	–	–	0.4/1/0.1	0.01	NE[8]	0.01	
Lindane	–	–	0.5/–/0.0005	0.02	2	0.002	0.9/1.4
Methyl bromide	–	–	110/10/0.01	–	–	–	–
Nonylphenol	–	–	–	0.3	–	0.3	1400/–
PAH[9]	NSL[4]	–	0.09/–/0.4	0.05	0.7	0.05	32/782
PBDEs[10]	–	–	–	0.0005	–	0.0002	–
PCBs	NP[6]	420/280/3200	1/–/–	–	–	–	34/277
Phenol	–	–	47000/–/5	–	–	–	–
Styrene	260	–	16000/1500/0.2	–	20	–	–
Tetrachloroethene	250	–	12/11/0.003	10	40	10	–
Toluene	260	610/120/4400	16,000/650/0.6	–	700	–	–
Toxaphene	–	–	0.6/89/2	–	20	–	0.1/–
Tributyltin		–	–	0.0002	–	0.0002	–
Trichloroethene	NSL[4]	–	58/5/0.003	10	20	10	–
Xylenes	–	250/160/2600[11]	160,000/410/9[11]	–	500	–	–

Sources:

a) World Health Organization: Regional Office for Europe. 2000. *Air Quality Guidelines for Europe*, 2nd Edition.

b) Environment Agency, UK. Soil Guideline Values (SGVs). Available at www.environment-agency.co.uk. For each contaminant, three SGVs are given for the following land uses: residential, allotment (domestic food growing) and commercial, respectively. The EA states: 'The SGVs should only be used in conjunction with the information contained [in the accompanying technical notes] and with an understanding of the [separately published] exposure and toxicological assumptions'.

c) Environmental Protection Agency, USA. 1996. *Soil Screening Guidance: Technical Background Document*, 2nd Edition. USEPA, Washington DC. The three values shown for each contaminant are soil screening levels, based on risks via human ingestion, human inhalation (where relevant) and migration to groundwater (assuming no dilution/attenuation), respectively.

d) European Union, Directive 2008/105/EC (relating to the Water Framework Directive). Figures shown are environmental quality standards for the protection of human health and the aquatic environment, based on chronic exposure (annual averages).

e) World Health Organization. 2011. *Guidelines for Drinking Water Quality*. Provisional guidelines.

f) Canadian Council of Ministers of the Environment. Interim Sediment Quality Guidelines, for the protection of aquatic life. Freshwater sediments. The first and second figures are, respectively, threshold effect levels and probable effect levels. Figures rounded in some cases.

1 Units: $\mu g\ m^{-3}$ for air; $mg\ kg^{-1}$ (dry weight) for soil; $\mu g\ L^{-1}$ for waters; $\mu g\ kg^{-1}$ for sediments (with the exception of dioxin, expressed as $ng\ kg^{-1}$ TEQ, see below).

2 Guideline level shown is for the sum of all cyclodienes (aldrin, dieldrin, endrin and isodrin).

3 The guideline level shown is for aldrin and dieldrin combined.

4 NSL = no safe level (for carcinogenicity). Concentrations giving an estimated excess lifetime cancer risk of 1:10,000 for benzene, PAH [using B(a)P as an indicator] and trichloroethene are 17 $\mu g\ m^{-3}$, 1.2 $ng\ m^{-3}$ and 230 $\mu g\ m^{-3}$, respectively. For chloroethene (vinyl chloride), the lifetime risk of exposure to 1 $\mu g\ m^{-3}$ is 1×10^{-6}.

5 Applies to polychlorinated dibenzo-p-dioxins ('dioxins') and polychlorinated dibenzofurans ('furans').

6 NP = guideline values not proposed for dioxins/furans or PCBs because 'direct inhalation exposures constitute only a small proportion of the total exposure' [which is mainly via food].

7 Dioxin guidelines for sediments expressed as $ng\ kg^{-1}$ toxic equivalent (TEQ).

8 NE = not established; guideline unnecessary because concentrations in drinking waters well below those of health concern.

9 PAH = polycyclic aromatic hydrocarbons. In all cases shown, benzo(a)pyrene is used as the indicator species.

10 PBDEs = Polybrominated diphenyl ethers.

11 SGVs and SSLs shown are for *o*-xylene. The corresponding SGVs for *m*-xylene and *p*-xylene are, respectively: 240/180/3500 and 230/160/3200 $mg\ kg^{-1}$. The corresponding SSLs for *m*-xylene and *p*-xylene are, respectively, 160,000/420 and 160,000/460 $mg\ kg^{-1}$.

Table A3 Air quality guideline levels proposed by the World Health Organization for priority air pollutants ($\mu g\ m^{-3}$).[1]

	Averaging period				
	10 min	*1 hour*	*8 hours*	*Daily*	*Annual*
Nitrogen dioxide	–	200	–	–	40
Ozone	–	–	100[2]	–	–
Sulfur dioxide	500[3]	–	–	20[4]	–
$PM_{2.5}$	–	–	–	25[5]	10[6]
PM_{10}	–	–	–	50[7]	20[8]

1 For ozone, sulfur dioxide and particulate matter (PM), interim targets are also published by the WHO (see below); these are intended for areas with high levels of air pollution as a stepwise means of ultimately achieving the guideline values. See Tables A1 and A2 for guidelines relating to other air pollutants.

2 Interim target of 160 $\mu g\ m^{-3}$.

3 Recommendation only; to protect against respiratory damage from short bursts of physical activity by asthmatics.

4 Interim targets (in two steps) of 125 and 50 $\mu g\ m^{-3}$.

5 Interim targets (in three steps) of 75, 50 and 37.5 $\mu g\ m^{-3}$.

6 Interim targets (in three steps) of 35, 25 and 15 $\mu g\ m^{-3}$.

7 Interim targets (in three steps) of 150, 100 and 75 $\mu g\ m^{-3}$.

8 Interim targets (in three steps) of 70, 50 and 30 $\mu g\ m^{-3}$.

Appendix 3. Periodic table of the elements of environmental pollution

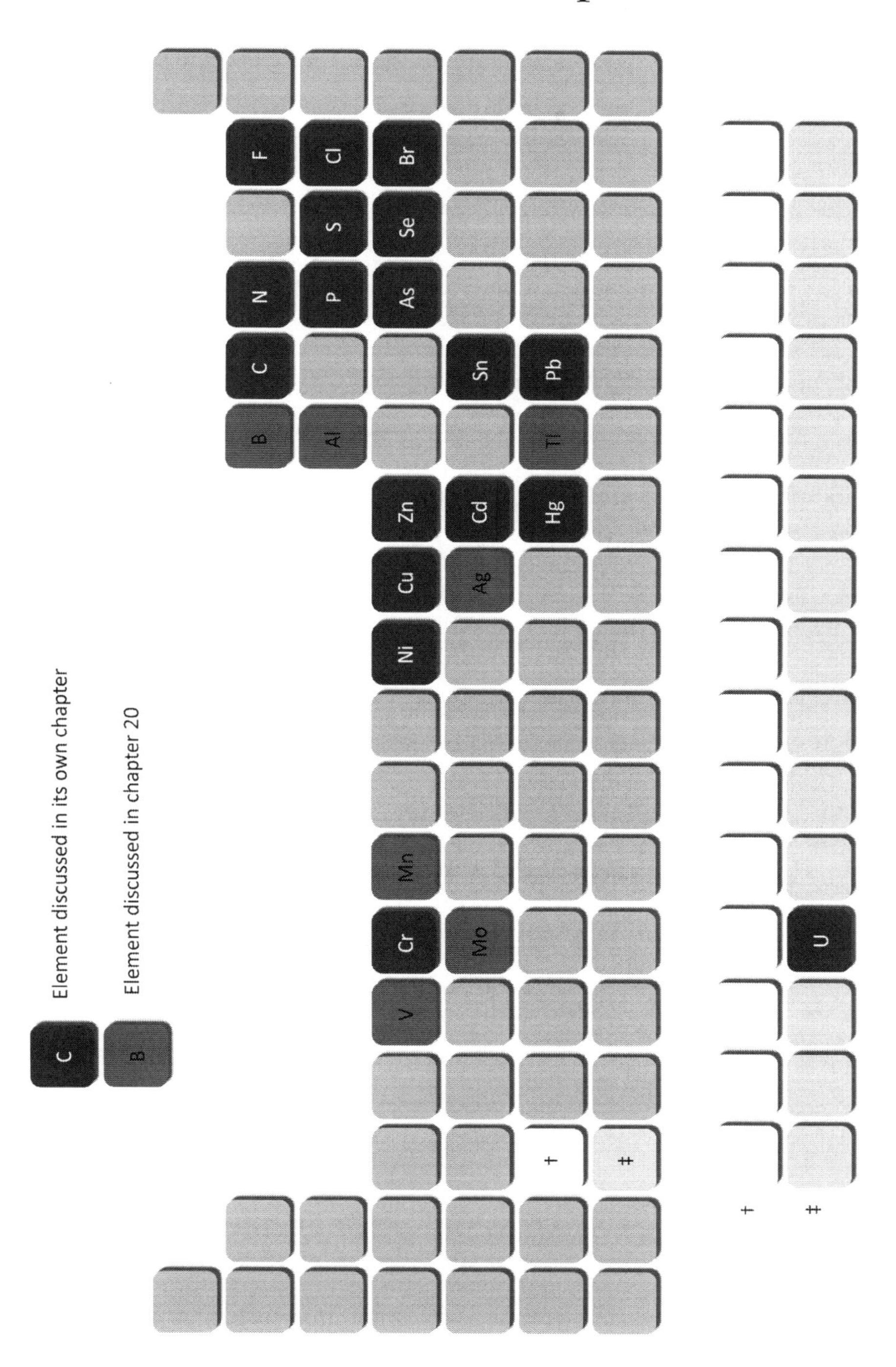

Glossary

Note: Many terms and concepts that are commonly used in relation to pollution are defined in Chapter 1 and not repeated here (see words shown in bold type in Chapter 1).

Acid See also →base and →alkali. Water dissociates into acidic H^+ and basic OH^- ions; in pure water the concentration of each (at 25°C) is 1×10^{-7} mol L^{-1} and the sum total of both is 1×10^{-14} mol L^{-1}. Acidity (and alkalinity, or basicity) is measured on the pH scale, from 0 to 14, pH being the common logarithm of the reciprocal of the H^+ concentration or 'activity'. For example, a concentration of 1×10^{-6} mol L^{-1} gives a pH of 6, an acidic solution, a concentration of 1×10^{-7} mol L^{-1} gives a pH of 7, a neutral solution, and a concentration of 1×10^{-8} mol L^{-1} gives a pH of 8, an alkaline solution. An acid is defined as any chemical species that can donate a 'proton', i.e. a hydrogen ion, H^+ (essentially a nucleus of a H atom) or any substance that, upon dissolution in water, increases the concentration of H^+ ions above 1×10^{-7} mol L^{-1} (0.0000001 mol L^{-1}), and therefore decreases the pH below 7. For example, an increase in H^+ concentration to 1×10^{-2} mol L^{-1} (0.01 mol L^{-1}) would give an acidic solution with a pH of 2.

Aeolian transport Movement of particulate matter from one location to another by the action of wind.

Aerobic Typically applied to organisms that live in →oxic environments and to their oxygen-based metabolic processes.

Aerosol A suspension of particles (liquid and/or solid) in a gaseous medium, typically air.

Alkali A water-soluble →base.

Anaerobic Typically applied to organisms that live in →anoxic environments and to their (non oxygen-based) metabolic processes.

Anhydride What remains when water is removed from a compound; in natural environments the latter is often an acid. For example, the anhydride of phosphoric acid, H_3PO_4, is phosphorus pentoxide, P_2O_5. See also →hydration.

Anion See →Ion.

Anoxia An absence of oxygen. An environment with anoxia is said to be anoxic. See also →Hypoxia.

Anthropogenic Caused by human activity.

Aqueous Of or containing water. Aqueous species are those that have dissolved in water. For example, the table salt, sodium chloride (NaCl) dissolves in water to yield aqueous Na^+ and Cl^- ions. Water is a strong solvent and is ubiquitous on planet Earth, so aqueous chemistry is an important aspect of pollution science.

Aquifer A water-permeable geological mass, either of fractured and/or porous rock or unconsolidated (loose) sediments. Aquifers act as sub-surface reservoirs of water and are often tapped for public supply.

Atmospheric lifetime See →Residence time.

Atmospheric stratification The atmosphere can be considered as having distinct horizontal layers defined by changes in temperature. The lowest layer, present at ground level and extending to variable altitudes of between approximately 9 and 17 km, is called the troposphere and is characterised by a steadily decreasing temperature with altitude (except in temperature inversions – see chapter 1). The layer above is the stratosphere, which has a higher temperature because of the absorption of solar radiation by abundant ozone molecules (the stratosphere being the home of the 'ozone layer'). There are two higher layers: the mesosphere, where the temperature profile reverts to a decreasing trend with altitude; and the thermosphere, which is the most rarefied of all the layers (atmospheric density decreasing with altitude) but has a relatively high temperature, because of the absorption of intense solar radiation, this time by molecular nitrogen and oxygen. Because of the very gradual rarefaction of the atmosphere with increasing distance from the Earth's surface, there is no clear boundary between the atmosphere and outer space; a separate 'layer', the exosphere, is sometimes used to define this indistinct zone above the thermosphere.

Base Any chemical species that can accept a proton, i.e. a hydrogen ion, H^+ (essentially a nucleus of a H atom), or any substance that, upon dissolution in water, increases the concentration of OH^- ions above 1×10^{-7} mol L^{-1}, giving a consequent decrease in the concentration of H^+ ions below 1×10^{-7} mol L^{-1} and a pH of >7 [→acid →alkali].

Biodegradation See →Respiration.

Biomethylation See →Methylation.

Blood–brain barrier A semi-permeable structure lining the blood capillaries of the brain that limits the passage of potentially harmful chemicals into the extracellular fluid of the central nervous system.

Boundary layer (or planetary boundary layer) In atmospheric science, a layer of air at or near to the Earth's surface that is affected by the latter in terms of friction and heat.

Catchment area (or drainage basin) In environmental science, the area within a →watershed, within which all atmospheric precipitation ultimately drains into a particular river system.

Cation See →Ion.

Chelation A type of →complexation whereby a single atom forms multiple bonds with a →ligand. Most chelators (chelating ligands) are organic compounds containing more than one atom able to form a bond. Chelation is pronounced with a 'hard c' rather than a 'ch' sound.

Chemical species The specific chemical (atomic, ionic and molecular) forms of an element. For example, the various chemical species of arsenic include arsenite, arsenate, arsenic trioxide, arsenious acid and so on.

Chemosynthesis The synthesis of carbohydrates from carbon atoms or carbon-containing molecules (usually CO_2 but sometimes methane) and other compounds by specialised micro-organisms using energy obtained from the oxidation of compounds such as ammonia and hydrogen sulfide (i.e. not from sunlight as in photosynthesis).

Chlorosis In botany, leaf discoloration (typically yellowing or whitening) resulting from a lack of chlorophyll, the molecule that absorbs light energy in photosynthesis. Leaves, or plants, suffering from chlorosis are said to be chloritic.

Cloud condensation nuclei Microscopic and sub-microscopic atmospheric particles that provide surfaces for the condensation of water. A number of microscopic water droplets thus formed may ultimately coalesce to form a raindrop.

Complex (or coordination complex) In chemistry, typically a metallic ion bonded to one or more →ligands. The central ion (or atom) is 'complexed' and the ligands are the 'complexing agents'.

Concentration The mass of a substance per unit mass or volume of the host matrix (e.g. soil, sediment, water, air, biomass). The units of concentration that are typically used to describe the extent of soil and water pollution are mg kg^{-1} (i.e. mg of pollutant per kg of the host matrix), μg g^{-1}, mg L^{-1}, μg kg^{-1} and μg L^{-1}; the first three units of this list equate to 'parts per million' (ppm) and the final two equate to 'parts per billion' (ppb). Concentrations of air pollutants (particulate and gaseous) are typically expressed in units of μg m^{-3}. Concentrations of pollutants may be reported for other matrices, such as biological tissues; for example, measurements of Pb in blood samples are often expressed in units of μg dL^{-1}. Very high concentrations of pollutants may be expressed in percentage terms. For example, a concentration of Cu in soil of 10,000 ppm may be expressed as 1%; i.e. 1% of the soil is comprised of Cu. Concentration may also be expressed in molar terms; for example, moles of a substance per unit volume.

Congeners Related elements or compounds. For example, the elements in the same group of the periodic table or the many different types of PBDE (see bromine chapter).

Conservative In oceanography, conservative ions are those that are stable over time and maintain a constant ratio to each other. They are (in descending order of concentration): sodium, chloride, sulfate, magnesium, calcium and potassium. In terms of pollution, the term 'conservative pollutant' is sometimes used to describe a toxic species that does not readily degrade or transform to a non-toxic species (e.g. toxic metals).

Covalent bond A chemical bond where each bonding atom shares one or more electrons with the other bonding atom.

Degassing The removal of gas from a body; for example a large amount of volatile mercury degasses each year from the Earth's crust. Degassing of pollutants from contaminated soils and waters largely occurs by simple →diffusion.

Diffusion In physics, the transfer of molecules from an area of high concentration to an area of low concentration, via random molecular motion.

Dissociation The splitting of a molecule, via physical or chemical means, into molecular fragments and/or individual atoms or ions.

Divalent See →Valence.

Dynamic steady state A characteristic of natural element cycles, whereby elements constantly move within and between environmental compartments but the amounts within each compartment stay relatively constant.

Eh A measure of →redox potential, measured in units of millivolts, mV. The zero point of the Eh scale is defined by the redox potential of the hydrogen ion, H^+. Substances that are more or less →electronegative than H^+ have, respectively, positive and negative Eh values; the former being substances capable of oxidising

H^+ and the latter of reducing H^+. In the natural environment, oxygen status is usually of overriding importance in these matters; positive Eh values are typically found in well-aerated environments (e.g. turbulent rivers, free-draining soils) while →anoxic or →hypoxic environments (e.g. waterlogged soils, bottom sediments) are likely to register negative Eh values.

Electronegativity The tendency of one atom in a bond to attract a bonding pair of electrons. For example, oxygen is a particularly electronegative element, strongly attracting electrons; in the water molecule this gives rise to polarity, an important factor in the power of water as a solvent, whereby the 'oxygen side' of the H_2O molecule has a partial negative charge and the 'hydrogen side' has a partial positive charge. This enables a number of water molecules to collectively overpower the attraction that holds one atom to another, in a rock for example, allowing the gradual dissolution of the rock. See also →water.

Enthalpy change The amount of heat energy absorbed or released in, respectively, endothermic and exothermic chemical reactions (or phase changes).

Environmental quality standard The target maximum concentration of a pollutant, usually in air or water, that is set to protect environmental and human health. Such standards are sometimes called 'limit values' or 'guidelines' and are set by governments and health agencies (including the World Health Organization), based on known effects concentrations (see also description of 'LOAELs' and 'NOAELs' in chapter 1).

Epidemiology In terms of pollution, the study of: (i) the distribution and occurrence of pollutants, or pollutant sources; and (ii) disease or adverse health outcomes and the statistical likelihood of causal links between the two.

Fluvial transport The movement of materials by water.

Flux In biogeochemistry, the mass of a substance transported from a particular environmental compartment (e.g. a →[catchment area]) over a defined period of time.

Free ions Unbound ions in aqueous solution, e.g. in water bodies or soil solution; i.e. ions that are not part of the inherent structure of solid materials, adsorbed to the surfaces of solid materials or present within aqueous →complexes.

Free radical In chemistry, a highly reactive atom, ion or molecule with one or more unpaired electrons. A commonly encountered example is the hydroxyl radical (OH).

Global warming potential (GWP) A relative measure of the capacity of a greenhouse gas to warm the Earth's atmosphere by absorbing infra-red (thermal) radiation from the Earth's surface. Carbon dioxide, CO_2, is used as the baseline. For example, the greenhouse gas methane, CH_4, has a 100-year GWP of 25, meaning that over this period an equivalent mass of CH_4 will trap 25 times more heat in the atmosphere than CO_2.

Haem synthesis An enzymatic process that produces the haem molecule, a component of haemoglobin and other proteins.

Hexavalent See →Valence.

Hydration In chemistry, the addition (generally reversible) of one or more molecules of water to a chemical species. In hydration, the addition of water does not involve the splitting of the water molecule(s) or the hydrated molecule(s) (see →hydrolysis). The absorption of water by a hydrating solid increases its volume and may cause internal stresses that promote weathering.

Hydrolysis A chemical reaction involving the ionic components of water, H^+ and OH^-, where the hydrolysed compound is decomposed.

Hydrophilic Literally, 'water loving' (cf. hydrophobic=water hating). Hydrophilic substances are those that tend to be readily dissolved in water (and other polar substances).

Hydrosphere The hydrosphere comprises the sum total of the water occurring upon and within the Earth's crust and in the atmosphere. The main parts of the hydrosphere are the oceans, sea ice, estuaries, rivers, lakes, glaciers, groundwaters and atmosphere moisture. The collective mass of sea ice, glaciers and other forms of frozen water is sometimes referred to separately as the cryosphere.

Hypoxia A low level of dissolved oxygen in water; defined in some cases as ≤2 mg L^{-1}. See also →Anoxia.

Ion An atom or molecule that has lost or gained electrons. An atom or molecule that has lost or donated one or more electrons has a positive charge and is called a cation (e.g. H^+, Na^+, Ca^{2+}). An atom or molecule that has gained one or more electrons has a negative charge and is called an anion (e.g. OH^-, Cl^-, SO_4^{2-}).

Ionic bond A chemical bond formed by the transfer of one or more electrons. One atom in the bond loses one or more electrons to become a cation and the other atom gains one or more electrons to become an anion (see also →ion).

Isotope The various isotopes of an element have the same number of protons (the number of protons being the defining characteristic of an element) but differing numbers of neutrons. For example, there are three stable isotopes of oxygen: O-16 (8 protons and 8 neutrons); O-17 (8 protons and 9 neutrons) and O-18 (8 protons and 10 neutrons).

Ligand A complexing agent. See →complex. A molecule or anion that donates one or more pairs of electrons to the central atom in a complex.

Limiting nutrient The nutrient in an environment that is in shorter supply (relative to nutrition requirements) than other nutrients, thus constraining the potential growth and reproduction of organisms inhabiting that environment. Nitrogen or phosphorus are typically limiting because they are not always readily available in the large amounts required by organisms.

Lipophilic Literally, 'lipid (fat) loving'. Lipophilic substances are those that tend to be readily dissolved in fats (and other non-polar substances). Therefore, they cannot be broken down in water (a polar substance) and tend to accumulate in fatty tissues.

Lithification The transformation of unconsolidated sediments to solid sedimentary rock by the processes of compaction (by the pressure of overlying sediments) and cementation.

Lithogenic Derived from the lithosphere. For example, elements or substances incorporated into soils from weathered parent material are described as lithogenic, to differentiate them from other sources (e.g. atmospheric deposition).

Lithosphere In geology, the Earth's crust and upper mantle. In some cases, the →pedosphere may also be included as an associated part of the lithosphere. In the context of biogeochemical cycling, the lithosphere is the major, long-term reservoir of most elements, nitrogen being a notable exception.

Metallothionein A protein that has the ability to bind toxic trace elements via the abundant →thiol groups in one of its constituent amino acids, cysteine. In biological systems, it may act as a defence (detoxification) mechanism against toxic metals and as a regulator of essential trace elements. →Phytochelatins are a similar class of structures.

Methanogenesis Generation of methane, CH_4, typically by →anaerobic microorganisms in →anoxic environments.

Methylation The addition of a methyl group (CH_3) to an atom, ion or molecule. In biological systems the process is referred to as biomethylation but abiotic methylation also occurs. Demethylation is the reverse process.

Mineralisation There are various meanings, more than one of which is pertinent to the current text. In biology, mineralisation of the major elements (C, N, etc.) refers to the release of these elements from the biodegradation of organic matter. In geology, the term relates to the formation of metal ore bodies in the Earth's crust. In physiology, bone mineralisation refers to the process of bone formation (see fluorine chapter).

Monovalent See →Valence.

Necrosis The premature death of cells in living tissues.

Ombrotrophic Fed by nutrients and water from the atmosphere. A typical ombrotrophic environment is a peat bog isolated from incoming surface drainage water.

Order of magnitude Put simply, a factor of ten. The number 520 (5.2×10^2) has an order of magnitude of 2, while the number 615,000 (6.15×10^5) has an order of magnitude of 5; the second number is approximately 3 orders of magnitude larger than the first.

Osmoregulation Regulation of the water and electrolyte (dissolved substance) content of cells.

Oxic Oxygenated; i.e. an environment characterised by the presence of free oxygen.

Oxidation The removal of one or more electrons from an atom, ion or molecule in a chemical reaction. For more detail, see →Redox.

Oxidation state Indicates the number of electrons gained, lost or shared by an atom as a result of chemical bonding; also called the oxidation number. Oxidation state increases with →oxidation and decreases with →reduction (see also →Redox).

Oxidative stress The damage caused to living cells by excess →[free radicals] and other 'reactive oxygen species' produced by the metabolism of oxygen. Free radicals (and therefore oxidative stress) are normally kept in check by antioxidants including certain vitamins and enzymes.

Oxyanion An anion containing oxygen. An example is arsenate, AsO_4^{3-}.

Parent material The geological materials underlying soils. The geological materials upon which soils form.

Pedogenesis The formation of soil, from weathered →[parent material] and decomposing organic residues.

Pedosphere The pedosphere comprises the sum total of all the soils upon the Earth's surface.

Pentavalent See →Valence.

pH See →Acid.

Photolysis The splitting (lysis) of chemical compounds into atoms and/or smaller molecules by photons (light). Also called photodegradation and photodissociation.

Photoreduction Chemical →reduction induced by photons (light).

Phytochelatin A protein that binds potentially toxic metals via →chelation. Similar to →metallothioneins, phytochelatins play an important role in detoxification.

Despite the 'phyto' suffix, phytochelatins also occur in micro-organisms and nematodes.

Phytotoxicity Toxicity to plants.

Placental barrier A semi-permeable structure in the placenta that limits the transfer of potentially harmful chemicals from the mother to the foetus.

Planetary boundary layer See →[Boundary layer].

Polarity See →Electronegativity →Water.

Radical See →[Free radical].

Redox A redox reaction involves →*red*uction and →*ox*idation and results in a change in the →[oxidation state] of each reactant. Redox potential, or reduction potential, is a measure of the tendency of a substance or a solution to lose or gain electrons upon addition of a chemical species and is measured on the →Eh scale.

Reduction The addition of electrons to an atom, ion or molecule in a chemical reaction. For more detail, see →Redox.

Reserve The amount of a →resource that can be practically and affordably extracted by available technologies.

Residence time The average amount of time spent by an element, or chemical species, in an environmental compartment (e.g. an individual lake or the whole atmosphere). Residence time can be defined as the total amount of a chemical species being held in an environmental compartment divided by its rate of flux, as measured by its rate of removal for example. Consider a simple example of a polluted lake that is estimated to contain 100 kg of mercury and has an annual efflux of 1 kg mercury; the residence time of mercury in the lake is 100 years (100/1).

Resource The estimated total amount of a commodity present; e.g. the total amount of a particular fossil fuel in the Earth's crust. Resources are estimated from measurements and observations. See also →Reserve.

Respiration The cellular metabolism of organic molecules, which provides energy to the respiring organism. In aerobic respiration this is achieved by the use of oxygen as the oxidising agent, yielding carbon dioxide and water; in anaerobic respiration, chemical species such as nitrate, NO_3^-, and sulphate, SO_4^{2-}, are used in the absence of oxygen.

Salinity The salt content of a water body, typically expressed in parts per hundred (i.e. percentage) or in parts per thousand (approximately, g L^{-1}), particularly for seawater. The most common cations included in measurements of salinity are sodium, magnesium, calcium and potassium; the main anions are chloride and sulfate, and to a lesser extent, bicarbonate and bromide. The main salt in seawater is sodium chloride, accounting for >85% of the overall salinity.

Sequestration In biogeochemistry, typically a mechanism by which an organism keeps potentially toxic chemical species away from tissues it could damage. Deciduous plants, for example, sometimes sequester toxins into old (senescent) leaves that remove the toxin from the plant when they fall. In some cases a toxic material is sequestered to the →vacuole. In chemistry, sequestration normally involves →chelation.

Sesquioxide Any molecule containing three oxygen atoms and two atoms of another element. Typically used in descriptions of the geochemistry of soils and sediments; the most common sesquioxides in these contexts include amorphous (i.e. non-crystalline) oxides of iron and aluminium.

Soil horizon A surface or sub-surface layer of soil approximately parallel to the soil surface, which is distinguished by colour and/or physico-chemical characteristics. Soils tend to have a distinct organic ('O') horizon at the surface underlain by increasingly mineral horizons termed (in downwards order) A, B and C, which may be further sub-divided (e.g. A_1, A_2, etc.), depending on subtle differences within the horizon.

Species and speciation See →[Chemical species].

Standards See →[Environmental quality standard].

Steady state See →[Dynamic steady state].

Stomata The openings (pores) on leaves and some other plant structures that allow gas exchange. Plural form of the singular 'stoma' or 'stomate'.

Stratosphere See →[Atmospheric stratification].

Subclinical effect A disease or other physiological effect without visible (i.e. clinical) symptoms.

Sulfhydryl A simple chemical group composed of a sulfur and a hydrogen atom. It is sometimes referred to as a →thiol, an organosulfur compound that contains the sulfhydryl group. The amino acid cysteine is particularly rich in thiols, and therefore sulfhydryl (–SH) groups. Pairs of –SH groups form disulfide bridges, giving structure to certain types of protein, in particular keratin. Sulfhydryl groups bind strongly to various metals and are abundant in metal-sequestering structures like →metallothioneins. However, the binding of toxic metals to –SH groups in biological tissues can also lead to enzyme disruption and protein damage.

Tetravalent See →Valence.

Thermocline See →Water stratification.

Thiol An organic molecule, similar to alcohol, with the general formula R-S-H, where R represents an alkane or similar chemical group. The –SH part of the molecule is the →sulfhydryl group.

Trivalent See →Valence.

Troposphere See →[Atmospheric stratification].

Upwelling The rising to the surface of deep, cold, nutrient-rich seawater, often as a consequence of surface water movement by strong winds.

Vacuole A structure present in the cytoplasm of many cells, which can store nutrients and harmful substances and isolate the latter from other parts of the cell.

Valence In chemistry, defined as the number of hydrogen atoms that a chemical species could theoretically bond with; in another sense this could be thought of as the number of electrons that could be donated, accepted or shared by an atom to achieve electronic stability upon chemical reaction. Many elements have more than one valence number and valency can be denoted by a suffix, the first six valence states being written as monovalent, divalent, trivalent, tetravalent, pentavalent and hexavalent.

Vapour pressure The pressure applied by the vapour phase of a substance when in equilibrium with its solid and/or liquid phases; i.e. the pressure applied by the vapour that is sublimating or evaporating, respectively, from a solid or liquid substance. A substance with a large vapour pressure is described as →volatile.

Volatile A substance that readily evaporates at normal temperatures. See also →[Vapour pressure].

Water Essential for life and ubiquitous in physical and biological environments. The hydrological cycle transports water, heat, aqueous ions and solid particles. The polar nature of water (see →electronegativity) means that it is a strong solvent and, as such, is an important factor in the weathering of rocks. See also →hydration →hydrolysis.

Water column The full depth of a water body, from the water surface down to the bottom sediments.

Watershed The boundary of a →catchment area. Rain falling on one side of a watershed will drain into one catchment area, while rain falling on the other side will drain into the adjacent catchment area.

Index

Numbers in *italic* refer to figures; numbers in **bold** refer to tables.